Airline Operations an

MW00827103

Airline Operations and Management: A Management Textbook presents a survey of the airline industry, with a strong managerial perspective. It integrates and applies the fundamentals of several management disciplines, particularly operations, marketing, economics and finance, to develop a comprehensive overview. It also provides readers with a solid historical background, and offers a global perspective of the industry, with examples drawn from airlines around the world.

Updates for the second edition include:

- Fresh data and examples.
- A range of international case studies exploring real-life applications.
- New or increased coverage of key topics such as the COVID-19 pandemic, state aid, and new business models.
- New chapters on fleet management and labor relations and HRM.
- Lecture slides for instructors.

This textbook is for advanced undergraduate and graduate students of airline management, but it should also be useful to entry and junior-level airline managers and professionals seeking to expand their knowledge of the industry beyond their functional area.

Gerald N. Cook is Adjunct Professor at Embry-Riddle Aeronautical University. He obtained his Bachelor of Science and Master of Science in Management from Purdue University and Doctor of Business Administration from Nova Southeastern University. He enjoyed a long airline career as a pilot and flight operations manager.

Bruce G. Billig is Adjunct Assistant Professor at Embry-Riddle Aeronautical University. He holds a Bachelor of Science in Electrical Engineering from the US Air Force Academy and a Master of Aeronautical Science from Embry-Riddle. His experience includes over 20 years as an Air Force pilot and 22 years as an airline pilot, retiring as a 17-year Captain at Southwest Airlines in 2019.

Airline Operations and Management

A Management Textbook

Second Edition

Gerald N. Cook and Bruce G. Billig

LONDON AND NEW YORK

Designed cover image: © Getty Images

Second edition published 2023
by Routledge
4 Park Square, Milton Park, Abingdon, Oxon, OX14 4RN

and by Routledge
605 Third Avenue, New York, NY 10158

Routledge is an imprint of the Taylor & Francis Group, an informa business

© 2023 Gerald N. Cook and Bruce G. Billig

First edition published by Routledge 2017

British Library Cataloguing-in-Publication Data
A catalogue record for this book is available from the British Library

Library of Congress Cataloging-in-Publication Data
Names: Cook, Gerald N., author. | Billig, Bruce G., author.
Title: Airline operations and management: a management textbook /
Gerald N. Cook & Bruce G. Billig.
Description: Second edition. | Abingdon, Oxon; New York, NY: Routledge, 2023. |
Includes bibliographical references and index. |
Identifiers: LCCN 2022054235 (print) | LCCN 2022054236 (ebook) |
ISBN 9781032268736 (hardback) | ISBN 9781032268729 (paperback) |
ISBN 9781003290308 (ebook)
Subjects: LCSH: Airlines–Management.
Classification: LCC HE9780 .C59 2023 (print) |
LCC HE9780 (ebook) | DDC 387.7068–dc23/eng/20221109
LC record available at https://lccn.loc.gov/2022054235
LC ebook record available at https://lccn.loc.gov/2022054236

ISBN: 9781032268736 (hbk)
ISBN: 9781032268729 (pbk)
ISBN: 9781003290308 (ebk)

DOI: 10.4324/9781003290308

Typeset in Bembo
by Newgen Publishing UK

Access the Support Material: www.routledge.com/9781032268736

Contents

Figures

Tables

Goal of Airline Management and Operations

This textbook is a survey of the airline industry from a managerial perspective. It does not include other sectors of the broader aviation industry, such as airport operations or general aviation. The primary audience is senior and graduate students of airline management, but it should also be useful for entry and junior-level airline managers needing a broad knowledge of the industry extending beyond their functional area. A background in the various management disciplines typically covered in an undergraduate degree program in business or management is helpful, but not essential. The text incorporates fundamental concepts from several management disciplines, particularly, economics, operations, marketing, and finance in developing an overview of the industry. The focus is on tactical, rather than strategic, management that is specialized or unique to the airline industry. Human Resource Management, for example, is not addressed, not because it is not crucial to an airline's success, but because the skills required are not unique to the industry. The manager with a broad understanding of the industry and its competitive environment is better equipped to work in interdisciplinary assignments and, ultimately, to succeed and progress in an airline career. This is the goal of this text.

Revisions in the Second Edition

This second edition maintains the basic structure of the original textbook. Airline examples, operational and financial data and statistics are updated throughout. This textbook was written in the first half of 2022 as the industry was recovering from the COVID-19 pandemic. As data in several chapters of the textbook illustrate, the pandemic severely disrupted airline operations. Data from 2019 are occasionally provided rather than the latest available information to provide valid historical comparisons.

The new labor relations Chapter 6 consolidates labor issues into one chapter. Chapter 8 covering airline fleet selection and acquisition is also new; however, much of the content was relocated from the first edition chapter on finance and economics. Each chapter concludes with a case study that illustrates the application of many concepts in a current industry setting. We appreciate the suggestions from several reviewers.

September 2022

Introduction

The dynamic and rapidly expanding airline industry is essential to the function of the world economy and for continued economic growth. It's also an industry often in turmoil, sometimes as a result of myopic management focus on market share but more often due economic crises beyond the control of airline executives. The first decade of the 21st century was especially cruel to the industry as it endured two severe economic recessions, the so-called Great Recession beginning in 2008 being the most devastating since the Great Depression of the 1930s. As a direct result of falling demand, several airlines failed, including ATA, Maxjet, Aloha, Sterling, and Oasis. Others were forced to par routes and frequencies and reduce costs in restructuring that continues today. This book examines the history, structure, and functions of the airline industry from a management perspective with the goal of providing students and junior managers with a broad overview of airline operations and management. Although the tools of several management disciplines are employed, previous study of business administration is not required.

The text begins with a summary of airline history from the emergence of earliest passenger airlines immediately following the end of World War I to the present. Chapter 1 emphasizes the role of government economic regulation, ownership, and subsidies in the development of the early industry. Rapid technological progress, especially in aircraft speed, reliability, and comfort, increased demand and lower costs to the point that government policy makers, beginning in the United States, freed airlines from regulation of routes and fares leading to an increasingly free market in much of the developed world.

Chapter 2 turns to the forces that drive airline demand and supply and the rapid growth in airline passenger and cargo transportation, which is now most evident in Asia. Airline product offerings are the subject of the next two chapters beginning with an airline's choice of route structure to connect the destinations it has chosen to serve. An airline has a range of product amenities it can provide with its core transportation product from the Spartan service characteristic of low-cost-airlines to a choice of amenities from first class to economy coach provided by most the world's largest airlines. The first half of the book ends by exploring the process of developing and managing a flight schedule.

Chapter 6 begins with a discussion of airline labor relations focusing on what is often the largest category of airline expenses. Chapter 7 dives into airline economics and finance. Fleet selection, renewal, expansion, and financing are included in Chapter 8. The complex and frequently misunderstood process of airline pricing and revenue management systems is the subject of Chapter 9. Sophisticated processes employed by airlines attempt to maximize revenue by charging different prices to different passenger segment based on willingness and ability to pay for air travel. Distribution of airline product information and ticketing is covered in Chapter 10. This chapter covers the reservation and ticketing process throughout history – from early hand-written paper ticket coupons to present day smart-device technology. Chapter 11 shifts to focus from domestic to international operations. Bilateral Air Service Agreements that began after WWII are being replaced with more liberal Open Skies agreements as airlines form the Big-3 global airline alliances. The text concludes with Chapter 12 which reviews challenges for the future. Environmental constraints and the weakening of lucrative business travel through teleconferencing are only two of the issues facing airline manages in the future. Although strong demand will exist for air travel, managers will need to continually adjust business models to maintain profitability.

1   Historical Perspective

Upon completion of this module, you should be able to:

*   Analyze how various governments supported their early airlines to ensure the growth of the industry.
*   Understand the significance of The Contract Mail Act of 1925 and how it resulted in the birth of the airline industry in the United States.
*   Using the airline industry as an example, explain how technology improves the material standard of living.
*   Interpret the pros and cons of economic airline regulation and the reasons for deregulation.
*   Compare and contrast the economic deregulation of the airline industries in the United States, Europe, and China.

1.1 Introduction

Several scheduled passenger airlines emerged in Europe shortly after the end of World War I in 1918. Economic regulation of routes and prices followed shortly thereafter and prevailed for much of the 20th century. Late in the century, deregulation of the US airline industry began a trend followed by the European Union and then by most countries worldwide. This history provides a foundation for understanding today's complex, competitive, and global airline industry. This chapter traces the industry from the era of economic regulation to the increasingly free-market competition allowed in most of the world's largest markets. After outlining the growth of the early airlines in Europe, the focus shifts to the United States beginning in 1918 with airmail operated by the US Post Office. Using this early airmail endeavor as our starting point, the chapter traces the fledgling industry through the years of economic regulation under the Civil Aeronautics Board until economic deregulation in 1978, emphasizing the economic, regulatory, and technical evolution that shaped the industry. This historical perspective then continues with post-deregulation developments, including the emergence of low-cost carriers, bankruptcies, mergers, and buyouts. Airline deregulation next occurred in Europe, then spread to several other world regions. To gain a broader perspective on the deregulation process

DOI: 10.4324/9781003290308-1

and outcome, the US experience is compared with that of Europe, China, and India, leading to the conclusion that competitive results are remarkably similar. A short characterization of the industry concludes the chapter.

1.2 Transportation and Commerce

Air transportation is just one element of the world's transportation infrastructure essential for economic growth and prosperity. Sailing ships have traded between countries for at least two millennia making those cities and states that controlled trade wealthy. The past riches of Venice, still evident today in magnificent buildings, were based on trade. Today, ever larger and more efficient container ships reduce the cost of ocean shipping facilitating extensive trade between China, Europe, and the Americas.

Rivers were the first and most efficient means for inland transportation of goods and people. The construction of canals to improve and supplement natural waterways dates to at least the sixth century BC when a canal linked the Nile River with the Red Sea. Extensive waterway building in China began in the third century BC. The famous Venetian canals were extensions of the natural waterways among the islands on which Venice is built and date from the city's founding in the fifth century. Elsewhere in Europe, the Netherlands began canal construction to foster economic development in the 12th century with construction then spreading across much of Europe. The British canal system played an important role in the industrial revolution beginning in the 18th century. In contrast, economic development in Africa has been stunted by the lack of navigable rivers to facilitate commerce. Waterway construction in the United States is more recent with the building of the Erie Canal connecting the Hudson River in New York with Lake Erie and the other Great Lakes completed in 1825. The alternatives to waterway transportation at that time, usually by horse and wagon train, were comparatively much more expensive, slower, and limited. Railroads, first introduced in Great Britain in the early 19th century, gradually replaced canals by offering higher speed over a much larger geographical area.

Transportation systems have often been financed and built by private firms, but market forces are generally insufficient to provide for the vast transportation infrastructure necessary to foster widespread economic growth. Therefore, governments frequently plan, build, subsidize, and regulate transportation systems.

In the United States, for example, the federal government encouraged and subsidized the development of both water and rail transportation. Railroads were often joint ventures of federal and state governments and private firms. The great western expansion of railroads following the Civil War, most notably the completion of the transcontinental railroad in 1869, allowed for farm products from the Midwest to reach the heavily populated East Coast cities beginning in the early 1870s. Midwest farmers, in turn, were able to readily purchase goods from East Coast manufacturers which vastly improved their standard of living. The Great Plains was transformed from a wasteland to a breadbasket. Transportation enables trade which improves the overall standard of living.

The federal government remains essential in supporting transportation systems, including highways and roads, waterways, railways, pipelines, airways, and airports. The US Army Corps of Engineers still operates the series of locks on the Mississippi and Ohio rivers over which Midwest grains are transported for export worldwide. Likewise, governments have also played a critical role in the development of commercial aviation. The Middle Eastern countries – especially the United Arab Emirates and Qatar – have recently invested heavily in airports and state-owned airlines to promote economic growth and diversity from dependence on petroleum exports.

1.3 The Earliest Airlines

Just over a decade after the Wright Brothers achieved powered flight, aviation entrepreneurs and military strategists around the world were finding new opportunities to use the innovative capability. For example, in 1914, people traveling from St. Petersburg to Tampa, Florida were faced with a circuitous train ride of over 11 hours around Tampa Bay. Seeing an opportunity, promoter Paul Fansler started an airboat service from St. Petersburg to Tampa across Tampa Bay. The Tampa – St. Petersburg Airboat Line (Figure 1.1, left) began the first-ever scheduled air passenger service using a Benoist XIV airboat (Figure 1.1, right). The flight took only 23 minutes and could carry one passenger seated next to the pilot. The airline flew twice daily round-trips Monday through Friday from

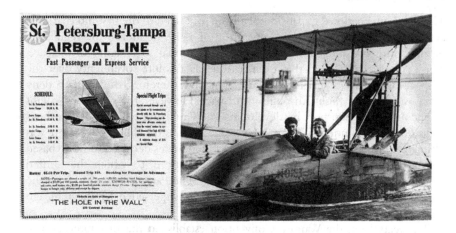

Figure 1.1 Tampa – St. Petersburg Airboat Line – 1914.
Source: Left photo from Public.Resource.Org. (2004) CC BY 2.0. Right photo from Florida Memory: State Library and Archives of Florida (1914).

January through March 1914, ending with the tourist season. The fare was $5 and the 90% average load factor would make today's airline executives jealous.

It wasn't until after World War I, however, that passenger air transportation companies began to develop. Enabled by the rapid wartime advances in technology and a post-war surplus of aircraft, many airlines were founded in the immediate aftermath of World War I in 1918. While there were a few short-lived ventures, many current European airlines trace their origins to that immediate post-war era. The British, French, and Dutch saw aviation as a method of linking together their distant colonial empires. KLM, founded in 1919 and claiming the title of the world's oldest airline under the same name, was "born out of the need to connect Amsterdam to what was then the East Indies" (EAS Barcelona, n.d., p.2). Scheduled flights between London and Amsterdam began in 1920 and continue to this day (Air France/KLM, n.d.). KLM's claim notwithstanding, British Airways initiated the first daily international scheduled service with its forerunner airline. Air Transport and Travel Limited started London to Paris service in August 1919. Similarly, the airlines that would later become Air France trace their origins to this period. Australia's Qantas dates from 1920 when two WWI veterans envisioned an air service connecting Australia to the rest of the world (Qantas, n.d.). In the Americas, Columbia's Avianca is the oldest airline (second worldwide behind KLM), where it revolutionized transportation, replacing riverboats for cross-country travel. Although Germany had fewer distant colonies, it was also home to several early airlines. Today's Lufthansa is a post-World War II creation, however, its predecessor, Deutsche Luft Hansa, dates to 1926. Asia is now the fastest-growing airline market, but its airlines generally trace their origins to the post-World War II era.

1.4 Regulation

The rapid international expansion of airline routes led to the first international aviation conference, the Paris Convention of 1919, to address the regulation of air commerce and conflicting claims of national sovereignty of airspace. The conferees concluded that each nation has absolute sovereignty over the airspace overlying its territories and waters and, therefore, the right to deny entry and regulate flights through its airspace. However, the conferees sought to encourage air transport by developing rules applying equally to all airlines and affording as much freedom of passage as possible. The convention was ultimately ratified by only 11 states. The United States declined to ratify based on the Convention's link to the League of Nations, to which the United States objected (De Gouyon Matignon, 2019). The Paris Convention built the foundation for later regulatory expansion until superseded by the Chicago Convention of 1944.

Ten years later, the Warsaw Convention established the first international airline liabilities and passenger rights. Airlines of the ratifying countries were required to issue passenger tickets and baggage claim checks for checked luggage. The Convention recognized the right to compensation for loss of cargo or luggage and injury or death, but it limited the airlines' liability. The Warsaw Convention has been modified several times but remains the basis for airline liability worldwide.

1.5 Government-Supported Airlines

Early airlines, regardless of their homeland, were generally not financially viable. The cost to carry passengers in aircraft with only a few seats was very high. While early European airlines began as private companies, most could not survive independently and quickly became reliant on state subsidies. All British airlines, finding themselves unable to compete with subsidized French companies, ceased operations for a few months in 1921 until the British government implemented subsidies, first temporary, and then permanent. Thus began an era of state control and often direct ownership of airlines. The United States and Australia also subsidized their first airlines with contracts to carry mail. Two examples, British Airways and Qantas Airways, exemplify how these two national flag carriers emerged in their respective countries.

1.5.1 British Airways

The evolution of British Airways is illustrative of the growth of European aviation. From 1916 to 1922, five airlines began operations in Great Britain: Air Transport and Travel (AT&T), Handley Page Transport Limited, Instone Airline, Daimler Airway, and British Marine Air Navigation. In 1920 AT&T ceased operations and its assets were bought by Daimler. In 1921, all British airlines suspended flying until government subsidies were implemented. In 1924, the government decided to combine the four subsidized airlines into one carrier. Imperial Airways was anointed as the "chosen instrument" of the British government with the mission of developing routes across the vast British Empire (British Airways, n.d.). By 1935 (with the help of Indian Trans-Continental and Qantas), Imperial Airways' routes extended from London to Cape Town, South Africa, and across Europe, the Middle East, India, and Asia terminating in Brisbane, Australia (see Figure 1.2 top).

Imperial Airlines operated a diverse fleet of aircraft, including four-engine flying boats. One of their land planes was the ungainly and aerodynamically inefficient Handley Page H.P. 42 biplane, pictured in Figure 1.2 (bottom). The H.P. 42 entered service in 1930 flying Imperial Airways eastern routes. Although cruising at only 100 mph, it had a range of 500 miles. Typical seating was for 24–26 passengers in surprisingly ornate first-class cabins split fore and aft. In 1934, Imperial Airways joined Qantas Limited to form Qantas Empire Airways (QEA). Jointly, Imperial Airways and QEA completed weekly London to Brisbane flights. Although the flight was primarily used for mail, a few hardy passengers did fly the 12,654-mile, 12-day trip (Eames, 2018).

The name British Airways (BA) first surfaced in 1936 with the consolidation of three airlines: United Airways, Hillman's Airways, and Spartan Airlines.

Figure 1.2 Imperial Airways Route System and Handley Page H.P.42.

Source: From Wikimedia Commons. Top photo: Imperial Routes April 1935 (Imperial Airways, 2020). Bottom photo Matson (G. Eric and Edith) Photograph collection at the Library of Congress.

British Airways and Imperial Airways operated independently for three years until the start of World War II in 1939 when they were combined to form British Overseas Airways Corporation (BOAC). BOAC operated under the direction of the British Secretary of State for the duration of the war.

With the War's conclusion, the new Labour government decided that air services would be provided by three state corporations – BOAC would continue to operate routes to the Far East and North America; British European Airways (BEA) would serve domestic and European destinations, and British South American Airways (BSAA) would begin new services to South American and Caribbean destinations. Following the unexplained disappearance of two aircraft over the Atlantic in 1948 and 1949, BSAA was absorbed by BOAC. The British Airways brand was resurrected in 1974 when BOAC and BEA merged. The modern era began in 1979 when the Conservative government under Margaret Thatcher declared that British Airways would no longer receive state support or interference in decision-making by offering it for privatization. Soon after, however, BA was reporting major losses due to a recession, so the privatization was delayed. BA was fully privatized in 1987, concurrent with European airline deregulation (British Airways, n.d.).

1.5.2 Qantas

The Queensland and Northern Territory Aerial Services (Qantas) Limited was started in 1920 when Australian Flying Corps World War I veterans Paul McGinness and Hudson Fysh envisioned an air service connecting Australia to the rest of the world (Qantas, n.d.). They started with just two biplanes offering air taxi and sightseeing services; regular airline service was initiated by 1922 (Hayward, 2019). Like British Airways, it was not long before Qantas needed government assistance to survive. Weekly subsidized domestic mail service started in 1922 followed by other contracts in 1924 and 1927. Additionally, in 1928, Qantas contracted with the government to provide on-demand medical flights, called the Australian Medical Service, flying 225 patients in its first year of operation (Qantas, n.d.). After flying domestically for 15 years, the founders' international vision was brought closer to fruition. In 1934, Qantas joined with Britain's Imperial Airways in a 50–50 venture to jointly form the subsidiary Qantas Empire Airways Ltd (QEA). Later that year, Qantas transferred all its operations to QEA. QEA started flying mail domestically, but in 1935 it began flying mail from Australia to Singapore as part of the UK's Empire Air Mail Route (Gunn, 1985; Hayward, 2019). These first overseas Empire flights from Brisbane to Singapore were in the De Havilland DH-86 (Figure 1.3), an aircraft De Havilland designed especially for Qantas (Qantas, n.d.; BAE Systems, n.d.). This airmail route was so popular for mail that often there was little room for passengers (Eames, 2018).

During World War II, Qantas continued service within Australia as well as the international flights to Singapore even though some of its aircraft were temporarily transferred to the Royal Australian Air Force. However, the Singapore service was shut down in 1942 after two Qantas Empire flying boats were shot down (Qantas, n.d.). Following the British lead with the nationalization of Imperial Airways, the Australian government began nationalizing QEA in 1947 by first purchasing BOAC's (formerly Imperial's) 50% share, then Qantas holding company's 50% share. The company continued to grow after WWII,

Figure 1.3 Qantas De Havilland DH-86s in 1935 at Charleville Airport, Queensland, Australia.

Source: From Wikimedia Commons (The Queenslander, 2020).

rapidly expanding internationally. The first round-the-world service began in 1958. The following year, the Boeing 707, Qantas' first jet aircraft, began replacing the propeller-driven fleet. In 1967, QEA placed an order for four Boeing 747s, beginning a transition from the 707 to the 747. Also in 1967, Qantas Empire Airways dropped the "Empire," changing its name to its current version, Qantas Airways Limited (Qantas). By 1979, Qantas became the world's first all-747 airline (Gale, 2019). In 1992, needing a capital infusion that the government was unwilling to provide, Qantas was privatized, with British Airways buying a 25% share. The partnership went further in 1998 with the two airlines forming the Oneworld Alliance. Ending a 50-year era with the 747 in 2020, Qantas now headlines its fleet with Boeing 787 and Airbus 380 aircraft.

1.6 US Air Mail

Early US airlines did not follow the pattern of carriers overseas. The US entered World War I late, did not produce any large aircraft, and generally did not enjoy the technological advances common in Europe. An extensive railway system served all major US cities offering more speed, range, and reliability than aircraft of the early 1920s. And, unlike several European countries, the US did not have colonies scattered across continents that could be reached only by ship or tortuous overland routes. Still, the US government recognized the importance of aviation, especially after WW I conclusively demonstrated the power of aircraft as weapons of war and the potential for passenger transportation. Military aircraft production was mostly dismantled after the War, but the US understood the need to retain an aircraft production capability in the event of a future war, a prospect that unfortunately materialized within 25 years. The limited passenger air transportation available after WWI wasn't economically viable. The aircraft were too small, slow, unsafe, uncomfortable, and expensive. The train offered a superior alternative for long-distance travel. Start-up airlines, unable

Figure 1.4 Curtis JN-4HM Jenny at the inauguration of US Airmail service in Washington, D.C., 15 May 1918.

Source: From Smithsonian Institution (2008b).

to compete with the railroads, failed. Rather than attempt to build a system focused on passenger travel, heavily subsidized airmail service was chosen as a vehicle to begin air transportation system development.

The Post Office had been interested in airmail as early as 1912, but Congress didn't appropriate funds until 1918 when $100,000 was made available for experimental airmail service to be conducted jointly by the Army and the Post Office on a route between Washington and New York with an intermediate stop in Philadelphia. The initial aircraft fleet consisted of six new Curtiss JN-4HM Jenny biplanes originally designed as pilot trainers but slightly modified for airmail service (Figure 1.4).

After three months of flights flown by the Army Air Corps, the Post Office bought their own Standard JR-1B biplanes designed specifically for mail service, hired civilian pilots, and assumed full control of the operation.

Over the next eight years, the Post Office operated an ever-expanding air mail system developing procedures and facilities for reliable night flying, including sequential beacon lights to guide pilots along the transcontinental airways, weather reporting, and new airports. It experimented with several types of aircraft, but the converted ex-Army DeHavilland DH-4s were widely used.

In the mid-1920s, Congress decided that the growing airmail service should be transferred from the Post Office to private companies, a move that contrasts with the strategy of government-owned airlines typical of most of the world at that time. In 1926, the Post Office began auctioning its routes to the lowest bidders. Figure 1.5 shows the system of Contract Air Mail routes (CAM) awarded to private airlines.

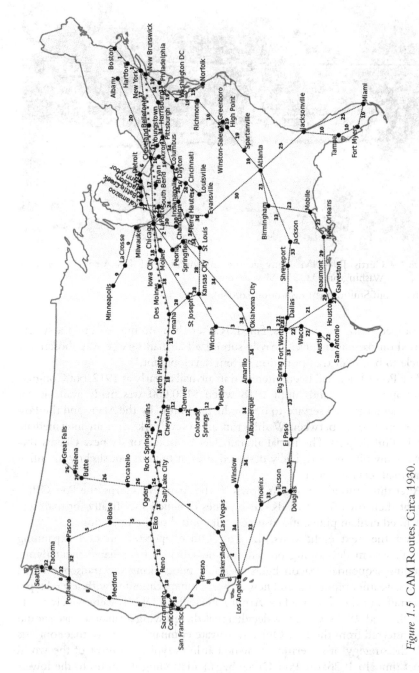

Figure 1.5 CAM Routes, Circa 1930.
Source: From Wikimedia Commons. Photo by Rlandmann (2020), CC BY-SA 3.0.

By 1928, 44 small, financially weak, airlines were transporting mail under government contracts but often operating recklessly using obsolete, but cheap, aircraft. These airlines were either unable or unwilling to invest in facilities and equipment, so the vision of a viable air transportation industry remained unfulfilled. Intending to make airlines profitable and end subsidies, the Hoover administration engineered the consolidation of several carriers, eliminated competitive route bidding, and designated new airmail routes in a series of closed-door, semi-secret, meetings that became known as the "Spoils Conference." The "Big Four" airlines – American, Eastern, United, and TWA – trace their origins to this conference and emerged with the spoils – the new routes. Gone were smaller carriers, such as Robertson in St. Louis for whom Charles Lindberg had flown, and Ludington in Philadelphia.

Under the airmail system, aircraft technology slowly improved. In 1925, Boeing produced the Model 40 specifically for transcontinental mail routes (Figure 1.6). Although primarily designed to fly the mail, the Model 40s also had an interior cabin for two passengers marking an early attempt to combine passenger carriage on scheduled mail routes.

A leap in technology came two years later with Ford Tri-Motor. Although slow and loud, it was designed for passenger service (Figure 1.7). The Tri-Motor, affectionately called the "tin goose" was flown by many European airlines as well as carriers in Africa, Asia, Australia, South America, and the United States.

Figure 1.6 Boeing Model 40. Note. Wikipedia Commons (Libby Photography Studio, 2022), CC BY 2.5.

Note the passengers' window in the fuselage between the two wings.

Figure 1.7 The Ford Tri-Motor.
Source: Photo from Wikimedia Commons (Royal Air Truck Ford Tri-Motor, 2021).

It was an important milestone in the growth of air transportation because it "helped convince the traveling public that safe, reliable-albeit loud-air travel was now a possibility" (Van der Linden, 2012, p. 207). The Tri-Motor carried 13–15 passengers, depending on the model, but at only 90 miles per hour, it still couldn't match the convenience and comfort of long-distance train travel.

1.7 Economic Regulation

In 1929, the US stock market crashed leading to a worldwide depression that further compounded the financial struggles of nascent airline industry. As a minor part of the new Roosevelt administration's sweeping New Deal legislation, the Air Mail Act of 1934 split the responsibility for airline regulation among the Post Office, Bureau of Commerce, and the Interstate Commerce Commission, a plan that soon proved cumbersome and unproductive. Within a few years, regulatory reform was again before Congress.

1.8 CAB Economic Regulation 1938 to 1978

In a series of acts beginning with the Civil Aeronautics Act of 1938 and culminating in 1940, airline regulation was transferred to two separate entities. The Civil Aeronautics Board (CAB) was given responsibility for economic regulation. The Civil Aviation Administration, the forerunner of today's Federal Aviation Administration, assumed responsibility for non-economic regulation, specifically for air traffic control, certification of aircraft and airmen, safety enforcement, and airway development.

The CAB was given a mandate to "promote the development and safety and to provide for the regulation of civil aeronautics" ("Civil Aeronautics Act," 1938, title). It was to develop adequate, economical, and efficient service "without unjust discrimination, undue preferences and advantages, and unfair or destructive competitive practices," but to preserve "competition to the extent necessary to assure the sound development of an air-transportation system" ("Civil Aeronautics Act," 1938, Section 2; Meyer & Oster, 1981). These conflicting objectives proved difficult to reconcile in practice. For most of its history, the CAB severely limited both route and price competition.

The CAB's mandate reflected the experiences of the Great Depression from which the nation had not fully recovered. The unregulated free market was judged destructive. The CAB regulation of the fledgling airline industry was informed by and patterned after railroad regulation. Evidence of the railroad legacy remains today as airline employee labor relations are still governed by the Railway Labor Act.

CAB regulation addressed three broad categories of airline operation: (a) initial approval to operate; (b) routes flown; and (c) fares charged.

The CAB approved airlines seeking to operate in interstate commerce. Airlines already operating in 1938 received grandfather rights to fly their existing markets. For new applicants, however, the CAB judged not only whether the prospective carrier was "fit, willing, and able," but also whether the new service was necessary for "public convenience and necessity." A prospective airline had to demonstrate that it had the financial, managerial, and technical capability to safely operate an airline and that its proposed service was needed to fill a void in the market that would benefit the public. This proved a high standard. In the years after World War II, the CAB approved many smaller airlines, but no new, large airlines, called trunk carriers, were approved during CAB's 40-year reign (Borenstein & Rose, 2007).

The CAB approved airlines to operate specific interstate routes with separate CAB approval needed to add a new route or to drop an existing route. New route requests originated for a variety of reasons such as a request from a city for new or expanded service or from the CAB's determination that more service was needed. Once a new route case was opened, all airlines could apply for the new service and, in a quasi-legal proceeding lasting a year or more, all arguments were heard. The CAB would eventually render its decision, but by the time the winning airline was announced, the market may have changed, and the winner often found the new route did not fit well into its existing structure. On all but the largest routes, only a single carrier was permitted. Trunk carriers wishing to expand on routes already served by another airline had to show that the competition would not harm the incumbent carrier. As late as 1958, there was no competition on 23 of the 100 largest domestic routes.

Most routes were operated in point-to-point or linear service often along the former airmail routes which themselves paralleled railroad lines. Enough flights operated from the largest airports to allow some passenger connections, but the CAB generally prevented airlines from optimizing their routes to improve service or reduce costs.

Finally, the CAB set the prices charged by all airlines in scheduled service. Before World War II, fares were set to prevailing first-class railroad fares but were later computed on a mileage basis and set to allow a reasonable return (12%) on investment (Bailey, 2002). Airlines were effectively required to cross-subsidize shorter routes with profitable, often longer segments. When fares were changed, increases applied to all flights; route-specific changes were not adopted. Charter flight fares were not subject to the same regulation that later became an avenue for price competition, but the CAB discouraged price competition and rarely approved discounting from the formula rate. Consequently, airlines attempted to compete and differentiate their products with inflight service. Even here, however, the CAB attempted to limit non-price competition by restricting first-class and sleeper-seat configurations.

Even under the protective shield of economic regulation, some carriers struggled financially. The CAB responded by awarding lucrative new routes to the weakest airlines. If this was insufficient to curb losses, the CAB orchestrated mergers. For example, Delta absorbed Northeastern with CAB encouragement in 1972.

1.9 Advances in Aircraft Technology

While the CAB was mandated to promote the airline industry, rapid advances in aircraft technology spurred the growth of passenger air travel. As a result of incentives provided by the Post Office, Boeing developed the Model 247, often called the "world's first modern airliner," in 1933 (Van der Linden, 2012, p.208). It was the first airliner to utilize efficient engine streamlining to reduce drag and improve the speed necessary to compete with the train (Redding & Yenne, 1983; Van der Linden, 2012). Although it needed several refueling stops, the 247 could fly from New York to Los Angeles in under 20 hours – seven hours less than any other airliner (Lombardi, 2008; Boeing, n.d.-a). Boeing took an order from its affiliate United Airlines for the first 60 247s; consequently, other airlines had to wait until United's order was filled. This would prove to be a strategic error as Transcontinental and Western Air (later TWA) persuaded Donald Douglas to design and build a competing airliner. The resulting DC-1 turned out to be superior to the 247. Douglas quickly improved on the DC-1 with the DC-2. Then, American Airlines president C.R. Smith convinced Douglas to design a sleeper version of the DC-2. The result was the fabulously successful DC-3, often proclaimed as the first airliner capable of making a profit from passenger service. The DC-3 in Figure 1.8 is pictured in American Airlines livery. American introduced the DC-3 in transcontinental sleeper service in 1935. With the popular DC-3 in production, the Model 247 sales fizzled, and only 75 were sold.

Figure 1.8 Douglas DC-3.
Source: From Wikimedia Commons (Jon Proctor collection).

The DC-3 incorporated the advanced technology introduced on Boeing's 247 – stress-carrying skin, two engines (with the capability for high altitude single-engine flight), controllable pitch propellers, retractable landing gear, and deicing for wings and propellers. Additionally, the DC-3 was slightly faster and much larger than the 247, capable of carrying 21–28 passengers in daytime and 14 in night sleeper configuration versus the 247's 10 (Boeing, n.d.-b). Within a year of its introduction into service, United began replacing its Model 247s with the more economical DC-3. The DC-3 soon dominated the airline fleets in the years immediately preceding World War II. By 1939, 90% of all US airline flights were operated with the DC-2s and -3s (Boeing, n.d.-b).

The DC-3 was larger, faster, and more reliable than the aircraft it replaced. These technological and economic improvements can be quantified with a *productivity index*. The productivity index is a measure of available seat miles per year, the product of three factors: (a) seat capacity; (b) speed; and (c) utilization measured in flight hours per year. A fourth component, load factor or the percentage of seats occupied, is sometimes included for economic comparisons. But load factor is a measure of market demand, competition, and management decisions rather than aircraft design productivity, so it is not included in the index. To compute the productivity index, multiply seat capacity, speed, and yearly utilization. For the DC-3, the index equates to 30 seats X 180 mph X 1,500 hours/year or 8.1 million available seat miles per year (ASM/year) per aircraft.

World War II, which the United States entered in late 1941 following the attack on Pearl Harbor, interrupted the growth and development of the civilian air transportation system. Airline capacity was dedicated to providing military and government transport in support of the war. Intensive wartime production allowed many firms to return to profitability following many years of losses in the Great Depression. Airlines benefited throughout the war years.

Advances in aircraft technology required to produce large bombers were quickly incorporated in the post-war period to field a new generation of airliners. The Douglas DC-6, originally a military transport (the C-118 Liftmaster), was reworked after the war for long-range commercial routes. The DC-6 entered commercial service with United Airlines in 1947. By 1949, it was in service with United, American, and Delta, among others. The DC-6 cruised at 300 miles per hour and carried about 60 passengers depending on the model and airline configuration. By the mid–1950s, piston-engine technology reached its zenith with turbocharged, compound radial engines capable of 3,500 hp. The Douglas DC-7C, tagged the Seven Seas, and Lockheed Super Constellation represent the pinnacle of piston aircraft technology. Both were capable of transcontinental and trans-Atlantic non-stop service. The Super Constellation with its distinctive three-tail configuration is pictured in Figure 1.9.

Figure 1.9 Lockheed Super Constellation.
Source: From Wikimedia Commons by RuthAS (2020) CC BY 3.0.

As an example of the post-war increase in aircraft productivity, the productivity index for the Super Constellation is 85 passengers X 335 mph X 2,000 hours/year = 57 million ASM/year per aircraft. During the 20 years from the introduction of the DC-3 to the largest piston-powered airliners, aircraft productivity increased seven-fold.

1.10 Post-War Airline Growth

In the later years of WWII, the airlines and the CAB turned their attention to expanding the domestic route system. With the increased range and capability of post-war airliners, the CAB awarded new routes between major cities to the 15 established trunk airlines. But competition was still restrained with even the highest demand markets limited to two carriers.

Under public and political pressure, the CAB approved another type of airline in addition to the trunk carrier – the local service airline (LSA). LSAs were awarded routes connecting smaller cities previously without air service to larger cities served by trunk carriers thereby allowing for air travel between many more cities across the US. Such travel, however, was often inconvenient because the schedules of the LSAs and trunk airlines were poorly coordinated. Passengers would often have to change airlines two or three times on an itinerary connecting two small, geographically dispersed cities. This inconvenience was an important factor in the post-deregulation development of hub-and-spoke route systems. Large numbers of war-surplus DC-3s became the backbone of the LSA fleets. LSAs were mostly unprofitable leading to various CAB subsidies (Cook, 1996).

The third type of carrier only flew internationally. Before World War II, Pan American World Airways was the sole US airline allowed to operate international routes. In return for this government monopoly, Pan Am was not permitted any domestic routes. After industry deregulation, this lack of domestic feed traffic proved an insurmountable problem for Pan Am. In 1946, Trans World Airlines broke the monopoly beginning trans-Atlantic service making it the second US-designated flag carrier. But, unlike Pan Am, TWA retained its domestic route structure.

The fourth type of airline, the supplemental carrier, operated charter services. The CAB intended for supplemental carriers to offer charters during the peak seasons to "supplement" existing scheduled service. Rather, with high-density seating and low costs, charter services gradually began to compete with trunk carriers introducing some price competition into domestic airline routes.

A new element of competition emerged in 1948 when the trunk carrier Capital Airlines inaugurated high-density coach seating on its DC-4 aircraft operating between Chicago and New York. The CAB approved coach fares of approximately two-thirds of previously existing first-class fares – and traffic soared. This event is a significant first step to greater competition and, eventually, deregulation. Capital Airlines later merged into United Airlines.

1.11 The Jet Age

Early jet engine designs date to the late 1930s; the Germans were the first to fly a jet-powered fighter aircraft towards the end of WWII. During the mid-1940s, Boeing engineers were designing a swept-back wing jet bomber they called the B-47, but the British were first to incorporate this new technology for commercial aircraft in the de Havilland Comet. The Comet entered service in 1952, first with BOAC, and then with other international airlines. Tragically, the aircraft experienced airframe metal fatigue, a phenomenon not then fully understood or appreciated, resulting in two in-flight breakups (Withey, 2001). Although de Havilland deserves considerable credit for starting the jet age, sales of the redesigned Comet never fully recovered.

After producing two military jet bombers, the B-47 and B-52, Boeing used similar design concepts in their first jet transport prototype – the Model 376–80. Affectionately called the "Dash Eighty," the aircraft first flew in 1954. Seeing the potential of the prototype, the US Air Force ordered an air refueling tanker version of the Dash Eighty – the KC-135. Perhaps the best example of military-civilian technology transfer, the KC-135, was introduced in 1956 and is still flying. While the KC-135 was an enormous success as a military plane, its civilian counterpart, the Boeing 707 (Figure 1.10, top) would be even more successful after its introduction as the first US-built passenger jet.

Jet engines proved far more dependable than piston-driven engines. Lower engine vibration delivered a smoother ride, placed less stress on the airframe, and reduced maintenance expenses. Jet engines burn kerosene, which, at that time, cost half as much as the high-octane gasoline used in large piston transports. While questions remained about the technical and economic feasibility of commercial jet aircraft, Pan American World Airways was the launch customer for the B-707 with a record-breaking order of 20 aircraft. With Pan Am's successful introduction of the B-707 in 1958, concerns about the feasibility of jet aircraft were resolved; other large airlines lined up to buy the new jet aircraft. The Jet Age had arrived.

Aircraft productivity took another big leap with the 707 and other manufacturers' jet transports that soon followed. The 707 was larger, much faster, and more dependable than the piston aircraft it replaced. The productivity index for the Boeing 707 quantifies another significant improvement in technology: 160 seats X 600 mph X 3,000 hours/year = 288 million available seat miles per year.

The Boeing 747, which first flew in 1969, rivals the B-707 for success and popularity as a commercial jet transport. It also ushered in another leap in productivity. Buoyed by its success with the 707, Pan American was the first airline to order and operate the world's first jumbo-jet when the Boeing 747 began transatlantic service in 1970. A Pan American World Airways 747 is pictured in Figure 1.10 (bottom). At two and half times the size of the B-707 and the rival Douglas DC-8, the 747 offered yet another leap in productivity. The

Figure 1.10 (Top) Boeing 707 operated by Qantas and (bottom) a Pan American World
Airways Boeing 747.

Source: Both photos from Wikimedia Commons. Top photo by Barry Lewis (Lewis,
2022) CC BY 2.0. Bottom photo by Paul Nelhams (Nelhams, 2020) CC BY 2.0.

productivity index for the Boeing 747: 360 seats X 550 mph X 4,000 hours/
year = 792 million available seat miles per year.

From the introduction of the Douglas DC-3 in 1933 to the Boeing 747 in
1970, a period of less than 40 years, aircraft productivity measured in available
seat miles per year increased by a factor of 100. This is an astonishing technological feat, a pace of innovation we now associate with computing.

Airbus, a government-sponsored consortium, including aircraft
manufacturers from France, Germany, Netherlands, and Spain, produced the
first wide-body, twin-engine commercial aircraft. The Airbus A300 entered
service in 1974 but initially enjoyed few sales (Figure 1.11, top). Airbus scored

Figure 1.11 Airbus A300 at Paris – Orly operated by launch customer Air France (top)
      and Singapore Airlines flagship A380 (bottom).

Source: Both photos are from Wikimedia Commons. Top photo by Kambui (2020) CC
BY 2.0. Bottom photo from Aeroprints.com (2020) CC BY-SA 3.0.

a breakthrough in the US when Eastern Airlines ordered the A-300-B4. Other
US airline orders soon followed firmly establishing Airbus in the world aircraft
market.

Airbus scored another first with the introduction into service of the A-380
by Singapore Airlines in 2007 (Figure 1.11, bottom). The A-380 is the world's

largest passenger airliner typically configured for 525 passengers in three cabins but with an all-economy capacity of 853.

Since the Boeing 747, the increase in technologically-driven productivity has slowed. Neither speed nor capacity increased until the launch of the Airbus 380, but improvements in fuel efficiency and reliability have continued. The supersonic British-French Concord notwithstanding, economical cruise speed is limited by shock wave-induced drag.

Rather than increases in size and speed, technological developments since the design of the 747 have focused on operating economics. Composite structures like those seen in the Boeing 787 and Airbus 350 lessen weight, and advanced engine technology provides greater power and better fuel economy. As a result, most new aircraft designs are now two-engine versus four-engine. Increased engine reliability permits two-engine aircraft, once limited in operations, to fly virtually any route in the world. Enhanced component reliability and advancements in cockpit automation improved flight operations efficiency and safety.

The productivity index is a measure of the rapid improvements in aircraft technology, but this index, as with any economic metric, is imperfect. Should the productivity index for the A-380 be computed using the typical passenger seating configuration or is the all-economy seating more representative even though no airline has chosen that configuration? Because we began with the computations with the DC-3 in single-class configuration, we'll use it again here: 850 seats x 600 mph x 4,750 hours per year = 2.5 billion available seat miles per year (ASM/year) or some 300 times that of the DC-3. Improved aircraft technology reduces aircraft operating costs per passenger, savings reflected in lower fares. The productivity index, however, does not measure improvements in quality, a common deficiency of economic indicators. The passenger riding current-generation commercial jet aircraft enjoys much greater safety and more comfort with a pressurized and air-conditioned cabin and other interior amenities unimaginable to the DC-3 passenger.

1.12 US Deregulation

With the success of the Boeing 747 in the earlier 1970s, other aircraft manufacturers rushed to build competing jumbo jets, notably the Douglas DC-10 and Lockheed L-1011. US trunk carriers dutifully ordered jumbo jets for domestic service so as not to be left behind. This resulted in an unintended consequence: too many available seats for the market demand established by CAB-set fares.

The CAB did not believe in or allow much price competition, so airlines resorted to flight frequency and inflight service to lure passengers to empty seats. Continental installed a piano bar in its DC-10 aircraft. Further, the CAB consistently refused to allow airlines operating older, slower, and less comfortable aircraft to charge lower fares; the newest aircraft became a competitive necessity (Borenstein & Rose, 2007).

Fares remained high and passenger load factors dropped from 70% in 1950 to 50% by the mid-1970s. Airlines were losing money to which the CAB responded with a route moratorium. But financial pressures increased, so in 1974 the CAB allowed a 20% fare increase which further depressed demand. The CAB also sanctioned capacity limitation agreements among major carriers to slow continuing losses (Borenstein, 1992; Button, 1991).

As the futility of the CAB's actions became apparent, many academics, regulators, politicians, and an occasional airline executive began calling for drastic changes in CAB regulation. Several factors provided impetus: (a) capacity increased on many routes as jumbo-jets entered service; (b) the Middle-Eastern oil embargo in 1973 led to skyrocketing fuel costs and contributed to price inflation; and (c) an economic downturn put a severe strain on airlines as business travel demand fell at the same time capacity and fuel prices were rising.

In 1974, the Ford Administration began to press for government regulatory reforms in response to a growing public sentiment that government regulations were overly burdensome on US industry and contributing to inflation. Shortly thereafter, Senator Edward Kennedy chaired Senate subcommittee hearings concluding that airline prices would fall if government constraints on competition were lifted. Comparisons drawn with unregulated intrastate carriers Southwest Airlines, Pacific Southwest Airlines, and Air California exposed inefficiencies of the regulated airlines that were costly to consumers. A CAB staff report reached the same conclusion in 1975. The report said the industry was "naturally competitive, not monopolistic," (Civil Aeronautics Board, 1975, Executive Summary) and that the CAB itself could no longer justify entry controls or public utility-type pricing. On its own, the Board began to loosen its grip on the industry (Bailey et al., 1985).

The debate over airline industry economic deregulation was based, in large part, on economic studies and recommendations. Economic theory holds that under conditions of perfect competition, consumer welfare is maximized without government restriction or interference. In attempting to earn a profit under intense competition, suppliers develop and market products that consumers want while competition ensures that products are produced at the lowest possible cost and sold at prices that just cover the production expense.

However, contrary to the theory of perfect competition, airline markets are natural oligopolies in which only a few airlines compete. While economists recognized this important distinction, they generally argued that airlines were sufficiently competitive to provide the benefits of competitive competition without government regulation. Proponents offered several arguments in favor of deregulation.

- Deregulated airlines would be more efficient. Proponents argued that regulation resulted in inefficient and costly operations. With fares set by the CAB based on airline costs plus a profit margin, airlines faced few incentives to pursue lower costs. Unable to differentiate their products based on prices, airlines promoted costly and inefficient passenger amenities. Labor successfully pursued a strategy of ever-increasing common industry-wide wages and work rules. Because all airlines incurred similar labor costs, higher labor costs were incorporated into the base fare computations.

- Subjecting airlines to competition would result in lower fares. Economists pointed to intrastate carriers Pacific Southwest Airlines (PSA), Air California (AirCal), and Southwest that were not subject to CAB economic regulation. These airlines charged only about half that of CAB-regulated airlines on routes of similar length. Deregulation would allow new, more efficient airlines to enter markets with low fares and innovative service thereby disciplining the established carriers. Incumbent carriers would either become cost-competitive or fail.
- Critics of the CAB also argued that it no longer regulated in the interest of passengers but rather in the interest of its clients, the airlines. The CAB staff were often recruited from the airlines or hoped to join the airline management following their time with the CAB. Indeed, many of the prominent post-deregulation airline executives began their careers at the CAB.
- Expansion by existing carriers and new entrants would provide more service in response to passenger demand.

Most airline executives and labor opposed deregulation. Their arguments against deregulation were also persuasive.

- The strongest carriers would engage in predatory pricing, lowering prices on routes until weaker carriers were forced out.
- Having dispatched competing airlines, strong airlines would then monopolize the most profitable routes raising fares to monopoly levels. Thus, rather than lower fares for passengers, prices would increase in the long run.
- Airlines would abandon routes with low demand that were cross-subsidized under CAB regulation. Some cities would lose air service altogether.
- Faced with intense price competition, airlines would cut maintenance and other essential safety expenditures.
- Finally, some airlines would fail with employees losing their careers and livelihoods (Cook, 1996).

1.12.1 The Airline Deregulation Act of 1978

In 1977, recognizing the growing movement towards deregulation, the CAB consented to American Airlines' request for SuperSaver discounts of 45% below existing coach fares. When American's traffic swelled as much as 60% in response, the solution to overcapacity seemed at hand in an apparent vindication of the deregulation proponents. Other carriers quickly filed and received approval for similar discounts (Meyer & Oster, 1981).

Congress first freed cargo airlines from economic regulation in 1977 leading directly to the success of Federal Express (now FedEx). The same principle of free-market competition was next applied to the passenger carriers in the Airline Deregulation Act of 1978. The Act phased out domestic route and rate restrictions over four years. The CAB moved much more quickly. It began granting new route authority so readily that, within a year of the law's passage, carriers were able to launch almost any domestic service they wanted.

The CAB ceased to exist on January 1, 1985, with remaining Board functions shifted to other government agencies, primarily the Department of Transportation.

1.12.2 Post-Deregulation Evolution

The post–deregulation evolution of the airline industry proved both proponents and skeptics partially correct. The general and uncritically examined expectation was that the newly deregulated industry would evolve to resemble the unregulated intrastate airlines to which it had been favorably compared. These carriers operated simple point-to-point route systems. Fares were low and uncomplicated, often with a single higher fare for times of peak travel demand. The result for the greater domestic system, however, was much more varied and complex.

CAB Chairman Alfred Kahn, often credited as the architect of airline deregulation, quickly moved the CAB into the sunset. All the dormant airline routes were put up for auction to certified airlines. Braniff International Airways, betting that deregulation would not last, applied for over 100 routes and expanded rapidly. The formerly intrastate airline Southwest began to expand out of Texas, and, similarly, Air Florida out of Florida. As predicted by the proponents of deregulation, a host of new carriers formed and entered markets. Twenty-four new jet airlines started service between 1978 and 1985 with People Express perhaps best known.

Contrary to expectation, however, the former trunk airlines moved to develop hub-and-spoke route systems, often through acquisition or merger with former local service airlines. As one of many examples, here is a partial history of vertical integration that eventually led to the current Delta Air Lines:

- In 1968, Hughes Air West was formed by merging Bonanza Air Lines, Pacific Airlines, and West Coast Airlines.
- North Central Airlines merged with Southern Airways to become Republic Airlines in 1979; then Republic acquired Hughes Air West the following year.
- Next, Republic merged with Northwest Airlines in 1986.
- Finally in 2010, Northwest merged with Delta Air Lines following the bankruptcies of both carriers.

Today's American and United Airlines are the result of a similarly long history of mergers and acquisitions. Even Southwest Airlines has acquired other airlines, most recently AirTran Airways in 2010. Many of the small, independent commuter airlines that provided service from the smallest commercial airports under contract to both trunk and local service carriers were acquired by the major airlines to support the growing hub-and-spoke systems. American Airlines' subsidiary American Eagle Airlines, now rebranded as Envoy, itself a union of several commuters, is but one example.

Unfortunately, tough times quickly roiled the industry. By the mid-1980s, intense competition, an oil embargo, and the air traffic controllers' strike resulted in the failure of all but two post-deregulation start-ups: America West and Midwest Express. Nor was this first deregulation carnage restricted to new and small airlines. Several of the large, formerly regulated, carriers also failed. Notable are Braniff, Eastern, and the fabled Pan American International Airways (Borenstein, 1992; Cook, 1996).

1.12.3 US Deregulation Results

Perhaps not surprisingly, some results of deregulation were anticipated while others were unforeseen. As predicted, deregulation did lead to lower fares, primarily for leisure travelers. Business fares remained high, but frequency increased, although generally with a stop required at one of the newly emerging hub airports. Economists estimated that in the years shortly after deregulation consumers saved some $18.4 billion per year (in 1995 dollars) as a result of the Deregulation Act; 55% of the savings resulted from lower fares; 45% from increased service frequency, which helps reduce the number of nights travelers must spend on the road (Morrison & Winston, 1995). By 1986, 90% of passengers traveled on some type of a discounted fare (Bailey, 2002).

Also as predicted, the entry of low-cost new airlines offering much lower fares than incumbents increased competition. These carriers should have prospered, but, except for Southwest Airlines, which has been consistently profitable and grown to the largest US domestic airline, early new entrant carriers mostly struggled and failed. Much later, airlines established in the late 1990s were more successful. JetBlue and Spirit are prime examples.

Proponents anticipated point-to-point route systems with simple fares, similar to those developed by the intrastate airlines Southwest, PSA, and AirCal. Instead, hub-and-spoke systems developed rapidly. The hub-and-spoke system connects many cities using fewer aircraft than linear or point-to-point route systems and provides an airline a better opportunity to keep its passengers all the way to their final destination rather than handing them off to other carriers. Travelers enjoy the advantage of staying with a single airline. Hub-and-spoke systems can generate high load factors, employ larger aircraft, provide more flight frequency, and potentially lower fares. Route structures are addressed in Chapter 3.

Deregulated fares were expected to be simple, similar to the intrastate airlines on which much of the argument in favor of deregulation had been based. Instead, the largest airlines quickly developed complex fare structures offering deeply discounted fares to leisure passengers with higher fares imposed on business travelers who were unable to book far in advance. Initially, this fare structure was devised to compete with the low fares offered by the new entrant carriers, but then took on increasing sophistication as the revenue potential of price discrimination was fully realized (Kahn, 1988). Chapter 9 expands on pricing and revenue management.

Although some of the oldest airlines failed, those incumbent carriers that survived enjoyed advantages over new entrants that were not fully appreciated. The rapidly developing hub-and-spoke route systems allowed large airlines to control passenger feed to their own hubs. Incumbents also controlled access to the distribution system through ownership of computer reservation systems. These systems were highly biased in favor of the owner placing new entrants at a significant competitive disadvantage. Similarly, the development of loyalty programs for passengers and travel agents benefitted the largest airlines. In total, these advantages rendered the low-cost structure of the new entrant carriers less advantageous than envisioned (Levine, 1987).

Failure of both old-line and new entrant carriers forced many airline employees into unemployment and ruined careers. Employees at surviving carriers faced demands for wage concessions and increased productivity. Labor-management turmoil spread across the industry (Cook, 1996).

The failure of the new entrants would seem to have decreased competition and, indeed, industry concentration increased with the failures. However, because the surviving carriers rapidly expanded their route systems, these airlines competed on many more routes in sharp contrast to the regulated era when the CAB allowed little route competition.

One of the arguments against deregulation was that routes with low demand and cross-subsidized under CAB regulation would be abandoned and that some cities would lose air service altogether. To ease that concern and gain congressional support for the Act, a section was added establishing the Essential Air Service (EAS) program. The EAS program ensures smaller communities retain airline service even if a federal subsidy is necessary. Although there are strict criteria for inclusion in the program, in 2018, 174 communities in the US and Puerto Rico received annual EAS subsidies of more than $315M (Tang, 2018).

1.12.4 CAB in Retrospect

In the debate that preceded the passage of the Airline Deregulation Act of 1978, many argued the CAB had outlived its purpose and effectiveness, but it had also accomplished the vision for which it was created: to develop a nationwide, effective air transportation system. During the CAB era, the domestic air transportation network grew to include all large US cities. With the emergence of the local service airlines after World War II, many mid-sized cities also received air service. Some itineraries, however, required changing carriers along the route, known as *interlining*, with connections sometimes poorly timed. Data from Airlines for America show that shortly after World War II in 1948, about 15 million passengers traveled almost 8 billion revenue passenger miles (RPM) on US airlines system-wide. By deregulation in 1978, the passenger count grew to almost 275 million totaling over 226 billion RPMs – increases of 19-fold and 29-fold, respectively (Airlines for America, n.d.). Despite the CAB's strict control of airfares, the real price of air travel also decreased substantially over the same period. The average price to travel one mile, known as yield, decreased from approximately 15¢ in 1948 to about 8¢ in 1978 as measured in inflation-adjusted 1978 dollars (Air Transport Association as cited in Sherry, 2010). Over the 30 years, the real cost of airline travel had fallen by half. Because the CAB allowed for little price competition, airlines were not driven to cost efficiency. The decrease in the price of air travel resulted from the dramatic increase in aircraft technology (productivity index discussed earlier).

The lower real cost of air travel is one factor that drove the increase in passengers and miles traveled by air. Lower fares allowed more people to travel for both business and pleasure, thereby increasing business productivity and consumer pleasure, both of which increased the standard of living as measured by the amount of goods and services consumed by the average American.

Its overall effectiveness in promoting air transport notwithstanding, it's hard to appreciate the minutia into which the CAB once delved. Here is a particularly enlightening passage from Alan Greenspan, the former chairman of the US Federal Reserve Bank, commenting on airline regulation:

> Deregulation was the Ford administration's great unsung achievement. It's difficult to imagine how straitjacketed American business was then. Airlines, trucking, railroads, buses, pipelines, telephones, television, stockbrokers, financial markets, savings banks, and utilities – all operated under heavy regulations. Operations were monitored down to the tiniest detail. My favorite description of this was by Alfred Kahn, a wisecracking economist from Cornell University whom Jimmy Carter made head of the Civil Aeronautics Board and who became known as the Father of Airline Deregulation. Speaking in 1978 on the need for change, Fred couldn't resist riffing on the thousands of picayune decisions he and the board were called upon to make: "May an air taxi acquire a fifty-seat plane? May a supplemental carrier carry horses from Florida to somewhere in the Northeast. Should we let a scheduled carrier pick up stranded charter customers and carry them on seats that would otherwise be empty, at charter rates? ... May a carrier introduce a special fare for skiers but refund the cost of their ticket if there is no snow? May the employees of two financially affiliated airlines wear similar-looking uniforms?" Then he looked at the congressmen and said, "Is it any wonder that I ask myself every day: Is this action necessary? Is this what my mother raised me to do?
>
> (Greenspan, 2007, pp. 71–72)

Deregulation of the airline industry is important in its own right, but it was also the first in a series of industry deregulations. Trucking was freed from economic regulation by the Motor Carrier Act of 1980 followed by similar legislation for energy, communications, and finance. Faith in the benefits of free markets spread to most of the developed world as nationalized firms were privatized and industries deregulated in many countries, perhaps most famously by Margaret Thatcher in the United Kingdom.

1.13 Deregulation in Europe and China

Spurred on by the positive results of deregulation in the United States, many other countries followed suit. The early nationalization and/or subsidization of airlines often placed tight restrictions on the operations and ownership of airlines. Although these restrictions were often needed to protect the interests of the public and promote the growth of the airlines, they eventually outgrew themselves and instead became a hindrance to the overall economic growth of countries' aviation industries.

1.13.1 Europe

Europe followed the US in airline deregulation. After WWII, airline markets in Europe were regulated by reciprocal agreements between individual countries that designated the airlines allowed to offer service, the routes operated and fares charged as well as the flight frequency and capacity. Usually, only one airline from each country was designated, most often the state-owned national carrier. Rather than competition, these *bilateral agreements* were intended to split the traffic and revenues between the signature countries.

Moves to integrate the many European states into a single market began shortly after World War II and progressed gradually beginning with the European Coal and Steel Community in 1950. Over time, more goods and services were included and trade restrictions among the member countries were reduced. The European airline industry was an important piece of the integration project.

Unlike in the US, where deregulation was achieved with one congressional law in 1978 and implemented swiftly thereafter, European deregulation occurred in three phases or packages in a decade-long process, the first of which was completed in 1987. The third package, implemented in 1993, fully opened airline markets to competition among the member states (European Parliament, 2022). Any member state airline could operate any route to any airport where it could obtain facilities and charge whatever it chose. As important, airlines were free to merge or purchase other carriers within the European Union (Doganis, 2006).

The initial response to European deregulation was somewhat timid; most dense markets continued as high-fare duopolies. Established carriers did expand creating more competition on some routes, but fares remained high. However, Irish-based Ryanair, following the Southwest low-fare/low-cost business model, offered low fares that stimulated new traffic. Noting Ryanair's success, new low-cost carriers (LCC) appeared as both independent airlines and subsidiaries of incumbents. Debonair, EasyJet, and Virgin Express began as independent airlines; whereas, British Airways and KLM set up LCC subsidiaries Go and Buzz, respectively.

The greatest impact of LCCs in Europe occurred only after Ryanair and EasyJet emerged as the clear leaders. Each began rapid expansion in the early 21st century. By 2005, LCCs had captured 25% of the European market.

As in the US, low fares greatly expanded traffic that doubled and even quadrupled on many routes. Manchester-Dublin traffic, as an example, grew from 230,000 passengers annually to over 600,000 within three years of Ryanair's entry. As in the US, however, most new entrant LCCs did not survive. Many merged or were acquired – both Ryanair and EasyJet grew through acquisition – others failed and went out of business (Doganis, 2006; Hanlon, 2007).

Before deregulation, European airlines operated networks concentrated around the national airport, but flights were not scheduled to provide timely connections. Transfer opportunities were mostly random. Following European deregulation, large airlines developed hub-and-spokes systems similar to the earlier US experience (Burghouwt & de Wit, 2005).

1.13.2 China

China presents another informative example. During their first 30 years of operation from 1950 until 1980, airlines in the People's Republic of China were owned and operated by the Civil Aviation Administration of China (CAAC), a division of the Air Force (Wang et al., 2016). The tightly-controlled passenger service, restricted to high-level government officials and senior managers of large state-owned firms, was limited, inefficient, and generally of poor quality.

After the death of Mao Zedong in 1976, China gradually adopted mixed-market policies and began to embrace competition as the means of economic development and growth. As in Europe, China pursued a more gradual and cautious approach to airline liberalization.

The CAAC separated from the military in 1980, becoming a civilian authority directly under the State Council of China. Between 1980 and 1986, the CAAC acted not only as an industry regulator but also as the owner of the country's sole airline, engaging in its day-to-day operations. The CAAC controlled all aspects of the industry, including market entry, route authority, frequencies, fare, aircraft purchasing, funding, and even passenger eligibility for taking flights. From 1987 to 1991, six state-owned trunk carriers were established, but the CAAC's heavy hand continued to regulate the industry.

Similar in many respects to the 40 years of US regulation under the CAB, the CAAC retained regulatory control over airline approval, fares, frequency, and route entry and exit, making the carriers compete with each other based on service and safety records. The CAAC allowed and even encouraged the entry of new regional airlines to feed traffic to the six trunk carriers. This development is like the growth of local service airlines in the US following WWII except that the regional airlines were owned by provincial governments (Zhang, 1998). After 1994, the CAAC allowed limited foreign investment and ownership in state-owned airlines. Public shares began trading on major stock exchanges beginning with China Eastern Airlines in 1997.

Aggressive liberalization of air transportation markets began in 1997, although not always following CAAC plans and desires. To encourage airline growth and expand the passenger market, the CAAC allowed airlines to provide several layers of discount tickets. A fare war promptly ensued. The CAAC attempted to regain control of fares by limiting the range of discounts, revenue sharing schemes and punishing non-compliant airlines with route expulsions. Airlines circumvented all attempts at control.

Demand for air transportation fell during the Asian Financial Crisis in 1997; CAAC advocated airline consolidation to save costs. By 2002, all state-owned airlines were merged into the "Big Three" – Air China, China Southern, and China Eastern (Wang et al., 2016). Interestingly, by 2004, the CAAC essentially acknowledged defeat and gave up on efforts to control fares even as it remained nominally in control.

Privately-owned airlines were permitted following China's entry into the World Trade Organization in 2001. Several new private airlines, mostly following the low-cost carrier business model, entered service but most struggled financially. Of the more than 40 private carriers that entered the domestic market, after some initial success, most either merged with larger carriers or went out of business. The CAAC also gradually permitted free entry and exit from most routes and markets (Zhang & Round, 2008).

As we have already seen, fare wars erupted as soon as the CAAC permitted discount fares. Since de facto deregulation in 1997, China's airlines experienced an explosive growth of more than 15% per year until slowed by the global recession of 2008 and again by the pandemic of 2020. A combination of lower fares and expanded routes provided widespread, convenient, frequent, and affordable airline service.

Another wave of consolidation began in 2002. Today's Big Three dominate regions of the country based in the largest cities of Beijing, Shanghai, and Guangzhou with each having merged or absorbed three or more formerly independent airlines. The consolidation created carriers with sufficient geographical scope so that the linear route structures developed under CAAC regulation were rationalized in hub-and-spoke networks connecting cities across the entire country.

Finally, as in Europe and the US, deregulation brought inconsistent financial results. In 2008, the Chinese government provided huge bailout packages to keep the largest carriers aloft, but recovery from the Great Recession was quick with most carriers enjoying record profits by 2010 (Cantle, 2009, 2011; Zhang & Round, 2008). China's zero-Covid policy hobbled its airlines leading to further government support in 2022.

1.13.3 *Comparison of Deregulation Outcomes*

In the US, EU, and China, deregulation not only brought lower prices and increased service that benefitted passengers, but also caused turmoil, including new entrants, failure, and consolidation. Interestingly, the results are similar even though the economic systems differ significantly. More generally, liberalization and the adoption of free markets foster economic growth, more jobs, and higher incomes, but are accompanied by consolidation and corporate failures. Some turmoil and chaos, famously termed "creative destruction" by the economist Joseph Schumpeter, is the price of free competition.

As countries liberalized their airline industries and previously nationalized carriers privatized, bankruptcy became more likely. Under strict economic regulation, government subsidies, government ownership, and regulated fares, airlines are less prone to failure. In countries that have deregulated and privatized formerly state-owned airlines, governments are reluctant to provide ongoing financial support. Some airlines fail in newly competitive markets.

Bankruptcy laws vary throughout the world, but North America has the highest number of bankruptcies. In the United States, the years following deregulation saw numerous bankruptcies, mostly start-up carriers that either lacked capital or viable business plans. Additionally, the recession following the September 2001 attacks led to several larger carriers' bankruptcies. Since deregulation in the US in 1978, there have been over 200 airline bankruptcies in the US (Airlines for America, 2019). About 88% have entered Chapter 11 of the bankruptcy code that allows for reorganization under court supervision and potential exit from bankruptcy. The remaining 12% were liquidated.

1.14 Airline Industry Today

Although the global airline industry continues to evolve, some relatively stable characteristics are addressed in the following chapters:

• Demand for air travel is "derived" from passengers' wants and needs to travel to different locations for business or pleasure. In other words, passengers do not fly to enjoy the flight, but rather, to get to a destination for a particular purpose. Especially for business, the need to travel follows the business cycle, so air travel demand is closely tied to the general economy.

• The domestic coach seat is a commodity. A commodity product is one in which the consumer considers all producers' products equal – many agricultural products are examples. With low brand loyalty and little product differentiation, leisure passengers often choose among airlines based on price. Business passengers display somewhat more loyalty.

• Although freed from economic regulation, the airline industry in the US remains highly regulated by the Department of Transportation and the Federal Aviation Administration. Other countries have a similar regulatory structure. Gradually, the world's safety regulations and requirements are being harmonized. The United States and the European Union have agreed on many common standards.

• The airlines do not own or control many factors of production. Airport facilities are usually not owned by airlines, national airspace (NAS) is controlled by the states and the system has reached capacity around many hub airports. In Europe, major airports also operate at capacity with flights limited by rights, known as slots, granted to individual airlines. Infrastructure constraints are increasing in most industrialized countries.

- The industry is heavily unionized; because the product cannot be stored, pilots or other employee groups can shut down an airline and all revenue ceases. Harmonious labor relations are a key element in airline success.
- Fuel and many other costs are not controllable. Under pressure from growing low-cost carriers, major airlines have enjoyed some success in reducing labor costs, aircraft lease costs, and some other expenses once considered nearly uncontrollable.
- The airline product, a seat from one place to another, is consumed when produced – it is fully perishable. This contrasts with manufactured products, such as air conditioners and automobiles, that can be held in inventory until purchased, or stockpiled to fend off a possible labor strike.
- Once the flight schedule is finalized, most costs of operation are fixed. On the other hand, as long as an airplane is not full, the cost of carrying an additional passenger, the marginal cost, is very low. Low marginal cost leads to intense price competition, especially for leisure passengers.
- There is a relatively low market entry cost considering the size of achievable revenue. Equipment and airport space can be leased and free entry is permitted in the US, EU, and increasingly in many other world markets. New carriers still have to prove financial fitness to the Department of Transportation and the ability to operate safely under regulations of the Federal Aviation Administration in the US with similar requirements in other countries.
- The industry has historically suffered from chronic overcapacity for several reasons. New aircraft lead time is long, three years or more. As a result, aircraft are often ordered during the economic boom but arrive during the subsequent down cycle creating a drain on financial resources at an inopportune time. High flight frequency is valued by passengers and used as a competitive weapon. Finally, the capacity to meet peak demand leaves idle capacity at off-peak hours.

As Robert Crandall, the former CEO of American Airlines, once observed, "in many respects, conventional solutions to problems of inadequate revenues, excessive costs, and unsatisfactory profitability are not terribly useful to airline managers." This is a challenging and, in many ways, unique industry.

1.15 Summary

The earliest airlines emerged in Europe shortly after the end of World War I, often to provide service to colonial empires. Over time, most countries developed state-owned, flag carriers. Airline development in the US followed a somewhat different path. In the early 20th century, the US public and especially politicians harbored a profound distrust of free competition that led to 40 years of economic regulation of US airlines. The Civil Aeronautics Board controlled which airlines were allowed to operate, the routes they flew, and the fares charged. After the industry was deregulated in 1978, an often-traumatic transformation swept the industry with consequences that are still felt today. Many new airlines started and failed accompanied by the loss of some of the great old carriers. Other major airlines grew developing the hub-and-spoke route systems that predominate today. Price structures became increasingly complex even as passengers flocked to new low-cost carriers. Europe, China, and other countries followed the US in deregulating their airline industries. Though the paths to deregulation were quite different, the results were surprisingly similar.

1.16 Case Study: India

Commercial aviation in India traces its roots to Tata Air Services' first flight in 1932. With its independence in 1947, India adopted a socialist-oriented approach toward economic development that included taking a 49% stake in the renamed Air India. As the national flag carrier, Air India expanded domestically and internationally. By 1950, access to war-surplus aircraft enabled the rise of several regional airlines and intense competition. Airlines resorted to predatory pricing and soon losses grew. With several carriers going bankrupt, the government responded by taking control and ownership of all major airlines. Eight airlines merged into two, Air India and Indian Airlines. Indian Airlines flew domestic routes and Air India served international routes. Initially, the two airlines fared well economically, although Air India, operating in the international regime, was less able to manage fares and needed government subsidies to compete with foreign carriers. The 1970s were marred with war and domestic disputes. Indian Airlines struggled with an aging fleet and a government mandate to operate unprofitable domestic routes. In 1986, India's tourist industry, which was heavily dependent on air transportation, pressed for needed reforms. A proposal to replace and expand both airlines' fleets was upended by the Indian government's inability to provide the needed capital. With fleet replacement/augmentation not possible, the government's solution was to again allow privately-owned airlines.

The government first permitted private airlines to operate as air taxis to boost domestic service. Influenced the success of airline deregulation in other countries like the US, the UK, and the EU, in 1994, India repealed all laws regulating the formation of airlines. Under deregulation, early new entrant airlines, including Jet Airways, Air Sahara, and ModiLuft chipped away at the dominance of the government-run carriers. Competition intensified as IndiGo, SpiceJet, GoAir, AirAsia India, and several other low-cost carriers began flying domestic routes. Low fares spurred rapid traffic growth; the Indian domestic market was the world's third-largest by the early 2000s. Air India, Indian Airlines, and the new full-service carrier Jet Airways, however, suffered from the competition. The government merged Indian Airlines into Air India in 2007; Jet Airways failed in 2019 although it is planning a comeback. By 2021, Air India had a measly 9% of the domestic market. With its 55% market share, IndiGo is India's largest airline. Several other LCCs split the remainder of the market.

No longer willing to shoulder the burden of Air India's losses, the government tried to privatize Air India in 2018, but with the airline's overwhelming debt, it found no buyers. After the government bought back some of the debt and liabilities, it again offered Air India for sale in 2020. Coming full circle, the Tata Group bought the whole of Air India for $2.4B in January 2022.

Concepts: Government economic regulation, state-owned airlines, deregulation, privatization, low-cost carriers. Major sources: Devi & Rajoriya, 2020; Hooper, 1997; Mazumdar, 2009; Pande, 2022; Sharma, 2020; statista.

References

Aeroprints.com. (2020). Airbus A380 Singapore Airlines. *Wikimedia Commons, the free media repository*. Retrieved 22:56, July 31, 2022 from https://commons.wikimedia. org/w/index.php?title=File:9V-SKN_Airbus_A380_Singapore_Airlines_(1389 1475825).jpg

Air France/KLM. (n.d.). History. www.airfranceklm.com/en/group/history.

Airlines for America (2019). Data & statistics: U.S. airline bankruptcies. www.airlines. org/dataset/u-s-bankruptcies-and-services-cessations/

Airlines for America. (n.d.). Annual results: U.S. airlines/Systemwide operations.

BAE Systems. (n.d.). De Havilland DH-86 Express. www.baesystems.com/en/heritage/ dh86-express

Bailey, E. (2002). Aviation policy: Past and present. *Southern Economic Journal, 69*(1), 12–20.

Bailey, E. E., Graham, D. R., & Kaplan, D. P. (1985). *Deregulating the airlines*. Cambridge, MA: MIT Press

Boeing (n.d.-a). Historical snapshot: Model 247/C-73 transport. www.boeing.com/hist ory/products/model-247-c-73.page

Boeing (n.d.-b). Historical snapshot: DC-3 commercial transport. www.boeing.com/ history/products/dc-3.page

Boeing. (2022). Boeing history. www.boeing.com/history/

Borenstein, S. (1992). The evolution of the US airline industry competition. *Journal of Economic Perspectives, 6*(2), Spring, 45–73.

Borenstein, S., & Rose, N. (2007). How airline markets work...or do they? Regulatory reform in the airline industry. NBER Working Paper #13452.

British Airways. (n.d.) History and heritage. www.britishairways.com/en-us/informat ion/about-ba/history-and-heritage

Burghouwt, G., & de Wit, J. (2005). Temporal configurations of European airline networks. *Journal of Air Transport Management*, 11(3), 185–198.

Button, K. (1991). *Airline deregulation: International experiences*. New York, NY: New York University Press.

Cantle, K. (2009, June). An uneven game. *Air Transport World*, 54–57.

Cantle, K. (2011, February). A golden opportunity. *Air Transport World*, 40–42.

Civil Aeronautics Act. (1938). Declaration of policy. http://libraryonline.erau.edu/onl ine-full-text/books-online/CivilAeronauticsAct.pdf

Civil Aeronautics Board. (1975). Regulatory reform: Report of the C.A.B. special staff.

Cook, G. (1996). A review of history, structure, and competition in the US airline industry. *The Journal of Aviation/Aerospace Education and Research, 7*(1), 33–44.

Correll, J.T. (2008, March). The air mail fiasco. *Air Force Magazine, 91*(3), 60–65.

De Gouyon Matignon, L. (2019). The Paris Convention of 1919. *Space Legal Issues – 2021.* www.spacelegalissues.com/space-law-the-paris-convention-of-1919/

Devi, S., & Rajoriya, C. M. (2020). Recent change in air deregulation in India and its impact on airline competition. *International Journal of Advance Research and Innovative Ideas in Education,* 6(1), 1371-1377.

Doganis, R. (2006). *The airline business* (2nd ed.). London, England: Routledge.

Eames, J. (2018). Courage in the skies: The untold story of Qantas, its brave men and women and their extraordinary role in World War II. Allen & Unwin. https://ebook central.proquest.com/lib/erau/detail.action?docID=5144771

EAS Barcelona. (n.d.) The five oldest airlines in the world. https://easbcn.com/en/the-five-oldest-airlines-in-the-world/

Eney, J.A. (n.d.). US air mail service – 90th anniversary. Originally published in *Skyways Journal Magazine.* www.antiqueairfield.com/features/us_airmail.html

European Parliament. (2022). Fact sheets on the European Parliament: Air transport: Market rules. www.europarl.europa.eu/factsheets/en/sheet/131/air-transport-market-rules

Florida Memory: State Library and Archives of Florida. (1914). Men sitting in the flying boat "Benoist" – St. Petersburg, Florida. www.flickr.com/photos/floridamemory/4010850579/in/photolist-77qE9c-cRGymN-KDYAJ-bJCNy8-e6CKNY-j3fJ91-smU5Bn-2kqyyra

Gale. (2019). Qantas Airways Ltd. *Notable Corporate Chronologies Online.* Gale. *Business Insights: Essentials.* http://bi.gale.com.ezproxy.libproxy.db.erau.edu/essentials/article/GALE|I2501151401?u=embry

Greenspan, A. (2007). The age of turbulence: Adventures in a new world. New York, NY: The Penguin Press.

Gunn, J. (1985). The defeat of distance: Qantas 1919 to 1939. University of Queensland Press.

Handley Page HP42 (cropped). (2022, June 29). *Wikimedia Commons, the free media repository.* Retrieved 19:35, July 31, 2022 from https://commons.wikimedia.org/w/index.php?title=File:Handley_Page_HP42_(cropped).jpg&oldid=669821251

Hanlon, P. (2007). *Global airlines: Competition in a transnational industry* (3rd ed.) Oxford, UK: Elsevier.

Hayward, J. (2019). A look At Qantas' history – How the airline became the Australian flag carrier. https://simpleflying.com/qantas-history-australian-flag-carrier/

Hooper, P. (1997). Liberalization of the airline industry in India. *Journal of Air Transport Management,* 3(3), 115–123.

Imperial Airways. (2020). Imperial routes April 1935. (2020, October 6). *Wikimedia Commons, the free media repository.* Retrieved 19:28, July 31, 2022 from https://commons.wikimedia.org/w/index.php?title=File:Imperial_routes_April_1935.jpg&oldid=482967648

Kahn, A. (1988). Surprises of airline deregulation. *American Economic Review,* 78(2), 316–322.

Kambui. (2020). Air France Airbus A300B4–203. *Wikimedia Commons, the free media repository.* Retrieved 22:53, July 31, 2022 from https://commons.wikimedia.org/w/index.php?title=File:Air_France_Airbus_A300B4-203_F-BVGT_(25815524052).jpg&oldid=485419217

Larkins, B. (2022). Douglas DC-6 AA N90739. *Wikimedia Commons, the free media repository*. Retrieved 22:28, July 31, 2022 from https://commons.wikimedia.org/w/index.php?title=File:Douglas_DC-6_AA_N90739_(8574037818).jpg&oldid=660856346

Levine, M. E. (1987). Airline competition in deregulated markets: Theory, firm strategy, and public policy. *Yale Journal on Regulation*, 4, 393–503.

Lewis, B. (2022). Boeing 707 – Qantas. *Wikimedia Commons, the free media repository*. Retrieved 22:39, July 31, 2022 from https://commons.wikimedia.org/w/index.php?title=File:Boeing_707_-_Qantas_(17342297349).jpg&oldid=667226198

Libby Photography Studio. (2022). Boeing Air Transport B-40. *Wikimedia Commons, the free media repository*. Retrieved 21:50, July 31, 2022 from https://commons.wikimedia.org/w/index.php?title=File:Boeing_Air_Transport_B-40.jpg&oldid=671489681

Lombardi, M. (2008). Meeting the need for speed. *Historical Perspective: Boeing Frontiers*. www.boeing.com/news/frontiers/archive/2008/april/i_history.pdf

Mazumdar, A. (2009, Apr–Jun). Deregulation of the airline industry in India: Issues, causes and rationale. *Indian Journal of Political Science,* 70(2), 451–469.

Meyer, J. R., & Oster, C.V. (1981). Airline deregulation: The early experience. Boston, MA: Auburn House Publishing Company.

Morrison, S., & Winston, C. (1995). *The of the airline industry*. Washington, DC: The Brookings Institution.

Nelhams, P. (2020). N729PA. *Wikimedia Commons, the free media repository*. Retrieved 22:42, July 31, 2022 from https://commons.wikimedia.org/w/index.php?title=File:N729PA_(8110221016).jpg&oldid=494125087

Pande, P. (2022, June 11). A brief history of commercial aviation in India. *Simple Flying.*

Public.Resource.Org. (2004). St. Petersburg-Tampa Airboat Line (Poster, Timetable). www.flickr.com/photos/publicresourceorg/494018870

Qantas. (n.d.). Our history. www.qantas.com/us/en/about-us/our-company/our-history.html

Redding, R. & Yenne, B. (1983). Boeing: Planemaker to the world. Crescent Books.

Rlandmann. (2020). Contract Air Mail routes. *Wikimedia Commons, the free media repository*. Retrieved 21:02, July 31, 2022 from https://commons.wikimedia.org/w/index.php?title=File:Contract_Air_Mail_routes.svg&oldid=505676834

Royal Air Truck Ford Tri-Motor. (2021, August 15). *Wikimedia Commons, the free media repository*. Retrieved 22:01, July 31, 2022 from https://commons.wikimedia.org/w/index.php?title=File:1927_-_Royal_Air_Truck_Ford_Tri-Motor.jpg&oldid=581454550

RuthAS. (2020). Lockheed L-749A. *Wikimedia Commons, the free media repository*. Retrieved 22:30, July 31, 2022 from https://commons.wikimedia.org/w/index.php?title=File:Lockheed_L-749A_PH-TDK_KLM_RWY_07.07.53_edited-2.jpg&oldid=486098585

San Diego Air and Space Museum Archives. (2022). LWF J-2 Twin DH rear. *Wikimedia Commons, the free media repository*. Retrieved 20:58, July 31, 2022 from https://commons.wikimedia.org/w/index.php?title=File:LWF_J-2_Twin_DH_rear.jpg&oldid=649341714

Sharma, J. (2020). How the Air Corporations Act, 1953, came to haunt Indian aviation. *Spontaneous Order.* https://spontaneousorder.in/how-the-air-corporations-act-1953-came-to-haunt-indian-aviation/

Sherry, L. (2010). Sustainability of the air transportation system: Successes, challenges & opportunities. Fairfax, VA: *George Mason University*.

Smith, R. (2021). American DC-3. *Wikimedia Commons, the free media repository*. Retrieved 22:24, July 31, 2022 from https://commons.wikimedia.org/w/index.php?title=File:American_DC-3.jpg&oldid=613738851

Smithsonian Institution. (2008a). JR-1B mail plane. www.flickr.com/photos/smithsonian/2550375657/in/album-72157605338989538/

Smithsonian Institution. (2008b). Curtis J-4HM biplane. www.flickr.com/photos/smithsonian/2550219215/in/photolist-Hyj4Um-4TmxjP-2jdDwuf-ouZzG5-REQhGM-viMLHb-drass-wMcXSQ-9S9rXe/

Tang, R. (2018). Essential Air Service. Congressional Research Service Report R44176. https://fas.org/sgp/crs/misc/R44176.pdf

Taylor, F. (1951). *High horizons: Daredevil flying postman to modern magic carpet – the United Air Lines story*. McGraw-Hill Book Company.

The Queenslander. (2020). StateLibQld 1 115436 Planes in front of the Qantas hangar at Charleville airport, 1935. *Wikimedia Commons, the free media repository*. Retrieved 19:40, July 31, 2022 from https://commons.wikimedia.org/w/index.php?title=File:StateLibQld_1_115436_Planes_in_front_of_the_Qantas_hangar_at_Charleville_airport,_1935.jpg&oldid=510932643

The rise of airlines. (n.d.). Century of Flight. www.century-of-flight.freeola.com/Aviation%20history/coming%20of%20age/airlines2.htm

Van der Linden, F. R. (1991). *The Boeing 247: The first modern airliner*. University of Washington Press.

Van der Linden, F. R. (2012). Government, business, and technology: US airliner development, 1927–2012. *World Neurosurg* 78(3/4) 2006–213.

Wang, J., Bonilla, D. & Banister, D. (2106). Air deregulation in China and its impact on airline competition 1994–2012. *Journal of Transport Geography, 50*, 12–23.

Wensveen, J. G. (2007). *Air transportation: A management perspective*. Burlington, VT: Ashgate Publishing.

Withey, P.A. (2001) Fatigue failure on the de Havilland Comet I. Science Direct www.sciencedirect.com/topics/engineering/de-havilland-comets

Zhang, A. & Round, D. K. (2008). China's airline deregulation since 1997 and the driving forces behind the 2002 airline consolidations. *Journal of Air Transport Management*, 14, 130–142.

Zhang, A. (1998). Industrial reform and air transport development in China. *Journal of Air Transport Management*, 4, 155–164.

Review Questions

1. Give an example of how trade increases people's standard of living.
2. Why do national governments often build or subsidize transportation infrastructure? Provide an example.
3. How and why did the early development of the airline industry in the US differ from that in Europe?
4. Early European airlines carried passengers. Why weren't passengers carried on the first US commercial flights?
5. What was the greater objective in federal subsidization of the airlines?
6. What was the Spoils Conference and was it an appropriate use of government power?
7. What two, sometimes conflicting, goals were established by Congress for the Civil Aeronautics Board (CAB)?

8. What regulated industry was the CAB patterned from?
9. Why did Congress not trust the free market to discipline the US airline industry?
10. What did the CAB regulate? Why?
11. Did the CAB allow airlines to compete on price?
12. Airline fares have declined in real terms since the introduction of the Douglas DC-3 throughout the regulated era. What has enabled this decline?
13. What are the three factors used to compute the Productivity Index? What is the relationship between productivity and standard of living?
14. Describe the differences among Trunk, Local Service, Flag (international), and Supplemental carriers in the post WWII period.
15. How did the introduction of wide-body aircraft (B-747, DC-10, L-1011) in the early 1970s contribute to industry losses?
16. What factors led to airline losses in the early to mid-1970s? How did these losses contribute to the move toward deregulation?
17. What were the arguments for and against deregulation? What other industries have been deregulated?
18. Compare the process for economic airline deregulation in the US, European Union, and China.

2 Supply and Demand for Air Transportation

Upon completion of this chapter, you should be able to:

- Explain the importance of air transportation in the world economy.
- Analyze the factors that drive global air transportation demand and supply.
- Explain the challenges that airlines face in matching supply with demand.
- Compare projected air transport growth across the world regions.
- Contrast the airline choice criteria of passengers traveling for business versus leisure.
- Discuss the methods for assessing the potential of a new route and market share.

2.1 Introduction

The air transport industry is a vital component of the global transportation network. Aviation provides the only means of rapid, long-distance travel making it essential for the conduct of global business and tourism. A study produced for the International Air Transport Association asserts that, "By facilitating tourism and trade, air transportation generates economic growth, provides jobs, improves living standards, alleviates poverty, and increases revenues from taxes" (Air Transport Action Group, 2012, p. 6). This is a bold statement, but these effects are most evident in developing countries. Africa, for example, has long been plagued by inadequate or non-existent surface transportation that hindered economic growth. Expansion of airline service provided a springboard for growth after decades of stagnation. By allowing executives, engineers, and other critical employees to travel between widely dispersed business locations, aviation supports business operations and efficiency, improves productivity, encourages investment, and allows companies to attract high-quality employees. Aviation is also essential to tourism, an important component of many national economies, again often in developing countries. Finally, aviation offers a vital link to communities that lack adequate road and rail infrastructure. This contribution is most apparent when aircraft deliver life-saving supplies in the immediate wake of natural disasters.

DOI: 10.4324/9781003290308-2

This chapter begins by examining the size and scope of the aviation industry and then turns to the factors driving the demand and supply of air transportation. Long-range forecasts of air travel demand are essential for commercial aircraft manufacturers who sell their products worldwide. The 20-year forecasts developed by the two largest commercial aircraft manufacturers, Airbus and Boeing, examine many of the factors driving global and regional demand. Airlines are concerned not only with macroscale forecasts for long-term strategic planning but also with shorter-term and smaller-scale forecasts down to the individual city-pair forecasts for their near-term planning and management. Macroforecasting is addressed first followed by micro and route-level forecasts.

2.2 Size, Scope, and Economic Importance

The COVID-19 pandemic devastated the airline industry. World revenue passenger kilometers, a standard measure of airline production, was down 94% from April of 2020 versus 2019. The airline industry will slowly recover; however, using recent data to assess the size of the industry provides a distorted picture. Therefore, this assessment uses 2019 data. That year total airline revenue was $841 billion earned by transporting over 4.5 billion passengers and 61.5 million metric tons of freight. The aviation industry directly employed some 11.3 million people. Including the indirect jobs, many of which are in tourism, aviation supports some 74 million jobs. This last figure is impressive; however, some caution is warranted when including indirect and induced effects. Tourism provides millions of jobs worldwide that wouldn't exist without aviation. On the other hand, tourism boosters would argue that tourism supports airline jobs. Certainly, there's merit in both assertions.

For perspective, airline revenues were similar to world restaurant ($868 billion) and pharmaceuticals ($1 trillion) sales but much less than spending on apparel ($1.9 trillion), automobiles ($3.6 trillion), and food ($5.9 trillion).

Table 2.1 lists other measures of the world air transport industry (Air Transport Action Group, 2020). Of course, the data are estimates, and figures from other studies vary, but these data are representative of the industry size and reflect the importance of the aviation industry in the world's economy.

By facilitating tourism and trade, aviation contributes to globalization. Trade allows people to purchase goods and services that would otherwise be unavailable. Consider, for example, that most fruits and vegetables would be unavailable in colder regions of the world were it not for international trade. Trade enlarges markets allowing for increased specialization. The rapid advance in personal electronics would slow if products were not sold worldwide. Further, trade across international markets has lifted millions of people from abject poverty, much more so than all the world's humanitarian programs combined. China is the best example. While the benefits are widespread, international trade and globalization are not without controversy. International trade is disruptive bringing ruin to some industries unable to compete with imports. Textile

Table 2.1 Airline Industry Size and Scope in 2019

Passengers traveling by air	4.5 billion; the largest markets are the US, EU, and China
	58% of international tourists travel by air.
Air cargo	61.5 million metric tons; $6.5 trillion in value or 35% of international trade by value. Air cargo benefitted from the pandemic with 66.3 million tons transported in 2021.
Jobs (in millions)	11.3 direct jobs in the airline industry
	18.1 in the aviation industry supply chain
	44.8 in the tourism industry
Airlines	1,478 airlines
	33,299 commercial aircraft in service
	3,780 airports with scheduled commercial flights
	$841 billion in total revenue
Contribution to global GDP	$3.5 trillion or 4.1% of world economic activity

Source: Air Transport Action Group, 2020.

manufacturing has mostly disappeared in the US, Europe, and other developed countries because it cannot compete with imports from low-wage countries, mostly in Southeast Asia. Just as airline deregulation created turmoil in the US, including the loss of jobs at several long-established carriers, expanding international trade creates both hardships and opportunities. These disruptions have led some countries to move away from international trade. In the US, despite the objections of many economists, the Donald Trump administration imposed tariffs on many goods imported from China intending to increase domestic production and reduce imports. China responded in kind, resulting in decreased trade between the countries. More recently, Russia's invasion of Ukraine in 2022, may further reduce global trade.

As Figure 2.1 left panel shows, in the 60 years between 1960 and 2020, world air traffic grew exponentially before collapsing during the pandemic in 2020. World economic recessions (shaded periods) have had a substantial, but relatively short-term, adverse effect on traffic. The depressing effect of the combination of the terrorist attacks of 2001 and the severe acute respiratory syndrome (SARS) pandemic that followed in 2002 is visible in the graph as a downturn in revenue passenger kilometers (RPK). The financial crisis of 2008, often termed the Great Recession, produced a similar, short-term decline in traffic. The fall in traffic in 2020 resulting from the COVID-19 pandemic is unprecedented. Airbus revises its 20-year forecast of traffic growth annually and publishes the projections in its Global Market Forecast (GMF). Figure 2.1 right panel compares the GMF projections for 2019 and 2021. Airbus believes the pandemic set back growth about two years. By June 2022, this forecast appears accurate as the world scheduled capacity was 15% below the 2019 level (OAG, 2022). Under the revised forecast, RPKs are projected to double by 2040.

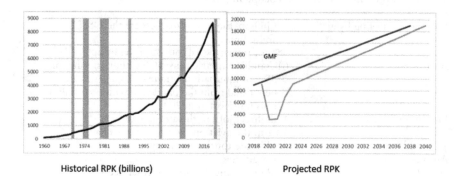

Historical RPK (billions)   **Projected RPK**

Figure 2.1 World Air Traffic from 1960 to 2040.

Notes: Traffic is measured in billions of revenue passenger kilometers (RPK). One RPK represents one paying passenger flying one kilometer. In the US, the more common statistic is the revenue passenger mile (RPM). Economic recessions are shaded.

Sources: A4A World Airlines Traffic and Capacity; Federal Reserve Economic Data; Airbus 2021.

2.3 Factors Driving Global Air Transportation Growth

Many factors that drive the growth of world air travel can be conveniently grouped into four categories: (a) globalization; (b) demographics; (c) regulation; and (d) factors of production. The first two primarily affect demand for travel whereas the latter two affect supply. The following summary provides a foundation for a more extensive exploration of methods for forecasting the future growth of demand and supply.

2.3.1 Globalization

Globalization broadly refers to the increasing integration of the world economies and societies. While the cultural, political, and environmental dimensions are important, the emphasis here is on the economic aspects of globalization. Long-distance trade has existed for thousands of years and shaped much of world history. The modern era of globalization began with the end of World War II accelerating with the breakup of the Soviet Union in 1991 as countries reduced tariffs, subsidies, and other restrictions on trade in goods and services. Impediments to foreign investment and the movement of capital were gradually relaxed. Shipping costs fell, especially with the introduction of containerized ocean and rail cargo transport.

Countries that embraced the global economy enjoyed faster growth and reduced poverty. Many East Asian countries, among the poorest in the world 50 years ago, grew rapidly by focusing on trade. Beginning with economic reforms introduced after the death of Mao Zedong in 1976, China grew over the next 40 years to become the "world's factory."

The International Monetary Fund (2010) identified four aspects of globalization: (a) trade; (b) capital movements; (c) migration; and (d) knowledge and technology. The global expansion of each requires extensive international travel by executives, managers, engineers, a host of other business specialists, and laborers. In the European Union, for example, laborers commonly travel on Europe's low-cost airlines to work in other countries but return regularly to their homeland. As the world has outsourced much of its manufacturing to China, India, and other low-wage countries in Asia, managers need to travel to manage their far-flung businesses.

Components for the Apple iPhone are manufactured in Australia, China, Europe, Japan, Latin America, Singapore, South Korea, Taiwan, and the United States and are mostly assembled in China by the Taiwanese companies Foxconn and Pegatron. Firms throughout the world are suppliers for the Boeing 787 Dreamliner with the main components built in the US, Japan, Italy, Korea, Germany, the United Kingdom, Sweden, and France all of which are shipped to Everett, Washington, and Charleston, South Carolina for assembly. The production of these and countless other products is dependent on global air transportation. To meet these needs, the largest airlines expanded their route systems and entered into alliances to serve all of the world's major business destinations.

Globalization allowed rich countries to benefit from cheaper products manufactured in low-wage countries – notably China – and for poorer countries to develop quickly. However, manufacturing regions in developed countries lost jobs and economic stability. Globalization, however, seems to have peaked around 2010 with recent trends that will reduce or at least realign world trade. Globalization slowed following the global financial crisis, The Great Recession, of 2008. Trade disputes between the US and China, the two largest economies intensified during the Trump administration. Then the COVID-19 pandemic upended the global supply chains. Finally, Russia's invasion of Ukraine in 2022 threatens to divide the world into competing trading blocs. Corporations responded by diversifying supply, often away from China and closer to home. The pandemic also profoundly altered business practices as teleconferencing replaced in-person meetings. The travel restrictions intended to mitigate the spread of COVID will slowly fade, but Zoom and competing platforms may have permanently reduced the need for business travel. The effects on demand for air travel are uncertain but likely negative. Full-service network carriers (FSNC) may find fewer opportunities for international growth and be forced to revamp route networks.

2.3.2 Demographics

Demographic factors – population growth, urbanization, diaspora, and per capita income – are perhaps the most important factors driving the demand for air travel. Population growth, of course, expands the potential market for air travel. The current world population of approximately 8 billion is forecast to grow to 9.7 billion by 2050. Of these billions, about half are middle class with the ability to purchase air travel. Airbus (2021) projects the middle class will be 63% of the world's population by 2040. The number of people who have flown commercially is less certain, but 20% is widely cited (e.g., Gossling & Humpe, 2020). The growth of cities also affects air travel demand. In 1800, 97% of the world's population was rural and mostly comprised of subsistence farmers scattered across immense reaches of land. The industrial revolution spurred migration to cities as people strove for an improved standard of living. By 2020, more than half of all people lived in urban areas with two-thirds expected to be urban dwellers by 2050 (United Nations, 2019). Cities have grown apace with migration. In 2020, 34 cities had populations of more than 10 million with the number expected to grow rapidly. The growth of the middle class, urbaniza-tion, and populations concentrated in very large cities create a vast potential for future air travel demand. Indeed, this rapid urbanization, and the growth of megacities, in particular, was the major rationale for Airbus' flagship A-380 aircraft.

As Booz and Company (2011) note, "few sectors will benefit as much from globalization and economic development as air travel." In Figure 2.2, per capita annual airline trips increase with per capita gross domestic product (GDP) but at a decreasing rate as the market reaches saturation. Although the high variance in the data suggests that other factors affect the propensity to fly, the dotted regression curve is useful for macroforecasting (see Morphet & Bottini, n.d.).

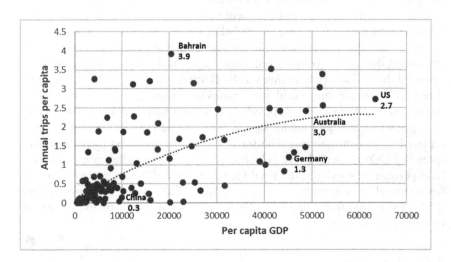

Figure 2.2 Air Travel and Per Capita Gross Domestic Product.
Note: Outliers, mostly city-states and small island countries, were deleted. Regression line R-squared = 0.41.
Sources: CIA World Factbook 2020, The World Bank.

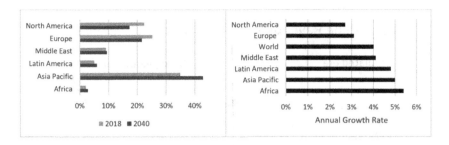

Figure 2.3 Left panel: World Market Share 2018 and 2040. Right panel: Annual
Estimated Passenger Air Traffic Growth from 2019 to 2040.

Data sources: IATA World Air Transportation Statistics, 2019 & Boeing Commercial
Aircraft market outlook, 2014, 2021.

The per capita GDP in the two most populated countries, China and India, is
low relative to the developed countries suggesting rapid growth in airline travel
with GDP growth.

The population in the developed economies of North America and Europe
is just 17% of the world's total but accounts for nearly 40% of total airline
traffic (Figure 2.3 left panel). The airline industry in these countries is mature
and forecast to grow at 2.9% annually. Mature economies rely on productivity
gains, service industries, and consumer markets for much of their gains, whereas
expanding labor forces, increased manufacturing, and entry into global capital
and trade markets characterize emerging economies. These factors will drive
higher growth rates in the Middle East, Latin America, Asia Pacific, and Africa
over the next two decades (Figure 2.3 right panel). The differences in growth
rates will result in shrinking market shares in North America and Europe and
higher shares elsewhere. By 2040, the Asia Pacific region will account for 43%
of the global air travel market (Figure 2.3 left panel).

Discretionary travel from the expanding middle class with sufficient disposable income to choose air travel will outpace travel for business. Marketers identify two primary sources of discretionary travel. Vacation and holiday travel is
one component. The other is traveling to visit or reunite with relatives, known
in industry jargon as *visiting friends and relatives* (VFR), which increases as ethnic
groups disperse across the world.

2.3.3 Regulation

Globalization and demographics drive increased demand for air travel.
Liberalization, or the reduction in government rules and restrictions on airline
competition, on the other hand, allows airlines to expand flights and product
offerings. Large airlines reorganize their networks and hubs, often in coordination with partner airlines in global airline alliances, providing more flight

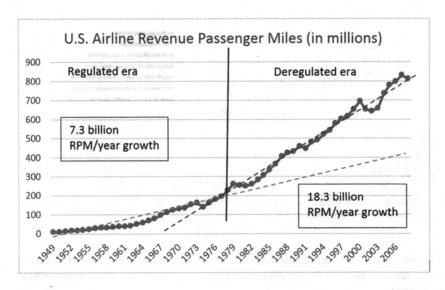

Figure 2.4 Growth in US Scheduled Revenue Passenger Miles.
Data source: Airlines for America.

frequency and improved connections. New airlines emerge, mostly following a
low-cost model, some as subsidiaries of existing network carriers and others as
independent private ventures. Expanded air service facilitates further economic
growth creating a virtuous circle that leads to more airline service.

The US experience is illustrative. Figure 2.4 shows that airline deregulation spurred air travel growth as airlines provided service according to market
demands rather than what the Civil Aeronautics Board believed was required
for public convenience and necessity. Increased competition, especially from
new entrant low-cost carriers, reduced prices inducing more people to travel by
air. In the 30 years following deregulation, revenue passenger miles grew by an
average of 18.3 billion per year, more than double the 7.3 billion RPM annual
growth in the 30 years preceding deregulation.

Led by US Open Skies initiatives begun in 1978, countries worldwide have
gradually liberalized restrictions on air travel allowing freer access to their
airways and markets. As competition increased and service expanded, fares have
fallen allowing more people to afford air travel. One example is the Open Skies
agreement between the US and EU first adopted in 2008 and extended in 2010.
The agreement allows any US or EU airline to fly between any point in the
European Union and any point in the United States. In what some critics view
as unfair, US airlines may also fly between points within the European Union.
The counterargument is that the US and the EU are comparable markets. Other
countries negotiated similar agreements. In 2019, several Latin American countries agreed to expand market access. The 10-member Association of Southeast

Asian Nations (ASEAN) agreed in 2009 to gradually liberalization of their aviation markets. In 2021, ASEAN and the EU concluded the first bloc-to-bloc air transport agreement that will allow the airlines of all member countries to operate unlimited flights between the regions (Association of Southeast Asian Nations, 2021).

Future liberalization seems less promising as the rise of populism and growing resistance to globalization may limit or even reverse aviation liberalization (Humphries, 2022). The history and growth of international aviation are the subjects of Chapter 11.

2.3.4 Factors of Production

Liberalization reduces the legal restrictions on airline services. Factors of production, on the other hand, determine the cost to provide air transportation. The two largest airline expenses are fuel and labor. Fuel prices are volatile and difficult to predict. Jet fuel prices were relatively stable in the 1990s but have fluctuated wildly since. Prices jumped more than eightfold from 48¢ in December 2001 to $4.11 (nominal USD) in July 2008, then fell sharply back to $1.16 during the recession of 2008 (Federal Reserve Bank of St. Louis, 2022). Since then, the graph of fuel prices resembles a roller coaster ride, going up during periods of economic expansion and falling during recessions. When fuel prices jump, airlines lose money on previously profitable routes. Flights are reduced and some routes are dropped. The price of labor is less volatile but a worldwide pilot shortage driven by retirements and fewer people entering pilot training will increase long-term airline labor costs.

Lack of infrastructure may also restrict airline growth in many world regions. Although Europe has two new airports in Berlin and Istanbul, most of Europe's major airports were operating at or near capacity before the pandemic. India must invest in airports and other aviation infrastructure if it is to meet its airline growth potential. China is an exception having aggressively built new airports with plans for a 60% increase between 2020 and 2035 to 400 commercial airports (Bryan, 2021). Finally, increased environmental regulation and taxation will likely increase future airline costs.

While high fuel and labor costs, infrastructure, and environmental regulation could potentially impede airline growth, improved technology will partially mitigate all three. Between 1978 and 2013, airline fuel efficiency more than doubled. As airlines continue to replace older aircraft with the latest generation of aircraft such as the Airbus 320neo and Boeing 737 Max, improved fuel efficiency will reduce fuel consumption and lower emissions per flight. Similarly, technology and consolidation of air traffic control promise to reduce flight times.

2.4 Air Cargo

Growing world trade also increases the demand for air cargo. The historical growth in air cargo as shown in Figure 2.5 is similar to the growth of

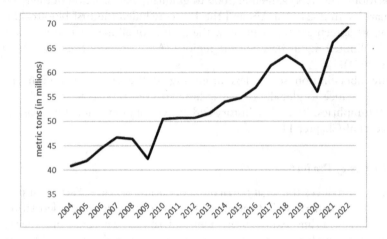

Figure 2.5 World Air Cargo Growth.
Note: 2022 metric tons are estimated.
Data source: Statista, Worldwide air freight traffic from 2004 to 2022.

passenger traffic (Figure 2.1); however, airfreight is closely tied to the world business cycle. For example, cargo volume dropped more than 10% beginning in 2008 and then recovered quickly and resumed growth in 2009 following the Great Recession that spanned December 2007 to June 2009. The response to the pandemic recession, however, is surprising. Trade disputes between the US and China caused world cargo volume to decline before the recession; then it fell precipitously as COVID-19 spread rapidly across the globe. In 2021, as consumers demanded more goods rather than services, air cargo quickly rebounded. Airfreight is much more expensive than competing forms of land and sea transportation, so air cargo is limited to high value-low volume, perishable, and emergency freight.

2.5 Forecasting Air Travel

Forecasts of demand and, to a lesser extent, supply are the basis for planning that is essential for aircraft manufacturers, airlines, airports, and other aviation-related firms. Long-term forecasts of air transport demand are critical for aircraft manufacturers, first because of the long lead-time and expense involved in developing and marketing new aircraft, and then because new aircraft will remain in service for 30 or more years. The manufacturers' horizon stretches out decades. Airports and airlines also depend on long-range forecasts to accommodate passenger and freight demand. Both Airbus and Boeing develop 20-year forecasts that are updated and published annually (Airbus, 2021; Boeing, 2021). As Boeing states, "The forecast has several important practical applications. It helps shape our product strategy and provides guidance for our long-term business planning. We have shared the forecast with the public since 1964 to help airlines, suppliers, and the financial community make informed decisions" (Boeing, 2014, p. 1).

Forecasting future sales is the first step in business planning. In most companies, this responsibility falls to the marketing department, which uses a myriad of inputs and information. In some airlines, the function resides in a separate airline-planning department. Whatever its source, the sales forecast and other statistics prepared with it form the basis for planning and budgeting by all other operational areas.

A typical annual planning and budgeting process begins with the preparation of passenger and market share forecasts by route. From these data, the marketing and finance departments jointly estimate yield, a measure of average price, and revenue. The marketing, operations, and engineering departments then prepare a proposed flight schedule and an estimate of flight hours (block hours). The proposed flight schedule drives more detailed departmental planning and budgeting. Here are a few examples of how various airline departments utilize the flight schedule:

- Reservations department begins selling flights, sometimes up to one year in advance.
- Flight Operations estimates required crewmember staffing, plans recruitment and hiring, schedules training, basing, and crewmember work schedules.
- Maintenance and Engineering establishes aircraft maintenance rotation, heavy maintenance schedules, staffing, and spare part requirements.
- Station Operations determines the staffing and airport facilities such as counter space, gates, and ground support equipment needed to support the planned flight schedule.

Each department prepares budgets that the finance department consolidates resulting in revised estimates of revenue, cash flow, capital expenditures, and internal and external financing for the upcoming year, usually with a less detailed plan extending out five years or more. This process may go through several iterations with inputs from all departments before the budget is finalized.

2.5.1 Macroforecasting

One of the easiest and least expensive methods to estimate long-run air travel growth is to assume that historical trends will continue years into the future. An exponential curve is a surprisingly close fit to historical RPK growth displayed in Figure 2.1. The regression equation for the curve can be used to estimate future RPKs. More sophisticated long-run demand forecasts usually begin with estimates of gross domestic product (GDP) which is highly correlated with air travel. Gross domestic product is the market value of all final goods and services produced in a country during a given period, usually one year. Aggregating

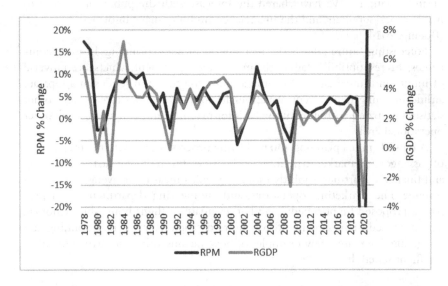

Figure 2.6 US Revenue Passenger Miles versus Real Gross Domestic Product.

Notes: Traffic volume measured in revenue passenger miles (RPM). Real gross domestic product (RGDP) is adjusted for inflation. RPM change for 2020 extends below the scale.

Data Sources: Federal Reserve Economic Data and A4A World Traffic and Capacity [databases].

country GDPs gives the gross world product. Figure 2.6 shows the relationship between US airline traffic and US real gross domestic product. The annual percentage change in passenger traffic, measured in revenue passenger miles (RPM), is shown on the left scale. The percentage change in real (adjusted for inflation) gross domestic product (RGDP) is displayed on the right scale.

The correlation between RPM and RGDP (r=.56) is evident in the figure. A graph of world air travel and gross world product shows a similar pattern but with a lower correlation. Notice that US airline traffic has grown at roughly twice the rate of GDP over the last four decades. Air travel is also more volatile than GDP and highly cyclical.

Because of the great interest that GDP growth holds for governments, corporations, and economists worldwide, it is subject to highly elaborate forecasts. GDP forecasts are available from many authoritative sources (e.g., International Monetary Fund, 2021). Manufacturers, airlines, and others in the aviation industry use these forecasts of GDP as one independent variable in deriving estimates of future air travel.

As explained earlier, population and per capita income are important drivers of air travel; both are subsumed in GDP, so the correlation between GDP and air travel is not surprising. Boeing concludes that 60% to 80% of air travel growth is the result of economic growth. The remaining 20% to 40% not directly

Figure 2.7 Airbus A-380.
Source: Photo courtesy of Airbus.

associated with GDP growth derives from other factors, primarily the fare level and service provided. As markets liberalize, airlines offer more flights and a variety of products, competition drives down fares. Travelers value a choice of arrival and departure times, routings, non-stop flights, choice of carriers, service class, and especially, lower prices, so they respond to the greater choice with more travel. Over time, this response to improved service and lower fares causes the share of GDP a country devotes to air travel to increase.

Long-range, macrolevel forecasting is essential for manufacturers' new aircraft decisions. Airbus' ill-fated decision to develop its flagship A-380 is an interesting example. The A-380 created the new class of very large aircraft (VLA) with a seating capacity of 850 passengers in an all-coach configuration, although a three or four-cabin configuration with 550 or fewer seats is more common. The upper and lower rows of windows seen in Figure 2.7 reveal the double-deck design. Even though Airbus and Boeing differed on the forecast for the size of the market for VLA, both understood that it could profitably support only one manufacturer. When Airbus committed to building the A380 in 2000, Boeing dropped plans for a VLA although it later upgraded its venerable B-747 introducing the somewhat larger B-747-8 that entered service in 2011.

How do manufacturers develop forecasts for market demand for the next 20 or more years? Airbus and Boeing outline the range of methodologies employed to develop their annual forecasts, the Global Market Forecast and the Commercial Market Outlook, respectively. For new product introductions, such as the A-380, the development and sales history of early similar products is helpful. Airbus certainly researched the introduction of the original Boeing 747. In addition to historical trend analysis, statistical methods generate equations for future demand by regressing historical air travel growth against independent variables, including GDP, population growth, urban population density, and oil prices, among others. Boeing and Airbus know every airline with the potential to purchase a new aircraft type. Each is asked for their requirements and projections for new aircraft acquisitions. Because relatively few airlines need an airplane the size of the A380, detailed discussions are feasible. Other methods include surveys of salesforce opinion as well as those of industry experts. Unfortunately, the demand for the A380 fell well short of Airbus' estimate. Production of both the A380 and Boeing's competing 747-8 ceased in 2022. For a detailed discussion of forecasting methods used in the airline industry, see Doganis (2019) and Vasigh et al. (2013).

2.5.2 Route Level Microforecasting

Whether evaluating the potential new routes or estimating growth in the existing network, airlines incorporate the factors that drive global and regional demand, but many other factors are important when estimating demand on a specific route. These include traffic history, the types of passengers, competition, and prevailing fares. With an estimate of total demand at the route level, the airline must then estimate the share of that total traffic, the *market share*, it can obtain versus that of competing airlines.

Historical data on the number of passengers and fares in a given market is usually the starting point for route-level demand estimation and planning. For its existing markets, of course, the airline will have internal data. Experience with previous entries on similar routes is particularly helpful. External data on passenger traffic and fares are available from several sources. The Marketing Information Data Transfer (MIDT) is a database that captures booking information from the major Global Distribution Systems (GDS). Actual sales data, in contrast to booking data, are available for transactions settled through the Airlines Reporting Corporation (usually referred to as ARC) and the Billing and Settlement Plan (BSP). Because direct sales through airline websites and reservation centers are not captured by any of these databases; the data represent only a portion of the total market. Data from each of these sources are processed and sold by several vendors, many of whom also provide estimates of total traffic and sales. For the US carriers, the Department of Transportation, Bureau of Transportation Statistics makes a 10% sample of all tickets available to the public without charge. Many airline economic studies utilize this database.

An airline considering a new route without existing service will have little or no historical traffic data on which to base its forecasts. It might rely on market research and expert opinion, but demand can be at least crudely estimated using the populations of the two cities and the distance between them. The *gravity model* has a long history first used in the late 19th century to estimate railroad demand and later by highway engineers. Demand is hypothesized to vary directly with the product of the populations and inversely with the square of the distance between the cities, hence the name because of its similarity to Newton's law of gravity. The model has been refined using other indicators of economic activity instead of populations and calibrated on existing airline routes (Doganis, 2019).

Forecast demand is a starting point for airline planning, but it's subject to considerable uncertainty and error. A common error has been the failure to differentiate between underlying demand and past traffic growth that was stimulated by declining yield (low fares). Falling prices will increase traffic, especially leisure travel. Airlines have frequently overestimated traffic growth and ordered new aircraft to meet the expected demand. When traffic failed to materialize, ticket prices fell which stimulated new traffic. Subsequent forecasts based on this traffic history could well be overstated but self-fulfilling; that is, extra capacity will force future yields even lower (Love et al., 2006).

2.5.2.1 Passenger Segmentation

Demand for air travel is derived from the purpose of travel. That is, passengers do not fly just to enjoy the flight, but because they need to be in another location for some reason, be it for business or pleasure. Economists term this *derived demand.* Some markets attract mostly business passengers whereas leisure travelers dominate others. The Chicago–New York market, for example, is predominately a business market whereas vacationing passengers mostly fill flights to Orlando and other Florida destinations. Similarly, in Europe passengers on holiday will predominate on routes to the Mediterranean cities. Passengers' wants and needs depend on the reason for travel. The airline must tailor its product and pricing to meet the desires of its passengers.

Airlines have traditionally segmented passengers into two major categories by the reason for travel, business and leisure, often with several subcategories (Doganis, 2019). Business passengers, for example, can be further classified as lower and higher end. Leisure passengers are universally categorized as those visiting friends and relatives (VFR) and those on vacation or holiday.

Passengers traveling for business tend to value flight frequency, non-stop flights, choice of cabin classes, in-flight service, and flexible, refundable fares. High-end business travelers are less price-sensitive and often book flights near the departure date while the lower-end business passengers are more price sensitive. Passengers visiting friends and relatives usually have some flexibility in travel plans, book far in advance, and often select the airline offering the lowest price. Flight frequency and amenities are less important. While having many of the same wants as those VFR, the destination appeal is the most important consideration for passengers traveling on vacation. Their preferred travel days are more likely constrained by work. To meet these varying wants and needs, airlines align their product strategy with flexibility for business passengers and lower prices for leisure and VFR passengers (Teichert et al., 2008).

Of course, the same passenger may be traveling for business one week and for leisure the next, so the passenger's wants and needs are tied to the purpose of the trip, not the individual. Passenger wants and needs also vary with the length of the travel. Many passengers will sacrifice cabin amenities, such as seat and legroom, for a lower price on a flight of three hours or less but will demand a higher quality product on longer flights.

2.5.2.2 Variation in Demand

The airline planner tries to match the demand in each market with capacity, flight frequency, and aircraft size. If an airline does not provide enough capacity to meet periods of peak demand, it risks losing passengers to other airlines, a loss in industry jargon known as *spill*. This concern tends to create excess capacity during lower demand periods. Unlike manufactured goods, airline capacity cannot be stored until demand is higher. On the other hand, too much capacity for the existing demand results in empty seats (known as *spoil*) or, more often, in lower prices. This challenge is daunting because demand in each market varies substantially by the hour of the day, day of the week, season, and business or economic cycle.

Business travelers favor morning departures and late afternoon or earlier evening return flights. In predominately business markets, a flight departing at 7 am may face heavy demand. However, when the aircraft operating this flight reaches its destination, little demand may exist for the next departure at, say, 10 am Similarly, business travelers also travel more frequently on Mondays and Fridays. In contrast, both VFR and vacationing passengers are less sensitive to the timing of flights but often wish to leave on a Saturday and return on Sunday. Because of these desires, demand mid-week on Tuesdays and Wednesdays is typically low.

Demand also varies by the season of the year. Using the North American market as an example, travel to Florida and other sun destinations is strongest in the winter whereas transcontinental and other east–west demand is strongest during the summer vacation season. Canadians flock to flights to the Caribbean islands during the winter, but this market dries up in the summer. European demand for sun destinations on the Mediterranean peaks in the summer with much less demand during the winter. Finally, holidays generate outsized, short-lived demand. For perspective, US domestic traffic in the weakest month of the year, usually in early February, is only 78% of the peak in mid-July.

Directional demand introduces another complication. This is easiest to visualize with special events such as big sporting attractions. Before the event, demand will be high for travel to the host city and strong leaving the city after the event. The result is that with sufficient capacity for the high demand, flights in the opposite direction will be wanting. Directional demand, however, is evident in some markets for reasons other than special events. Holiday travel to

Florida or other sun destinations is an example. Passengers travel to the sun at the beginning of holiday periods, returning at the end. Directional demand is even more pronounced in the world cargo market where demand from Asia, especially China, to Western Europe and North America greatly exceeds demand in the other direction.

The need for business travel also varies with the business economic environment; when business is good, managers and sales staff travel extensively. When the economy softens or contracts, business travel falls. Thus, business travel follows but lags behind the larger economic cycle. The business cycle has a lesser effect on leisure travelers. This effect is readily apparent in the earlier Figures 2.1 and 2.6.

Demand forecasting is further complicated by natural disasters; political upheavals, especially war; economic crises; and global pandemics – all of which are unpredictable. The impact of natural disasters from hurricanes, tsunamis, and volcanoes is regional rather than worldwide. Traffic falls because of a lack of demand; then, airlines reduce supply by canceling flights. The Japanese earthquake and tsunami of 2011 is one of many examples. The effects of war vary from regional to worldwide. Financial crises should, in principle, be predictable, but few economists foresaw the Great Recession of 2008. Pandemics, such as the COVID-19 that devastated the global airline industry, defy prediction.

2.5.3 Demand

Economists capture many of these concepts of demand with the demand curve that can provide additional insights into route-level forecasting. In the economics argot, demand is the quantity of goods or services that customers are willing and able to purchase in a certain time period at a given price. The demand curve is a graphical representation of the relationship between price and the quantity of goods or services demanded. Figure 2.8 shows a stylized demand curve for air travel. In the left panel, the average ticket price is on the Y-axis while the number of tickets sold during some period is on the X-axis. Not

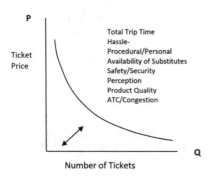
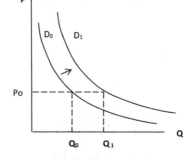

Figure 2.8 Demand for Air Travel.

surprisingly, the lower the price, the more tickets passengers are willing and able to purchase. This inverse relationship between price and the quantity of goods sold is the *law of demand,* one of the most robust principles of microeconomics.

The demand curve is not linear but is convex sloping steeply downward in the upper left corner and gradually flattening towards the lower right. At relatively high prices, few passengers are willing and able to purchase tickets. Those who do are typically traveling for business, although a few will be wealthy and traveling for leisure. This passenger segment is not especially sensitive to the price of travel. In the case of the business traveler, the potential gain from the trip far exceeds the cost of travel where the ticket price may be only a small portion of the total cost of the trip. As the price falls, many more people will purchase tickets. Most of these passengers will be traveling for leisure. They carefully consider alternatives such as driving, traveling to a different desirable destination, or may simply choose not to travel if the price is high. Leisure passengers are highly sensitive to price, or *price elastic,* and regularly choose the airline offering the lowest fare. The low-cost carriers target this price-sensitive passenger segment.

Low-cost carriers such as Europe's Wizz Air, which promotes its low price, can greatly expand a market as passengers who would not otherwise have flown flock to a perceived travel bargain. This concept is illustrated in Figure 2.9 which shows the dramatic increase in passenger traffic when a low-cost carrier enters a market. A 50% decrease in fares caused traffic to increase by 140% when JetBlue entered the New York-Burlington, Vermont market. Similarly, Southwest's entry into the Buffalo-Baltimore market resulted in a tripling of traffic on a 60% decrease in fares. A doubling or more of traffic in response to Southwest's entry into a market was common in the 1990s and dubbed the *Southwest Effect.*

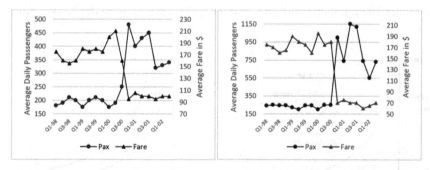

JetBlue enters New York – Burlington, VT market   Southwest enters Buffalo, NY – Baltimore, MD
Traffic increased by 140% while Fares fell by 50%   Traffic increased by 300% while Fares fell by 60%

Figure 2.9 The Southwest Effect.

Data source: Henderson, G. (2004). *The changing shape of the industry.*

Not all of the increase is new traffic; some is diverted from surrounding areas. Many leisure passengers will drive up to three hours to take advantage of lower fares.

The Southwest Effect illustrates the point that price is the most important variable in determining the amount of goods or services purchased in any given period, especially in the short term. However, many other variables, such as per capita income and population growth, also affect demand. The graphical depiction of the demand curve is conceptually helpful, but a two-dimensional graph shows the relationship between just two variables, price and quantity sold. To capture the effect of other variables, the demand curve shifts right or left. For example, if per capita income increases, then passengers will purchase more tickets at any given price, so the demand curve shifts right as shown in Figure 2.8, right panel. The demand curve shifts from D_0 to D_1. Taking any price, such as P_0, the number of tickets demanded increases from Q_0 and Q_1. The actual number of tickets sold depends on the supply of flights that airlines offer in response to the new demand. Figure 2.8 lists other factors that can cause a shift or change in demand. For example, in Europe, Japan, China, and the US northeast corridor between Boston and Washington, D.C., the high-speed train is a substitute for air travel. Some passengers who would have previously flown now choose the train. The demand curve for air travel shifts left. The Acela train between Boston and New York has gained more than 50% of the passenger market share. This is probably the first time since the 1950s that the train has a greater market share in a major US city-pair than the airlines. Consequently, the once large and lucrative airline shuttles between these cities have suffered. Figure 2.8 shows a single average price in the market. This is a needed simplification; in fact, airlines rarely charge a single price for a flight. Ten or more different prices for economy seating on any given flight are common. The lower prices usually come with several restrictions such as advanced purchase requirements. Airline pricing is the subject of Chapter 9.

2.5.4 Supply

Airline supply receives less emphasis than demand with the implicit assumption that airlines will expand the number of flights and seats offered to meet an increase in demand. As with changes in demand, several factors independently affect supply, most of which have already been considered. In general, any factor that increases the cost to operate flights will lead to a reduction in the supply of flights and vice versa.

2.6 New Route

An airline considering adding a new route will first assess the existing traffic and growth potential. Next, it must estimate the portion of traffic it can capture, its potential market share. A naïve estimate of market share is the percentage of total flights it intends to offer on the route. An airline with 25% of the flights (and seats) in a market might plan on capturing 25% of the market demand. Such naïve estimates are refined by careful consideration of the product offered by competing airlines. Variables considered include:

- flight frequency
- time of departures
- number of stops
- en-route time
- aircraft type
- product quality
- airline reputation
- competitive reaction

Of these variables, flight frequency has traditionally been considered critical. High frequency is an important choice criterion, especially with business passengers. The S-curve (Figure 2.10) illustrates the relationship between market share, measured in percent of revenue, and flight frequency.

In a market in which no carrier has a significant price or product advantage, the carrier with most frequency enjoys a disproportionately higher market share. For example, this S-curve shows that an airline with 70% of flights captures an 80% market share. While the precise shape of the curve would vary by market, the concept remains the same – high frequency gains more of the market.

There are several reasons for the S-curve phenomenon. An airline that offers more frequency will attract business travelers who will choose the airline with a flight closest to their desired time of travel. Because an airline can charge higher fares to passengers traveling for business, a greater share of business passengers

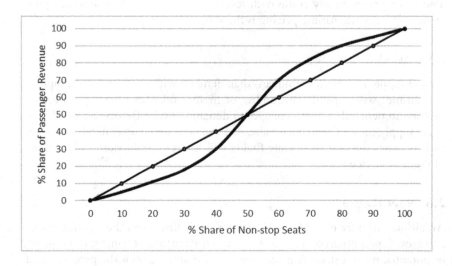

Figure 2.10 S-Curve.

results in higher total revenue. Further, passengers become familiar with the airlines offering high frequency, so search costs are lower and buying habits are established. Frequent flyer programs add incentives to stick with the dominant carrier in a market.

The S-curve's validity has been widely accepted but its relevance is fading. A study back in 2005 (Button & Dexler) found some past evidence for the curve but little recently. The growth of low-cost carriers is one reason. The curve implicitly assumes that all competitors charge the same price. With price being the first criterion of airline choice for leisure passengers, a carrier with low frequency may obtain a relatively high market share by offering a lower price. It follows that an LCC's entry into a market served by high-cost legacy carriers will reduce the S-curve advantage enjoyed by an incumbent airline offering higher frequency.

The existing service and competitive intensity are critical for estimating the market share a new entrant can obtain. Competition exists not just from airlines flying a route but also from those serving the same city pair with connecting flights through their hub airports. Five airlines, for example, fly non-stop between New York Kennedy and London Heathrow but at least as many others provide one-stop services connecting through their hub airports like Frankfurt, Paris, Rome, and Madrid. There's also competition from airlines flying between other nearby airports such as Newark to London Gatwick.

For success in a new market, the airline must offer a product feature(s) that will draw passengers from incumbent airlines or expand the market. This could be a superior product, perhaps more legroom, free internet, inflight entertainment, or full lay-flat seats in business class. For low-cost airlines, the attraction is a low price. Depending on the price sensitivity of the market, lower prices can generate more traffic expanding the existing market – the Southwest Effect.

Finally, a new entrant must be cognizant of the potential competitive reaction. Most airline markets are oligopolies – only a few competitors. In oligopoly markets, each airline is acutely aware of the product offered by competitors and new marketing initiatives. Competitors will usually match the lower fares of a new entrant, although perhaps only on a limited number of seats. Aggressive competitors may also increase flight frequency sometimes adding flights operating at the same time as the new entrant. If competitors view the new entrant as vulnerable, an aggressive competitive reaction is likely.

2.7 Summary of Factors Affecting Airline Demand and Supply

Although fraught with error, forecasting is an essential element of airline management. No airline, or other business, can effectively function without developing detailed forecasts. Nevertheless, even with extensive data and sophisticated planning tools, both of which large airlines employ, many circumstances are unforeseen. Table 2.2 summarizes the many variables that will change air travel supply and demand.

Table 2.2 Factors Causing a Change in Demand and Supply

Changes in Demand

Macro Factors	Effect
Tastes & Preferences	Most people want to fly to vacation destinations, visit friends and family, or support their businesses. Tastes and preferences, however, can change. A few people, for example, choose to limit flying for environmental concerns.
Population	Population growth increases the number of potential passengers.
Per capita income/ wealth	Higher per capita income, especially disposable income, increases peoples' ability to purchase air travel.
Business Cycle/ Economic environment	More passengers travel when the business cycle is positive and the economic environment is favorable.
Taxes/Fees	Higher taxes on travel decrease the ability of people to pay for travel. Higher passenger taxes effectively raise the price of travel.
Safety, Security	Concerns about safety and security decrease passengers' willingness to travel. Pandemics, wars, civil unrest, and similar events greatly decrease passengers' desire and willingness to travel.

Micro Factors	
Price	More people will travel at a lower price, especially those traveling for leisure. Low-cost carriers target this passenger segment.
Substitutes	More or improved substitutes, such as high-speed trains and video conferencing, provide passengers with greater choice. Some will choose a substitute rather than flying.
Competition	Increased competition, from both substitutes and other airlines, increases passenger choice and may improve the quality of service. Competition also lowers prices. More people will choose to fly. In general, competition benefits consumers but challenges producers.
Ease of travel/ Hassle	The greater the ease, the more people are willing to travel. New terminals, for example, increase ease. The hassle of security screening reduces the ease of travel.
Total trip time	Lower total trip time increases the willingness to travel. Non-stop flights; quick, easy connections, and reduced check-in and security times reduce total trip time.
Frequency of flights	Higher flight frequency allows passengers to choose a flight time that best meets their needs. Increased competition usually increases flight frequency.
Choice of service	Greater choice of service, both in the number of airlines and the services each airline offers, increases the appeal of air travel.

Table 2.2 Cont.

Changes in Supply

Factors	Effect
Input factors	Higher prices for the factors of production increase airline costs and decrease profitability. Some marginally profitable routes then operate at a loss. Airlines respond by reducing flights and may drop some routes. Fuel and labor are the two largest categories of operational expenses. Both are likely to increase over the long run.
Technology	Improved technology lowers the cost of producing flights. For example, the higher fuel efficiency of new aircraft lowers the cost of producing a flight, so airlines will add flight frequency and new routes. Some new aircraft have greater range capability so airlines can add long-haul routes that were not previously feasible. Technology affects many other components of the airline operation.
Government Regulation	Liberalization allows airlines to expand into previously restricted markets, so supply increases. More often, however, new regulations increase airline costs and, therefore, reduce supply. In the extreme, such as during the COVID-19 pandemic, governments prohibited flights.
Infrastructure	Governments provide many of the factors of production. Airports are often government-owned and a substantial source of government income. Too few airports and runways restrict flights as will higher fees for airport use. Air traffic control, whether private or government-operated, imposes substantial flight costs.
Subsidies and taxes	A government subsidy reduces an airline's cost of operation, so it operates more flights. National governments often directly or indirectly subsidize state-owned airlines. During the COVID-19 pandemic, many governments provided subsidies for airlines to continue flying. On the other hand, airlines are highly taxed. Higher taxes reduce supply.

The effects of shifts in supply and demand are no more dramatically elucidated than by the global COVID-19 pandemic that began in 2020. Many travelers were reluctant to fly fearing contracting the virus in confined aircraft cabins and crowded airports, so their preference for flying fell. The passengers who were willing to fly for vacation discovered many favorite spots closed. Similarly, employees who previously flew to promote their business found customers reluctant to meet and conferences canceled. Video conferencing emerged as a widely accepted substitute for in-person meetings. These factors substantially reduced demand for air travel.

To stop the spread of the virus, governments severely restricted or banned international flights. Late in the pandemic, a lack of crewmembers caused tens

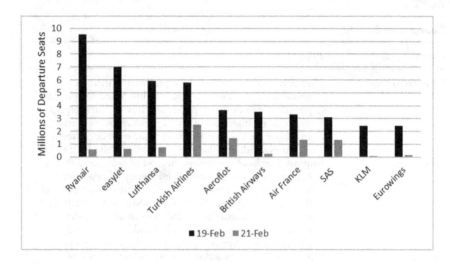

Figure 2.11 European Airlines Seats in February 2019 vs February 2021.
Data source: Aviation Week.

of thousands of flight cancelations, especially in the US. All these factors reduced supply. As airlines worldwide faced bankruptcy, governments provided billions of dollars in various subsidies to keep their airlines afloat, actions that modestly increased the number of flights offered. Figure 2.11 displays the unprecedented decline in European airline operations from 2019 to 2021.

The intersection of demand and supply determines the volume of air travel. At the lowest point in April–May of 2020, world revenue passenger kilometers dropped an astonishing 94% compared with the same period in 2019. Airlines grounded 64% of the global fleet and reduced flights by 72%. For all of 2020 compared to 2019, passengers carried and revenue passenger kilometers both fell 75%. Capacity measured in available seat kilometers was down slightly less at 64% (IATA, 2021). Recovery to 2019 levels is likely by 2023.

2.8 Summary

Aviation provides the only means of rapid, long–distance travel across the globe, which makes it essential to global business and tourism. Several factors drive worldwide air travel demand, including globalization and a growing, dispersed, and increasingly affluent population. Airline supply is positively affected by a reduction in government restrictions on airline service and the introduction of ever more efficient new aircraft. The wants and needs of airline passengers depend on the purpose of travel; business travelers value the convenience of schedule and service more than price whereas leisure passengers are often very price sensitive and choose the airline offering the lowest fare. The uneven and cyclical demand for air travel challenges airline managers who must design flight schedules and product features to satisfy anticipated customers' needs. Methods for estimating demand include various statistical techniques, historical comparison, and sales force opinion, among others. Other departments within the airline base their planning on initial marketing forecasts, thus estimating demand is critical to airline performance.

Airlines offer products ranging from the low-fare carriers who typically provide point-to-point service with a minimum of amenities to global, full-service network carriers whose inflight products range from first-class to tourist economy in three or more cabins. The next two chapters address this range of products.

2.9 Case Study: JetBlue to London

The transatlantic market is vital to US and European full-service network carriers. Before the pandemic, British Airways' annual revenues on the New York–London route exceeded $1 billion. In 2021, years after announcing its interest in starting European service, the US hybrid carrier JetBlue began flights between New York and London joining the US Big-3: American, Delta, and United. Figure 2.12 shows JetBlue's route map with the new flights from New York Kennedy to both London Heathrow and Gatwick airports. Initial service is only daily to Heathrow with less than daily service to Gatwick, but JetBlue intends to expand its European service.

Airlines have served the New York–London market for decades, so in planning its new service JetBlue could draw on extensive historical data on passenger volumes, prices, flights, airlines, and aircraft. British Airways, Virgin Atlantic, and the US Big-3 dominate the market. All are members of global alliances and have

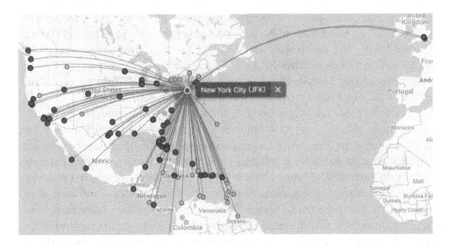

Figure 2.12 JetBlue's New York–London Route.
Note: Route map courtesy of FlightConnections.

other incestuous financial and marketing relationships, including one between JetBlue and American. While many of the incumbents will aggressively compete to maintain their market share, they also maintain complex relationships that affect their competitive responses. JetBlue could also benefit from studying past airline successes and failures. The market is littered with previous failures, including Laker Airways, People Express, WOW, and Primera, all of whom tried to undercut the major players. Paradoxically, Norwegian's withdrawal from the New York-London market and all other long-haul routes in early 2021 may have created an opening for JetBlue. The opening is short-lived as the new low-cost long-haul Norwegian airline Norse Atlantic Airways plans to begin flights in 2022.

JetBlue had to estimate the share of the market it could capture. The incumbents benefit from extensive feed traffic from their networks in the US and Europe, so the relevant market is much greater than just New York to London. As Figure 2.10 shows, JetBlue can also develop feed from its more limited network. The minimal flight frequency compares with competitors' hourly flights placing JetBlue at a severe disadvantage for the critical business fliers. JetBlue is touting low prices and its acclaimed, price-competitive business class product and superior offerings in coach class to entice travelers.

Both Kennedy and Heathrow are slot-controlled airports. JetBlue holds slots at Kennedy but had to obtain rights at Heathrow that have previously been difficult and very expensive to secure. Here JetBlue benefited from the pandemic by gaining temporary slots unused by incumbents at Heathrow. Its long-run access to these slots is uncertain.

JetBlue will be unique in operating a narrow-body aircraft between New York and London, the Airbus A321LR, which it believes will provide a cost advantage. However, with just 114 seats in coach and 24 Mint suites, as JetBlue brands its first class, there are a lot fewer seats than a typical configuration on which to spread the flight cost. Although British Airways doesn't use the A321 in the London–New York market, its international A321s are configured with 131 coach seats and 23 flatbed seats. Other airlines cram 198 seats, including a business class, in the same aircraft.

Not surprisingly, analysts disagree with JetBlue's potential for success. Paul Griffiths, CEO of Dubai Airports and independent airline analyst, is bullish. "A squeeze on corporate travel budgets, inflation and rising pressure on family finances mean almost every traveler will be looking for value for money for years to come. If JetBlue really can exploit the superior operating economics of smaller aircraft and offer a reliable, consistent service at a budget price, it stands to do very well." Henry Harteveldt of Atmosphere Research, holds the opposite view stating that JetBlue "can tout its 'low' business-class fares all it wants, but it's not quite true. I routinely observe that JetBlue's Mint fares here in the US are as high as and sometimes higher than, those of other carriers." The dispute illuminates the risk and uncertainty of new route decisions. JetBlue's early success in the European market will be known by the time this textbook is published.

Concepts: New route planning, demand forecasting, airline alliances, hub-and-spoke networks, S-curve, aircraft economics

Major sources: Arlidge, 2021; Bernie, 2017.

References

Air Transport Action Group. (2012, 2020). *Aviation benefits beyond borders.*

Airbus. (2021). *Global market forecast, 2021–2040.* www.airbus.com/en/products-servi ces/commercial-aircraft/market/global-market-forecast

Arlidge, J. (2021, August 15). Is JetBlue winging it from London to New York? The route has been a ticket to bankruptcy for low-cost airlines. Can the latest no-frills challenger succeed, asks John Arlidge. *Sunday Times.* https://go-gale-com.ezproxy.libpr oxy.db.erau.edu/ps/   i.do?p=AONE&u=embry&id=GALE%7CA672086161&v= 2.1&it=r&sid=summon

Association of Southeast Asian Nations. (2021, June 4). *ASEAN, EU conclude the world's first bloc-to-bloc Air Transport Agreement.* https://asean.org/asean-eu-conclude-the-wor lds-first-bloc-to-bloc-air-transport-agreement/

Bernie, N. (2017, January 3). JetBlue service to London. *Sabre.*

Boeing Commerical Aircraft. (2014, 2021). *Commercial Market Outlook.*

Booz and Company. (2011). Data points: Why air travel is bound to boom. *Strategy + Business.* www.strategy-business.com/article/11116?gko=5a99e

Bryan, V. (2021, October 28). How China's airport plans will make it the world's largest passenger market. *Aero Time Hub.*

Button, K., & Drexler, J. (2005). Recovering costs by increasing market share: An empirical critique of the S-curve. *Journal of Transport Economics and Policy*, 39(3), 391–410.

Doganis, R. (2019). *Flying off course: Airline economics and marketing* (5th ed.). Routledge.

Federal Reserve Bank of St. Louis. (2022, January 12). *Kerosene-type jet fuel prices: US Gulf Coast.* FRED Economic Data [data set]. https://fred.stlouisfed.org/series/ WJFUELUSGULF

Gossling, S., & Humpe, A. (2020). The global scale, distribution and growth of aviation: Implications for climate change. Global Environmental Change, 11(65).

Humphreys, B. (2022). Aviation's retreating liberalization. *Air Transport World*, 6–9.

IATA – International Air Transport Association. (2021). *World air transport statistics 2021.* www.imf.org/en/Publications/WEO/Issues/2021/10/12/world-economic-outl ook-october-2021

International Monetary Fund. (2010). *Globalization: Threat or opportunity.* www.imf.org/ external/np/exr/ib/2000/041200to.htm

International Monetary Fund. (2021). *World Economic Outlook.* www.imf.org/external/ ns/cs.aspx?id=29

Love, R., Goth, J., Budde, W., and Schilling D. (2006, June). Understanding the demand for air travel: How to compete more effectively. *Boston Consulting Group.* www.bcg. com/expertise_impact/Industries/Transportation_Travel_Tourism/Publication Details.aspx?id=tcm:12-14822.

Morphet, H., & Bottini, C. (n.d.) Propensity to fly in emerging economies: What do the trends mean for aviation infrastructure investment? *PwC.* www.pwc.com/ gx/en/capital-projects-infrastructure/pdf/pwc-propensity-to-fly-in-emerging-economies.pdf

OAG. (2022, June 7). *Steady aviation growth continues amid travel chaos.*

Teichert, T., Shehu, E., & von Wartburg, I. (2008). Customer segmentation revisited: The case of the airline industry. *Transportation Research*, Part A. 42, 227–242.

United Nations. (2019). *World Urbanization Prospects 2018.* https://population.un.org/wup/Publications/Files/WUP2018-Report.pdf

Vasigh, B., Fleming, K., & Tacker, T. (2013). Introduction to Air Transport Economics (2nd ed.) Ashgate.

Review Questions

1. How does the airline industry support global economic growth?
2. Provide some measures of aviation industry size and a comparison with other major industries.
3. What factors drive long-term global and regional air travel demand? Supply?
4. How did the COVID-19 pandemic affect worldwide air transportation supply and demand?
5. What is liberalization and how does it affect air travel? Provide an example.
6. Which world regions will have the highest air travel growth? What factors will cause these regions to grow more rapidly than others?
7. How does historical air cargo growth compare with that of passenger air travel?
8. What macroeconomic metric is most highly correlated with air travel growth? Why?
9. How does air travel demand vary by time of day, day of the week, season, and economic cycle?
10. If an airline planner could perfectly forecast future demand, would a problem in matching capacity with demand remain? Why?
11. The demand curve shows the relationship between what two economic variables?
12. What segment of the aviation industry is most concerned with macroforecasts? Microforecasts? Why?
13. What does the S-curve show?
14. Compare the wants and needs of the typical business passenger with that of leisure passengers.
15. What segment of the demand curve is targeted by the low-cost-carriers such as Ryanair?
16. Given the complexity and uncertainty in forecasting demand, why do airlines bother?
17. What are the steps an airline planner would follow to evaluate a potential new route?
18. What factors should be considered in estimating the market share a new entrant airline on a route with existing service can garner?

3   Route Structure

Upon completion of this chapter, you should be able to:

*   Describe and sketch each type of route structure.
*   Analyze the marketing and revenue advantages of alternative route systems.
*   Compare the cost of operation of hub-and-spoke with point-to-point route systems.
*   Explain why the hub-and-spoke system is the choice of the world's largest airlines.
*   Name a few carriers using each type of route structure and analyze their choice of route systems.

3.1 Introduction

An airline must provide a product that passengers value and are willing to purchase at a price from which the airline can profit. This chapter and the next explore the range of airline products. Among an airline's most critical strategic choices are the destinations served and the route structure connecting these cities. These decisions flow from even more fundamental strategic choices about the targeted customer segments and how the airline intends to obtain a competitive advantage in its product offering. These latter questions are the subject of business strategy and are not considered in depth in this text.

An airline can choose among three generic route system architectures that vary in complexity and cost of operation: point-to-point, linear, and hub-and-spoke. Route structure choice is examined in this chapter. Chapter 4 explores the type and number of product features to include with the core air transportation product.

3.2 History

Early airline route structures were mostly *linear* going from one city to the next and on to the next similar to a train. As we saw in the review of airline history, an airline passenger in the 1920s could leave London and, after about 10 days and many stops at cities along the way, arrive in Cape Town, South Africa. As aircraft

DOI: 10.4324/9781003290308-3

technology allowed for longer range, the number of stops on linear routes was reduced. Early air routes in the United States often followed the railroad lines connecting large cities. Indeed, in 1929, the first transcontinental passenger air service between New York and Los Angeles and continuing to San Francisco utilized airplanes by day and trains by night. Both plane and train made many stops so that passengers could travel between any two cities along the route, but usually with several intermediate stops. As airline economic deregulation spread to other countries, *hub-and-spoke* (H&S) systems were adopted by most of the world's largest airlines to the surprise of many airline economists. On the other hand, low-cost carriers that emerged following deregulation built *point-to-point* (P2P) systems. Competitive pressures have forced airlines to operate a mix of H&S and P2P routes.

3.3 Generic Route Structures

Route structure choice is the foundation of an airline's product. Point-to-point and hub-and-spoke architectures lie at the poles of a continuum. The linear system is in between but closer to the point-to-point pole. In practice, pure forms are rare; most airlines operate some hybrid form. For clarity, however, each is discussed in a pure form as depicted in Figure 3.1.

In a linear system, passengers may board at the origin or any intermediate stop and deplane at any subsequent stop along the route. A point-to-point system connects destinations with non-stop service. As shown in the center panel, all passengers board at the flight origin and deplane at the destination. In a hub-and-spoke system, flights operate from spoke cities (A through F in the right panel), a term adopted from a wheel analogy, to the hub airport (H). Passengers not destined for the hub, change planes for a connecting flight to their destination in another spoke city. Each system has advantages best suited for certain markets.

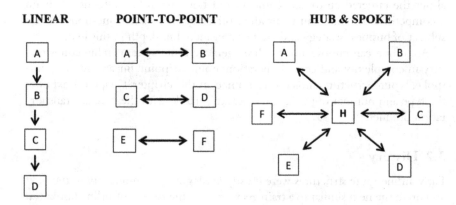

Figure 3.1 Generic Route Systems.

3.4 Linear

Linear route systems are interesting historically and as a comparison with point-to-point and hub-and-spoke routes. The linear system is a simple extension of the point-to-point structure. Rather than terminating at the first stop, the flight continues to one or several additional cities. This travel experience is familiar to anyone who has ridden a long-distance train and most commuter rail lines as well. Airline passengers traveling on a linear system board at the origin or other cities along the route and, likewise, deplane at their respective destinations. Some itineraries require several stops before reaching the destination.

The linear system has several advantages. Each route operates independently of the others. Scheduling and operational control are simplified; disruptions on one route do not affect another. This independence allows for high aircraft utilization. Flight frequency along the route can range from several per day to a few per week. However, because passengers traveling between several city pairs are consolidated on a single aircraft, the combined demand allows economical service to smaller cities. The multiple stops, however, are also the linear systems' major disadvantage resulting in its near demise. The several stops required to

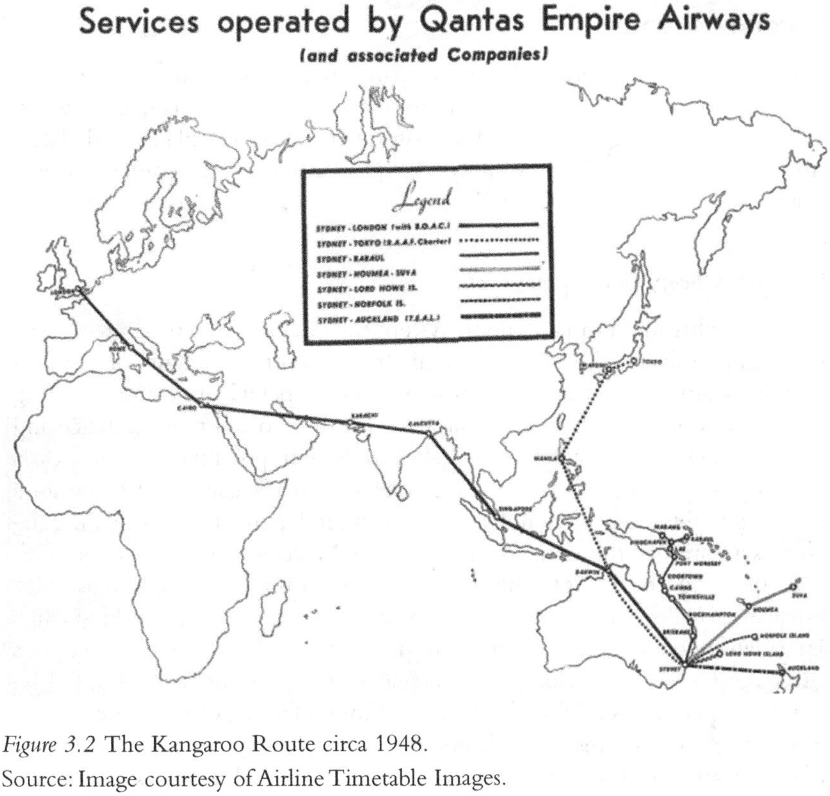

Figure 3.2 The Kangaroo Route circa 1948.
Source: Image courtesy of Airline Timetable Images.

serve many city pairs conflict with the passenger's desire for quick and convenient flights. Distance between stops is often short which increases the unit cost (cost per available seat mile) because the aircraft doesn't reach an economical cruising altitude and each stop adds airport costs. Passengers will choose a competing airline offering a non-stop or single-stop connecting service. The remaining linear routes are confined to thinly populated regions with little competition, such as Siberia and Alaska.

Although now rare and often not included in academic discussions, the linear system has a rich history. Before the introduction of DC-7 and Lockheed Constellation after World War II, the range limitations of early commercial aircraft required several stops along a route connecting distant cities. The famous Kangaroo route from Sydney to London as operated in 1948 by Qantas included six stops (Figure 3.2). At 10,600 miles, the route is within the non-stop range of the Airbus A350 and B787. Qantas completed research flights in 2019, but this and similar ultra-long-haul flights may not be profitable or, at more than 19 hours, even attractive to most passengers. Not to be deterred, Qantas announced in 2022 that it would acquire a fleet of Airbus A350–1000 jets and begin non-stop service between Sydney and London in late 2025.

3.5 Point-to-Point

The point-to-point system is the simplest means to connect the cities an airline chooses to serve. Each origin and destination city, or *city pair*, is served by non-stop flights. All passengers who board at the origin deplane at the destination. In a pure point-to-point system, passengers do not connect to any other flight.

3.5.1 Fast, Cheap, and Independent

Non-stop flights in a point-to-point system provide the least total travel time from origin to destination. Subject to air traffic control restraints or national government airspace restrictions, non-stop flights operate on a direct routing between the origin and destination allowing for the shortest flight distance and the most efficient and cost-effective flight profile with potential savings of 30% over competing service in a hub-and-spoke system (Doganis, 2019). Because connecting flights are not offered, flights operate independently eliminating the need to synchronize the entire flight schedule. Moreover, irregular operations on one flight do not disrupt other flights operated on other aircraft. This independence also allows flights to be spread across the day and night. Flight frequency can be scheduled depending on demand and type of market, business versus leisure. High utilization of aircraft, gates, and personnel is attainable. The schedule is easily adjusted for seasonality without affecting other traffic.

Point-to-point systems are well suited to high-density vacation and leisure markets because of low production costs. Passengers value non-stop service, a benefit for which business travelers will often pay a premium, so point-to-point service can be used to compete with full-service network airlines operating hub-and-spoke systems, a strategy that has been particularly effective for Ryanair, EasyJet, and other European LCCs.

3.5.2 Limited to Large Markets

The demand for any flight in the P2P system is strictly limited to the two cities served by the route. The inability to consolidate traffic bound for many destinations limits the number of city pairs that can profitably support P2P flights. The demand from most small and mid-sized cities is sufficient to support non-stop flights to only a few destinations. Of more than 400 airports in the contiguous US with commercial service, the top 10% account for 80% of all passenger enplanements. The remaining 90% are highly dependent on connecting services from a network carrier. About one-third have daily non-stop service to cities other than a major airline hub, flights usually provided by a low-cost carrier (authors' computations based on Federal Aviation Administration and airport websites).

Low-density markets may be served with smaller aircraft. One role envisioned for the regional jet was in point-to-point service between mid-sized cities bypassing large hub airports, but the seat-mile-costs of regional jets are more than twice that of low-cost carriers operating mainline jets, usually the Boeing 737 and Airbus 320 series. The fares required to cover the operating costs of the regional jet are too high to stimulate traffic for expanded point-to-point service. Consequently, the role of the regional jet has been almost exclusively as a feeder to large network carriers.

Finally, demand varies significantly by day and season. Without connecting traffic from other cities that might offset some variation, the point-to-point carrier cannot avoid swings in route-specific demand. Only changes in flight frequency or aircraft size can match capacity and demand.

3.5.3 Example: Ryanair

Europe's largest low-cost carrier by passengers, Ryanair, is an example of a pure point-to-point airline. Figure 3.3 shows Ryanair's route map from Stockholm, Sweden. Every flight operates from Stockholm to one of the many destinations such as Milan, Italy, and then returns to Stockholm; there are no connections offered to other flights. So Stockholm is not a hub, but rather a base city for crews and aircraft. Ryanair has more than 80 base cities across Europe with every base sporting a similar point-to-point route structure. Aircraft and crews originate from the base each day and return every evening. Ryanair's European low-cost rival, EasyJet, employs a similar basing system and point-to-point route structure.

Ryanair, EasyJet, and Wizz have captured much of the intra-European market with point-to-point routes that bypass the hubs and connections required for flights with Europe's full-service network airlines.

Figure 3.3 Ryanair's Stockholm-based Routes.
Source: Map courtesy of FlightConnections.

In US post-deregulation history, Allegiant Air was the only example of a pure point-to-point system operating from more than 20 US bases. Allegiant initially focused on selling all-inclusive vacations to sun-soaked destinations in the Southwest and Florida but has rapidly expanded across the US. Allegiant connects mid-sized cities with non-stop flights to major destinations, a competitive advantage as larger competitors' service most often requires a connection at a hub airport. Frequency is low with many markets served only a few days per week. No connections are provided to any other Allegiant flight. Allegiant's success with this business model has drawn new entrants. In 2021, Breeze Airways and Avelo Airlines, both ultra-low-cost carriers, entered the US domestic market with point-to-point service. Breeze emphasizes the passenger benefits on its website proclaiming, "Straight to the point. Our point-to-point network means no connections for an easier, quicker, and nicer journey." (Breeze Airways, n.d.)

3.6 Hub-and-Spoke

The hub-and-spoke route system, the predominant route structure for the world's largest airlines, is much more complex to design and operate than either the linear or point-to-point systems. As route structure is the foundation of the airline's product, understanding the advantages, disadvantages, and operation of the hub-and-spoke system is essential for understanding the airline industry.

The H&S is optimized when providing air service to a wide geographical area and many destinations. Passengers departing from any spoke city who are bound for another spoke city in the network fly first to the hub airport where they transfer to a second flight to their destination. Thus, passengers not originating from or bound for the hub can usually travel between any two cities in the routing system with one connecting stop at the hub, or, as one author described it, "from anywhere to everywhere" (Hansson et al., 2002, p. 1).

3.6.1 Operation

A wheel analogy is helpful to visualize the H&S system. At the beginning of the day's flight operations, aircraft are positioned at each spoke city on the wheel. Aircraft depart these spoke cities to arrive at the hub in a short time window. In airline parlance, these inbound flights are a *bank* or *wave*. With all of the inbound flights parked at the hub airport gates, passengers whose destination is not the hub city, change to the aircraft that will depart shortly for their spoke city destination. Simultaneously, baggage and cargo are transferred to the appropriate outbound aircraft. After a sufficient connecting time, usually about 45 minutes for domestic operations and longer for international flights, aircraft depart, again within a tight time window, for the spoke cities. At the hub, this sequence of arrivals, connections, and departures is known as a *complex*.

A large full-service network carrier (FSNC) will operate several complexes per day. American Airlines, for example, operates eight complexes at its Dallas-Ft. Worth hub. After the day's operations, aircraft will again be positioned at the spoke cities ready to begin the next day's operations. Actual H&S operations are not as orderly as this. Some aircraft terminate at the hub from which they originate the next day's flying. Not all spoke cities are served in every complex; international destinations are usually served less frequently than domestic cities. There are many variations on the basic concept, some of which are explored later, but the defining characteristic of the hub-and-spoke system is the ability of passengers to make convenient connections to their destination at the airline's hub airport.

Some complexes for large airlines will integrate international and domestic flights with aircraft ranging from the largest wide-body aircraft to the smallest regional jets and turboprops. While tight arrival and departure sequences are desirable to minimize passenger connection times, airport and air traffic control cannot accommodate nearly simultaneous arrivals or departures of many flights, so some adjustments are required. The following sequence is typical:

* Flights from international cities arrive first allowing time for passengers to clear government immigration and customs before transferring to domestic flights. Sometimes international terminals are separate from domestic terminals so passengers will need additional connection time. Many international destinations are served by wide-body aircraft that required more time for ground servicing.

- Next to arrive are flights on regional jets. Terminals and gates for regional flights at many airports are less conveniently located so additional transfer time is needed.
- The domestic mainline jets are the last to arrive providing for the shortest connection times to spoke-city destinations. Occasionally arriving passengers will stay on their aircraft (called a thru-flight passenger), possibly for a flight to another hub or a large spoke. Having these aircraft arrive last minimizes the sitting time for those passengers.

For the departing bank, the sequence is reversed with domestic jets, regional jets, and finally international departures.

3.6.2 American Airlines Dallas-Ft. Worth hub

American Airlines is the world's largest carrier with 2019 revenues of $46 billion and a fleet of 900 mainline aircraft. American is headquartered in Dallas-Ft. Worth, Texas where it operates its largest of several US hubs.

Figure 3.4 (left panel) shows the Dallas-Ft. Worth route system with spoke cities of greatly varying size and distance from the hub arrayed in all directions. The longest international routes to Europe, Asia, and South America extend beyond the image. The right panel shows the daily scheduled arrivals and departures. The eight complexes are evident in the peak arrivals followed by peak departures. Each twin peak is a connecting complex that defines the hub-and-spoke system. Aircraft arrive in a bank, passengers and baggage transfer to flights for their destinations, and the outbound bank departs. There is also a departure bank at 7:00 am without a preceding inbound bank. Only a few connecting passengers feed this first outbound bank, so it depends on the substantial demand from the Dallas-Ft. Worth metropolitan area. Many passengers,

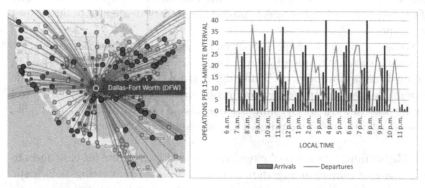

DFW Hub   Daily Arrivals and Departures

Figure 3.4 DFW Route System.
Source: Map courtesy of FlightConnections.

especially those traveling for business, demand the early morning departures provided in this first outbound bank. The next outbound bank from the first complex departs around 8:30 am, too late to meet the needs of business passengers on which all full-service network carriers depend. With the basic H&S concept in mind, the advantages and disadvantages of this route system architecture are now explored in detail.

3.6.3 Advantages

3.6.3.1 Minimizes Required Flight Segments

The H&S system connects an airline's destinations with the fewest number of aircraft and flights. As shown in Figure 3.5, providing service to five destinations consisting of four spoke cities and one hub, only four routes are needed. In contrast, 10 routes are required using the point-to-point system. Consequently, for any given level of frequency and number of destinations, the H&S requires the fewest number of aircraft. With the list price of a mainline aircraft beginning at around $90 million, minimizing the number of aircraft is a major financial consideration.

The H&S system combines the passengers bound for many destinations on a single flight to the hub airport. In Figure 3.5, a flight from spoke city #1 to the hub at #3 will include passengers destined to the hub and to spoke cities #2, #4, and #5. As the system grows, additional flight frequency and/or the use of larger aircraft positively affect both supply and demand. Passengers prefer to use a single airline for their entire journey, so the ability of an airline to offer service to many cities of varying sizes confers a competitive advantage.

3.6.3.2 City Pair Expansion

The number of city pairs served in a H&S system is given by the formula $[N(N-1)]/2$ where N is the total number of destinations (including the hub city). The five-destination network in Figure 3.5 comprises 10 city pairs: $(5 \times 4)/2 = 10$ city pairs, where the city pairs are 1–2, 1–3, 1–4, 1–5, 2–3, 2–4,

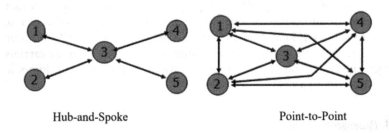

Hub-and-Spoke   Point-to-Point

Figure 3.5 Flight Segments Required.

2–5, 3–4, 3–5, 5–4. A large airline network serving 100 destinations from a single hub, such as American does from DFW, would be comprised of 4,950 city pairs. Adding a new destination strengthens the product offering. Adding one new city to this existing hub network creates 100 potential new city pairs available for sale. The additional (or marginal) cost of adding a new city is low in comparison to the number of new markets. However, the power of the additional city is not as great as the mathematics indicates because some city pairs involve such circuitous routings that passengers will opt for a competitor who offers less total travel time. For example, few passengers would choose to travel from city #4 to city #5 on the H&S system if a competitor offered P2P flights between #4 and #5.

In the United States, post-deregulation carriers were quick to realize the competitive advantages of expanded destinations and geographical coverage. Network expansion encourages travel, increases connectivity, and improves asset utilization; consequently, the former trunk carriers moved quickly to transform their route structures through growth, acquisition, and merger.

3.6.3.3 Consolidation of Demand

The hub-and-spoke's fundamental advantage is the ability to consolidate the demand from many city-pairs on each flight segment. Passengers on each flight from a spoke city to the hub are destined for many cities across the airline's route system. This consolidation allows for service to cities that would otherwise not be profitable. By combining the demand for many origins-destinations on a single flight, high frequency can be sustained if only two or three passengers are traveling to each of many spoke cities. Consider a hypothetical American Airlines flight from Albuquerque, New Mexico to its Dallas hub. Of the 123 passengers on board, 43 are going to Dallas, two will connect to Atlanta, three to Boston, two to London, and the remaining 71 to 28 other destinations. With a large network, a new destination or spoke city can be added if it generates only a few passengers per day to each of the many destinations in the network. Thus, a network carrier can profitably serve smaller cities that a point-to-point carrier cannot.

However, as low-cost airlines have captured more short-haul traffic with point-to-point service by-passing traditional hubs, the number of passengers connecting at hub airports has fallen. McKinsey and Company, a consultancy, estimates that in 2019, 83% of passengers worldwide traveled without a connection. Connections for intercontinental travel, on the other hand, increased from about 45% in 2005 to between 50% and 60% in 2019 (Bouwer et al., 2020). Even before the pandemic, Europe's full-service network carriers had largely given up on intra-European travel structuring their domestic systems to feed international flights.

3.6.3.4 Passenger Convenience

As destinations are added and more passengers funnel through the hub, flight frequency can be increased. High frequency allows the passenger to match flights with desired itinerary times. This choice of departure and arrival times is especially valued by business travelers. Passengers prefer to use a single airline for their entire journey, so the ability to service many cities of varying sizes confers a competitive advantage. As compared with a journey requiring a change of airline, passengers making a single-hub connection benefit from closely timed flights, single check-in, more convenient gate and facility locations, and reduced risk of lost baggage. Knowing that an airline likely serves the desired destination saves the passenger search and transaction costs. Familiarity with the airline's product lessens uncertainties and increases loyalty, particularly when linked to loyalty programs. In economic jargon, these benefits are *positive network externalities*.

On the supply side, economies of traffic density lower seat-mile costs. Larger aircraft can be utilized as the number of passengers per flight segment increases. Because seating capacity increases faster than operating and capital costs, the cost per available seat mile is reduced. Finally, to the extent that demand patterns of city pairs are not highly correlated, total demand is smoothed. A decrease in demand in one city pair may be offset by an increase in another. Both of these effects allow the network carrier to provide service over a wide geographical area that is only feasible with the H&S system.

3.6.3.5 Hub Dominance

As an airline's hub system grows, the airline comes to dominate its hub airport(s). Domination confers a degree of market power enabling the carrier to extract a price premium on flights to and from the hub. In the wake of US deregulation, this premium was estimated at 20% or more, although it has gradually decreased with the growth of LCC competition (Borenstein, 2011). The reason for the hub premium has been debated with many critics claiming quasi-monopoly pricing. It's also possible, however, that passengers are willing to pay a premium for the higher flight frequency and convenience that the hub carrier provides.

3.6.3.6 Competitive Strength

Hub airlines also exercise considerable influence over airport operators that can be used to impede competition through control of airport facilities. If faced with a new entrant to its hub threatening an existing market, the airline can defend its hub airport by adding flights and matching or undercutting the new competitor's fares. This competitive advantage results in the *fortress hub*. Aggressive competitive responses are potentially anti-competitive and illegal in many countries. In one flagrant example, in 1993 Reno Air, a small new entrant airline, started service from Reno, Nevada to Minneapolis-St. Paul (MSP), home to Northwest Airlines (now merged into Delta). Even before the service started, Northwest announced new service from MSP to Reno plus additional flights from Reno to Seattle, Los Angeles, and San Diego overlaying the heart of Reno Air's route system. In the face of Northwest's threat, Reno Air dropped its MSP flights after only three months.

In response to frequent aggressive tactics employed by Northwest and other legacy US carriers, the Justice Department sued American Airlines in 2001 for engaging in illegal predatory pricing intended to drive new entrant airlines from its markets. The judge dismissed the suit ruling that American did not price below its cost and did not undercut the new entrant's fares (Swododa, 2001).

3.6.3.7 Widespread Distribution

Advertising and distribution also benefit from the H&S system. The hub-and-spoke carrier can advertise flights to many destinations thus spreading its promotional expenses across many destinations and daily flights. A low-cost carrier utilizing a point-to-point system, on the other hand, offers relatively few destinations from any city. Large-scale advertising is often not economically feasible until the base of flights is large.

3.6.3.8 Attractive FFP

Finally, frequent flier programs (FFP) that encourage passenger loyalty are more effective when the airline serves many destinations, particularly a variety of desirable vacation areas. Business passengers value their mileage awards that allow for free travel – awards that would likely be viewed as kickbacks in other settings – and want to use the awards for vacation travel. As business travelers often are members of several airline FFPs, loyalty effects have gradually faded. However, FSNCs and other large airlines subsequently leveraged their FFPs by selling points to third parties substantially contributing to ancillary revenues.

3.6.4 Disadvantages

For all their marketing advantages, hub-and-spoke systems are expensive to operate compared with point-to-point and linear systems with higher costs incurred in both aircraft operation and passenger handling. The strength of the H&S system, the ability to channel many destinations through a hub airport, is also the source of its weakness. Typically, 40% of all H&S airline passengers have the hub city as their origin or destination. The remainder only passes through the hub to make outbound connections. This additional stop, as compared with point-to-point service, increases costs.

3.6.4.1 Infrastructure and Labor

Extensive facilities and personnel are needed solely to accommodate connecting passengers. The passenger service agents, gates, lounges, baggage facilities, and ramp personnel dedicated to passenger connections are not necessary for point-to-point flights. At times between the connecting complexes, many of these assets and personnel are idle.

3.6.4.2 Flight Operations Expense

The intermediate stop at the hub increases flight expenses. Each of the two flight segments connecting a passenger's origin and destination is shorter than a single non-stop segment. Short segments are more expensive to operate. The aircraft achieves lower block speeds because of additional taxi time and maneuvering for takeoff and landing. More time spent at lower altitudes increases fuel burn. Flight crew pay is based on block time as are some maintenance costs. Other maintenance costs vary with the number of takeoffs and landings (cycles), so these increase with the intermediate stop.

3.6.4.3 Pacing spokes and circuitous routing

Route system geography also drives higher costs. The wheel analogy implies that all spoke cities are at an equal distance from the hub, but this is misleading. The distance from the hub to the spokes varies greatly. If aircraft in a departing wave comprise part of the subsequent connecting complex, aircraft operating from the closer spoke cities must await the return of aircraft from the most distant cities before returning to the hub. The longest spoke flight in a complex thus becomes the *pacing spoke*. Consequently, aircraft and crews sit idle at many close-in spokes as the connecting complexes cycle through the day. Consider again American's DFW hub. El Paso, Texas (ELP), and Las Vegas, Nevada (LAS) are spoke cities but LAS is twice as far from DFW as ELP. For LAS and ELP flights to arrive in the same inbound bank at DFW, the ELP aircraft must lie idle in ELP allowing the LAS airplane to travel a much greater distance. Scheduling techniques, particularly for carriers operating many complexes per day, can mitigate this poor asset utilization, but limits remain.

Second, the hub is in line with only a few city pairs, so connecting flights between most cities entail circuitous routings to the hub lengthening the total mileage flown and flight time with an attendant increase in cost.

3.6.4.4 Lengthy complex times

As an airline grows its operations at a hub, airport capacity limits cause rapidly escalating costs. The most obvious cause is flight schedule imposed hub congestion. Arrivals and departures are limited by the available airspace, runways, and taxiways. Inclement weather further reduces airport capacity as greater aircraft spacing is required, particularly for arrivals. Arrival rates are halved at some airports when weather requires instrument approaches. If departing aircraft are delayed, gates are unavailable for inbound flights causing more delays. Terminal space, especially in older terminals, is taxed. Essentially, the hub carrier creates its own traffic congestion by scheduling ever more flights in each complex. Scheduling constraints sometimes cause airline planners to publish unrealistically short connection times that increase passenger stress and detract from customer experience. Addressing the problems by extending the complex time further lowers aircraft and crew utilization. The largest airline at a hub bears most of the cost of congestion delays, but other airlines are caught up in the queues as well. Critics often accuse airlines of scheduling more arrivals and departures than the airport can handle in the allotted time.

3.6.4.5 Mixed Fleet Requirement

The H&S system lends itself to serving cities of greatly varying size and demand. Typically, these include a range from small cities only a few hundred miles from the hub to the world's largest cities that may be thousands of miles away. Short segments to small cities are best served with regional aircraft or the smallest mainline jets whereas long-haul international flights are usually operated with wide-body aircraft. The fleet of a large network carrier usually consists of several aircraft types with seating capacities ranging from 50 to 500 or more. As fleet commonality decreases, costs increase to train pilots and mechanics, inventory varied parts, and acquire and maintain fleet-specific support equipment. Aircraft and crew scheduling are more difficult and constrained as fleet complexity grows. Learning curves are more slowly exploited.

3.6.4.6 Susceptibility to Delays

Finally, H&S spoke systems are highly susceptible to delays. A delay on one or a few inbound flights can spread as outbound flights are held for connecting passengers. Disruptions that affect the entire hub, particularly weather, but also radar or computer outages, often propagate rapidly through the entire flight operation. Multiple hubs provide some opportunities for mitigation.

3.6.5 Bottom Line

The cost to handle a connecting passenger in a hub-and-spoke system is 30% to 50% higher than with a point-to-point system, a disparity that costs H&S airlines billions of dollars annually (Doganis, 2019). To earn a profit, these carriers attempt to maximize revenues by offering a full array of services. The high frequency and geographical scope of the H&S system are particularly valuable to business travelers for which they pay a price premium. Without this fare premium, the large H&S carrier cannot compete with lower-cost rivals.

3.7 Hub Airport Requisites

Most airports are not well-suited for hubs. Successful hub airports share several characteristics, such as:

- The city must have a large population and a strong economic base. In addition to the service to spoke cities that the hub supports, the hub city must be a significant destination in its own right. Typically, 40% or more of the passenger traffic should have the hub city as either origin or destination. For example, about 40% of Lufthansa's long-haul passengers originate or terminate at its hubs in Frankfurt and Munich, while about 60% are transfer passengers (O'Connell, 2011).

- The hub must be centrally located so that routes to the spoke cities avoid extensive circuitous routings. For an airline with a large and geographically dispersed route, multiple hubs can reduce circuitous routings.
- Airport infrastructure and local transportation must be sufficient to support the hub operations. Many hub airports lack runways, taxiways, terminals, and gates to support future airline growth.

3.7.1 Legal, Financial, and Capacity Restrictions

Airlines must often compete for scarce airport facilities at the world's largest airports which are frequently the first choice for additional flights. The International Air Transport Association (IATA) provides guidelines for managing airport capacity restrictions. A slot is a legal right for an airline to land or takeoff at a restricted airport within a specific time period. Slots are only useful in pairs, one for a landing and the other for a takeoff. Slots are administered by the operator of the airport or a government aviation regulator ("Landing slot, n.d.). In 2021, IATA listed 156 airports worldwide that were slot restricted. Another 196 were subject to less stringent conditions encouraging airlines to coordinate their schedule through the airport operator. Most of Europe's airports operate at capacity and are slot restricted, including 17 airports in Greece and 18 in Italy. In China, 22 airports were slot-controlled (IATA, 2021).

Slots not used regularly used must be returned to the operator. Slots are valuable and can be traded between airlines, often for millions of dollars. In 2016, Oman Air, a tiny Gulf airline, paid $75 million to Air France-KLM for a pair of slots at London Heathrow, the highest known price (O'Connell & Collingridge, 2016).

3.7.2 Hub Failures

An airport may meet the hub requirements and still fail as a hub, often as the victim of airline consolidation. Redondi et al. (2012) examined 37 hubs that had been "de-hubbed" worldwide between 1997 and 2009, including St. Louis, Milan, Brussels, and London Gatwick. Only three subsequently regained hub status. Some airports that seem attractive have failed repeatedly. Kansas City was a hub for TWA, then Eastern, and finally Vanguard. Perhaps not coincidentally, these airlines also passed into history.

3.8 Hybrid Route Systems

Examining the three generic route systems in isolation is useful for illustration; however, pure generic structures are, in fact, rare. Most large airlines operate

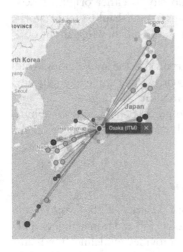 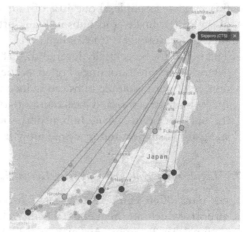

Osaka domestic hub-and-spoke   **Sapporo point-to-point**

Figure 3.6 Japan Airlines Route Systems.
Source: Maps courtesy of FlightConnections.

hybrid systems with a mix of hub-and-spoke, focus cities, and point-to-point systems. China Eastern Airlines operates five primary hubs, eight secondary hubs, and nine focus cities. Japan Airlines is a simpler example. From its primary hub in Tokyo, JAL operates international flights to Europe, the US, and Southeast Asia plus all large domestic markets. A second hub at Osaka serves domestic markets as shown in Figure 3.6 (left panel). Other cities, often served by JAL's several regional airlines, such as Sapporo on the northern island of Hokkaido, receive point-to-point service (right panel).

Low-cost carriers are not exclusively point-to-point airlines; Frontier Airlines is an example. Frontier (the original Frontier was absorbed in a series of mergers) began operations in 1994 with a hub-and-spoke system from Denver that continues to grow in direct competition with United Airlines' Denver hub operations. But it also operates many point-to-point routes throughout the States often with less than daily service. In 2021, it began developing another hub in Miami that competes with American Airlines. Surprisingly, the ultra-low-cost carrier Spirit Airlines, which is based next door in Ft. Lauderdale, also initiated extensive service from Miami at the same time.

3.9 Hub-and-Spoke Variations

As successful airlines grow and expand to serve more destinations often across many continents, several variations to the H&S system have evolved. This section examines the most common.

3.9.1 Multiple Hubs

The operation of a hub-and-spoke system with one hub is challenging, but large network airlines often have multiple hubs. The largest network airlines in China, Europe, and the United States all operate several hubs. The most important reasons for this additional complexity are (a) hub congestion; (b) geographical dispersion of destinations; and (c) operational flexibility (Wang & Wang, 2019). Before the COVID-19 pandemic, the International Air Transport Association projected that just four of the world's largest 100 airports would have sufficient capacity to meet passenger demand by 2028 (cited in Bouwer et al., 2020). Beijing opened a second airport in 2019 to relieve congestion; Air China operates hubs at both Beijing airports. European hub airports have long been at or near capacity. Lufthansa uses Munich as a secondary hub to alleviate congestion at Frankfurt. Following the merger in 2004, Air France/KLM adopted a twin-hub strategy developing both the Amsterdam and Paris hubs. British Airways operates a primary hub at London Heathrow. Heathrow, being the most valuable of the European hubs, has been at capacity for decades. International Airline Group, the holding company for British Airways, owns several other European carriers effectively providing a multiple hub system. Multiple hubs alleviate congestion but often serve many of the same city pairs reducing individual hub demand.

For an airline network that spans a large geographical area such as China and the US, destinations often cannot be efficiently served from a single hub. China Eastern Airlines operates five primary and several secondary hubs. Air China and China Southern also have multiple hubs although fewer than China Eastern. In the US, American Airlines, Delta Air Lines, and United Airlines currently operate 10, nine, and seven domestic hubs, respectively. Qantas, Australia's flag carrier, operates five hubs as shown in Figure 3.7.

Qantas serves all of Australia's major cities, but, because of the vast geographical expanse of the country, cannot do so efficiently from a single hub. Cities in the more densely populated east are connected by hubs in Brisbane (BNE), Sydney (SYD), and Melbourne (MEL); Adelaide (ADL) is needed to connect some central destinations while Perth (PER) is better situated to serve the sparsely populated west.

Travel between some smaller cities can only be accomplished with connections in two hubs. Suppose, for example, you wished to travel from Cairns, on Australia's northeast coast to Port Hedland in Western Australia. Of several options, all requiring two hub connections, the shortest total travel time is over 11 hours with connections in BNE and PER. The service, however arduous, is only possible because Qantas operates multiple hubs. There are many other examples worldwide.

Hubs are subject to operational disruptions, especially from adverse weather such as snow, ice, or thunderstorms causing extensive delays or closing the hub for hours throwing an airline's operation into chaos. If the airline operates just this one hub, recovery from irregular operations may require a few days.

Figure 3.7 Qantas' Five Hubs.

Multiple hubs, on the other hand, provide an airline with greater flexibility to reroute passengers, aircraft, and crew providing for better passenger service and faster resumption of normal operations.

3.9.2 Directional Hub-and-Spoke Systems

The hub-and-spoke system provides convenient passenger connections through tightly-timed complexes. With the omnidirectional hub, flights arrive at the hub from all spoke cities for each connecting complex. This system works best if spoke cities are evenly distributed around the hub. But the system is complex and expensive to operate. With the addition of new destinations, congestion and delays increase.

For many airlines, spoke cities are not uniformly dispersed around the hub but are instead aligned on opposite sides, usually east-west or north-south. For such a network, a directional hub-and-spoke system is more efficient and less costly to operate. In a directional hub, only flights from spoke cities on one geographical side of the hub airport arrive in the inbound wave of a complex. After the transfers of passengers, baggage, and cargo are complete, flights then depart for spokes cities on the opposite geographical side of the hub. This directional complex is then followed by a complex operating in the opposite direction. The series of alternating directional complexes may be repeated several times during the day. In contrast with the omnidirectional hub, the directional system requires about half as many airport facilities and personnel resulting in substantial cost reduction.

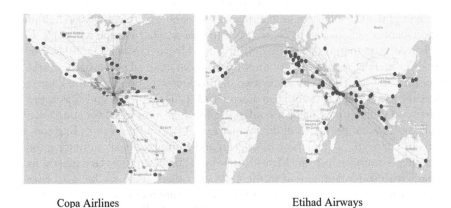

Copa Airlines   Etihad Airways

Figure 3.8 Examples of Hub-and-Spoke Systems.
Source: Maps courtesy of FlightConnections.

The route networks for two airlines operating directional hub–and–spoke systems are shown in Figure 3.8.

The first is Copa Airlines. Copa has a long history having been founded in 1947 by a group of Panamanian businessmen in partnership with Pan American World Airways. Ownership changed several times over the years, including 10 years when Continental Airlines held a 49% stake. Today, Copa is a public company listed on the New York Stock Exchange. In 2005, Copa acquired the Columbian domestic regional airline AeroRepublica, which it renamed Copa Airlines Columbia (Centre for Aviation, 2015). Copa operates a single fleet type with several Boeing 737 models.

Copa's only hub in Panama City, Panama is ideally located to connect cities in South America with those in the Caribbean, Central, and North America. For example, a passenger traveling from Santiago, Chile to Toronto, Canada can do so with a single connection in Panama City. However, few passengers would be willing to travel from Santiago to Rio de Janeiro, Brazil with a connection in Panama City because of the long, circuitous routing. Other airlines offer much more convenient flights. So, an omnidirectional hub would not attract passengers traveling between most cities in North America or between cities in South America. A better solution is a series of directional hubs. Copa's first daily complex is comprised of aircraft arriving in Panama City from South American destinations. After time for passengers and their baggage to connect, flights depart to northern spoke cities. This south-to-north complex is followed by a complex operating north-to-south. The series of directional complexes continues throughout the day. Of course, it's not this simple as connections to some potential destinations are not included in every complex. Some cities are served several times per day and others only once. Copa, for example, operates four daily flights to and from New York but only one to Toronto.

Etihad Airways, the national airline of the United Arab Emirates and the smallest of the three Persian Gulf airlines, is a second example. Its single hub in Abu Dhabi is well located to connect North America and Europe with Asia. As Etihad Airways then president and CEO James Hogan explained in 2013, "We are quite literally at the center of the world. We can fly to all points of the world nonstop" and thereby connect any two destinations via one stop in the United Arab Emirates' capital (quoted in Karp, 2013, p. 1). Emirates and Qatar, the other two Gulf airlines, make a similar case for their hubs in Dubai and Doha. Etihad operates most flights in directional, east-west complexes providing convenient connections from North America and Europe to Asia and return, for example, a round-trip from Madrid to Tokyo. Etihad's routes vary greatly in length with some destinations in the Gulf and others, such as Chicago, being more than 7,000 miles. Operating this route structure requires a mixed fleet of varying capacity and range capability. As of 2021, Etihad's fleet of 88 passenger aircraft comprised Boeing 787s, 777s, and Airbus 320s. During the pandemic, Etihad reduced the number of fleet types by eliminating Airbus 330s, 340s, and 380s as part of a restructuring to increase efficiency and reduce cost.

The Persian Gulf carriers operate two or three connecting complexes per day. Their rapid growth before the pandemic posed a serious competitive threat to other global carriers, particularly the European airlines whose valuable connections to Asia faced increased competition. US and European airlines complained to their governments and anyone else who would listen that the Gulf airlines were unfairly subsidized by their national governments. The Gulf carriers responded that their governments understood the importance of the airline industry and adopted policies that support growth whereas the US and EU impose heavy taxes and burdensome regulations.

3.9.3 Rolling Hub

During periods of economic recessions, airlines typically focus on reducing costs while waiting for economic recovery and a return of air travel demand. One approach to cost-reduction is a modification of the hub-and-spoke system known as the *rolling hub* or *depeaking*. In a rolling hub, arriving and departing flights are spread more evenly through the day with less emphasis on the closely timed connecting complexes. This results in fewer constraints on flight schedules, higher aircraft utilization, better employee productivity, and fewer hub infrastructure requirements, but at the cost of longer connecting times for passengers.

In response to the financial stresses that followed the recession of 2001, American Airlines and Delta Air Lines both experimented with rolling hubs. The flight frequency histogram (Figure 3.9) compares American's flight schedule at Dallas-Ft. Worth before and after the implementation of the rolling hub. In August of 2001, before depeaking, the ten daily peaks in flight frequency correspond to the connecting complexes. In March 2003, by spreading arriving and departing flights more evenly throughout the day, peak frequencies

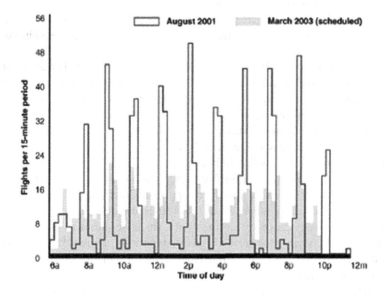

Figure 3.9 Effect of Rolling Hub at Dallas-Ft. Worth.
Source: Courtesy of Tam & Hansman, 2003.

fell by more than half. These results were due in part to implementing more pronounced directional flow at Dallas and in Chicago somewhat blurring the distinction between the rolling and directional hubs.

American gained the equivalent of 11 aircraft by improving aircraft utilization at Dallas and another five in Chicago. Flight crew, ground crew, and airport gate utilization increase. The process saved American an estimated $100 million annually (Goedeking & Sala, 2003); however, passenger convenience suffered. Some connections were lost entirely and the average passenger connection time at Dallas-Ft. Worth increased by approximately 11 minutes.

When Delta restructured under bankruptcy in the fall of 2004, depeaking played a central role. At Atlanta, Delta lowered flight operations from 80 to 66 per hour but increased the total number of daily flights from 970 to 1,051. Secondary hubs at Cincinnati and Salt Lake City were also depeaked. Both received more flights a day, but fewer per hour, with Cincinnati going from 590 to 619, and Salt Lake City from 318 to 376. Despite this improvement, as Delta's restructuring continued, Cincinnati was subsequently dropped as a hub ("Hubs Rejig," 2004).

Depeaking can improve productivity and efficiency, but it's not without downsides, especially in the presence of a strong competitor. Cost savings are relatively easy to measure, but the loss of connectivity adversely affects revenue. American's experience at O'Hare, where United Airlines also operates a hub, is instructive. As American's rolling hub freed up airport capacity, United further concentrated its connections. United's competitive response contributed to American losing four market share points in Chicago. In Dallas, where American has scant competition, it suffered a market share loss of only 1% (Goedeking & Sala, 2003).

As the American and Delta examples illustrate, a rolling hub is feasible at airports where an airline offers hundreds of daily flights because the high frequency ensures that most city pairs will continue to enjoy convenient connections. Still, some city-pair connections will be lost entirely and the quality of other connections will suffer. The schedule must protect high volume, high yield markets whereas less profitable markets can be compromised to optimize productivity. At smaller hub airports with less frequency, the loss of connectivity renders the rolling hub uneconomic. Hubs best suited to depeaking are characterized by highly directional, short-haul, high volume, and limited airport capacity whereas hubs that are multi-directional, long-haul, and have a high dependence on connectivity are least suited to depeaking. Most European hubs are not well suited to depeaking, although Lufthansa depeaked its Frankfurt hub and Scandinavian did so at Copenhagen and Oslo (Katz & Garrow, 2013).

As with most strategic decisions, depeaking involves a compromise between cost and revenue. Depeaking reduces costs but sacrifices some revenue from lost or unattractive connections. American and Delta both reversed course and *re-banked*, to use American's terminology, beginning in 2014 with American's Miami hub. Judging the Miami experiment a success, American then re-banked both its Chicago and Dallas hubs in 2015. United also reversed its earlier de-peaking strategy re-banking its Denver and Houston hubs in 2014 followed by the Chicago O'Hare hub in 2015 nearly in unison with American. In its new schedule, United shorten connection times but retained directional flows (Unnikrishana, 2015) demonstrating that rolling and directional hubs are separate, distinct strategies that can be implemented individually or jointly.

3.9.4 Focus Cities

Many airlines that do not operate traditional hub–and–spoke systems concentrate flights in several cities. In a *focus city*, a term that has gained wide acceptance, an airline operates so many flights during the day that convenient connections arise without specifically planning to do so.

At Southwest Airlines, an early adopter of the focus city, analysts determine and publish the available connections after the flight schedule is finalized. In his analysis, James Pearson (2021) found that in 2019 more than 38% of Southwest's passengers connected to other Southwest flights at its several focus cities. The percentage of connecting passengers is increasing and has become a key to Southwest's operations.

JetBlue is another example of an airline that initially operated a point-to-point schedule that increasingly relies on connections. In 2021, it entered an alliance with American Airlines promoting seamless connections in New York and Boston. It also partners with several other domestic and foreign airlines.

3.9.5 Tailored Complexes

Hub operations can be further tailored to meet specific objectives. For example, intense competition from low-cost carriers, especially Ryanair and EasyJet, has rendered intra-European flights unprofitable for Europe's FSNCs. Among the responses to this challenge is a redesign of the connecting complexes. Before the pandemic devastated international travel, Air France operated six daily complexes at its Paris-Charles de Gaulle (CDG) hub. Although there is significant short-haul flying built into the CDG hub, the complexes are designed so that intra-European connections are unattractive. Instead, short-haul European flights feed international flights. The first-morning complex includes inbound long-haul flights connecting to short-haul European destinations. The late-morning outbound wave is focused on North American flights whereas the early afternoon complex is planned so that inbound European flights feed long-hauls to North America and Africa (Flo, 2013). This design illustrates how the flow of aircraft to and from the hub supports the most profitable passenger connections.

3.10 Competing Hub and Spoke Systems

Airlines compete intensely in most of the world's largest markets, competition that benefits passengers with lower prices and expanded services. With each large airline merger come complaints that fewer carriers will reduce competition and harm consumers. The analysis often draws on fears of reduced competition because of a likely reduction or elimination of competition on overlapping routes where a route is understood to be a flight between two cities. This analysis, however, is too simple and misleading because airlines compete in city pairs, which may or may not be served by overlapping routes.

Demand for airline flights is derived from each passenger's desire to travel from one location to another, not from the airports or routes available to fulfill the passenger's travel needs. Except for short flights, passengers usually have a choice of airlines, schedules, and route structures. Figure 3.10 left panel shows two highly simplified hub-and-spoke systems, one centered at Denver (DEN) and the other at Dallas-Ft. Worth (DFW). Note that the only overlapping route is between the two hubs. Both airlines fly between DEN and DFW, but each airline's route system connects all city pairs and the airlines compete for the same passengers. Thus, competition may be intense despite the lack of overlapping routes. This competition among hubs creates many rivalries within the domestic US market. Similarly, but on a global scale, the three Persian Gulf carriers and Turkish Airlines compete with the largest European airlines long-haul international travel (right panel).

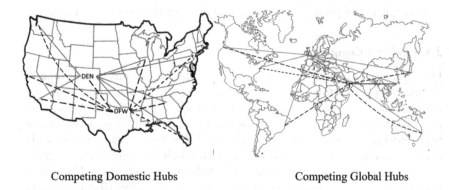

Competing Domestic Hubs   Competing Global Hubs

Figure 3.10 Competing Hub-and-Spoke Systems.

Many large cities are served by more than one commercial airport, so in analyzing competition between city pairs, it's important to consider all of the airports within the *catchment* area. London, for example, is served by six commercial airports with Heathrow being the most desirable for long-distance flights. Three major commercial airports serve Paris. In the US, the Washington DC area has three major airports – Washington Reagan, slot-and range-restricted, but closest to the centers of government in the capital; Dulles in Northern Virginia; and Baltimore-Washington in Southern Maryland. For a business traveler, a non-stop flight from Dulles may not be an acceptable substitute for a similar flight from Reagan as it is much closer to the city center and Federal government offices. However, the vast majority of passengers in the US travel for leisure so Dulles and Baltimore-Washington are substitutes.

3.11 Evolving Routes Systems

The airline industry is dynamic; route systems are not static but constantly evolving to meet changes in competition, travel demand, and passenger wants and needs. When Scott Kirby left American Airlines to become president of United Airlines in 2017, he expanded United's hub-and-spoke network by adding smaller cities most of which were already served by American (see case study below). This expansion may seem to run counter to passengers' preference for non-stop flights, but as Brett Snyder (2017) observed, low-carriers have expanded to compete with FSNCs in most major US domestic markets driving fares lower. There's much less competition in small cities, so the hub-and-spoke airline can charge much higher fares. In 2022, the lowest price for a midweek roundtrip on United Airlines from Chicago to Ft Lauderdale, a route with lots of LCC competition, was a bargain at $91. In contrast, a roundtrip originating in the nearby small city of Madison, Wisconsin connecting in Chicago to the same flight to Ft Lauderdale was priced at $239 roundtrip. The LCCs don't serve Madison providing United with the market power to extract a large fare premium.

3.12 Summary

An airline's most critical strategic choices are the destinations it serves and the route structure design connecting these cities. The three generic route system architectures, point-to-point, linear, and hub-and-spoke, vary in complexity and cost of operation. Point-to-point is simple and the least costly. The hub-and-spoke is complex to design and operate but allows for a broad geographical coverage to cities of greatly varying sizes. The linear system falls somewhere in between. Most large carriers, as well as many low-cost airlines, choose the hub-and-spoke system despite the higher cost of operation because it allows for many destinations to be served on each flight to one or more hub cities where passengers transfer to a second flight to their destination. This combined demand is sufficient for profitable service to cities too small to support point-to-point service. Because of the predominance of the hub-and-spoke route system in both domestic and international air travel, students and practitioners must appreciate the marketing advantages and operational complexity this route architecture poses. Table 3.1 compares the point-to-point and hub-and-spoke systems.

Route structure and the destinations an airline chooses to serve are the foundations of its product. The airline, however, may choose a product ranging from Spartan to expansive. This is the subject of the next chapter.

3.13 Case Study: United and American Airlines' Fickle Domestic Route Systems

When Scott Kirby left American Airlines in 2016 to become president – and later CEO – of United Airlines (UAL), he knew that United had a superior international network but lagged American in the scope of its domestic hub-and-spoke system. UAL lacked the connecting traffic that its rival, and Kirby's long-time employer, American Airlines enjoyed. Kirby set about correcting this imbalance in the spring of 2017 when United added dozens of new domestic flights, several on new routes. As Brett Snyder (Cranky Flier, 2017 March 2) noted at the time, "These new routes were clearly pulled from the American Airlines playbook that Scott was operating under until he headed north to Chicago last year. This week, United put out a big summer expansion plan, and guess what? It's all about connections."

Kirby's emphasis on building domestic connections at its several US hub airports was the result of several factors. International expansion, long the focus of US FSNC growth, became less attractive. Immediately before the onset of the Great Recession of 2008, fuel prices spiked at $140 per barrel, only to fall precipitously to $42 six months later. In the interim, demand collapsed. Airlines reacted by cutting flights to match the lower demand and reduce costs.

Table 3.1 Comparison of Hub-and-Spoke and Point-to-Point Systems

Attribute	Point-to-Point	Hub-and-Spoke
Scope	Each route serves a single city pair. Routes may be widely dispersed.	Optimized by connecting service to a wide geographical area and many destinations.
Connectivity	No connections provided (although some may arise randomly as focus cities develop).	Most passengers, typically about 60%, connect at the hub airport(s) for flights to their destinations.
Dependence	Flights operate independently.	Flights are highly dependent to provide passenger connections.
Market Size	Requires high-density markets with at least one city of the pair being a high-traffic destination.	Efficiently serves cities of varying sizes, including cities with too little demand to be profitably served with point-to-point routes.
Demand	Only varying frequency and pricing are available to counter demand variations.	Varying demand in one or more city pairs may be offset by demand from other city pairs.
Frequency	Supports a large range of frequencies from several per day to only a few per week.	Generally, high frequency is tied to the connecting hub complexes.
Pricing	Both business and leisure pricing depends on the city-pair demand characteristics. Passenger preference for non-stop flights will often support a price premium.	Dependent on business travelers at high fares with excess capacity filled with discounted fares to leisure travelers.
Asset Utilization	No network constraints on utilization usually result in high aircraft and crew utilization	Limited by network geography, the timing of complexes, and the capacity of hub airports.
Cost of Operation	Lowest cost per available seat mile per city pair.	Hub connections significantly increase the cost per available seat mile which is somewhat offset by the use of larger aircraft.
Fleet Requirement	Suited to a single aircraft type.	Range in city demand usually requires a mixed fleet.

Source: Adapted from Cook & Goodwin, 2008.

Small and midsize cities bore the brunt of route cuts. By 2017, with a robust and growing economy, airlines were again targeting expansion. Adding flights from smaller cities to major hubs boosted connecting traffic while avoiding intense competition from aggressive low-cost carriers that depressed prices.

Over several months, United added or upgraded service in 31 markets; some were previously American Airlines' monopoly routes.

As would be expected, American Airlines reacted aggressively to protect its markets. In January 2018, it announced plans to launch 49 new nonstop domestic flights. With its 2013 US Air merger completed, Vasu Raja, American's vice president for network and schedule planning, said American appraised its network, undoubtedly with an eye on United's intrusions. A new, more flexible pilot contract allowed American to outsource most new routes to its regional airline partners operating 50-seat jets.

These domestic expansions were followed shortly thereafter by the pandemic, which of course, forced airlines to prune their routes and reduce costs. United began a series of cuts to domestic cities in late 2021 announcing it was dropping unprofitable routes to a dozen cities, all of which were operated by its regional partners. In the Spring of 2022, the effects of the pandemic were compounded by the pilot shortage. United said it was exiting 29 cities because its regional partner, Skywest Airlines, was short of pilots. The airports are all part of the federal Essential Air Service program, an accommodation to airline deregulation in 1978 that ensured air service to small Communities. The Department of Transportation stayed SkyWest's exit until replacement airlines could begin service.

Concepts: Full-service network carriers, regional airlines, low-cost carriers, hub-and-spoke systems, recessions, pandemic.

Sources: Cameron, D., 2018; Cranky Flier, 2017 March 2; Cranky Flier, 2017 March 18; Griff, Z., 2021; Rains, T., 2022.

References

Borenstein, S. (2011). What happened to airline market power? UC Berkeley Working Paper.

Bouwer, J., Krishnan, V., & Saxon. (2020). Will airline hubs recover from Covid-19? *McKinsey and Company.* www.mckinsey.com/industries/travel-logistics-and-infrastructure/our-insights/will-airline-hubs-recover-from-covid-19

Boyd, M. (2010, Aug 16). Hot Flash. *Boyd Group International.*

Breeze Airways. (n.d.) www.flybreeze.com/home

Cameron, D. (2018, January 22). Large airlines flock back to midsize cities; lower fuel prices, competition from discount carriers increase appeal of flying to smaller markets. *Wall Street Journal.*

Cook, G. & Goodwin, J. (2008, Winter). Airline networks: A comparison of hub-and-spoke and point-to-point systems. *Journal of Aviation/Aerospace Research and Education*, 51–60.

Cranky Flier. (2017, March 14). *American and United fight to see who can lose more money in small cities.*

Cranky Flier. (2017, March 2). *United's big summer expansion confirms its focus on connecting traffic.*

Doganis, R. (2019). *Flying off course: Airline economics and marketing.* Routledge.

Federal Aviation Administration. (2021). *Passenger boarding (enplanement) and all-cargo data for US airports* [Data set].

Flo, J. (2013, July 22). Europe's large hubs struggle as the small ones disappear. *Aviation Daily*.

Goedeking, P., & Sala, S. (2003). Breaking the bank. *Airline Business*, 19(9), 93. http://search.proquest.com.ezproxy.libproxy.db.erau.edu/docview/204263745?accountid=27203

Grant, J. (2021, March 10). The power of connecting passengers – later guest arrivals to become the new normal. *OAG*. www.oag.com/blog/power-connecting-passengers-later-guest-arrivals-become-new-normal

Griff, Z. (2021, November 8). United Airlines is dropping 8 more US cities from its route map. *The Points Guy*.

Hansson, T., Ringbeck, J, &Franke, M. (2002). Fight for survival: A new operating model for airlines. *Strategy + Business*. Retrieved from www.strategy-business.com/article/22462?pg=0

Hubs rejig will open door for rivals. (2004). *Airline Business*, 20(10), 14. Retrieved from   http://search.proquest.com.ezproxy.libproxy.db.erau.edu/docview/204266708?accountid=27203

IATA. (2021) *List of slot coordinated and facilities airports*. www.iata.org/en/programs/ops-infra/slots/slot-guidelines/

Karp, A. (2013, April 5). Can Abu Dhabi, Dubai and Doha all be thriving global hubs? *AirKarp*. Retrieved from http://atwonline.com/blog/can-abu-dhabi-dubai-and-doha-all-be-thriving-global-hubs

Katz, D., & Garrow, L. A. (2013). Depeaking schedules: Beneficial for airports and airlines? Journal of the Transportation Research Board, 2325. http://dx.doi.org/10.3141/2325-05

Landing slot. (n.d.) *In Wikipedia*. https://en.wikipedia.org/wiki/Landing_slot

Low-cost line-up. (2015, June). *Airline Business*, p. 26–27.

O'Connell, D. & Collingridge, J. (2016, February 14). Oman breaks Heathrow record with deal for slots. *The Times*. www.thetimes.co.uk/article/oman-breaks-heathrow-record-with-deal-for-slots-5mhdlzs23mn

O'Connell, J. F. (2011). Airlines: An inherently turbulent industry. In J. F. O'Connell (Ed.), *Air Transport in the 21st Century*(pp. 59–96). London: Ashgate Publishing.

Person, J. (2021, April 30). Transit passengers are key to Southwest's Airlines' operations. *Simple Flying*. https://simpleflying.com/southwest-transit-passengers/

Rains, T. (2022, February 28). United is cutting 17 routes and leaving one US city entirely. *Business Insider*.

Redondi, R., Malighetti, P., & Paleari, S. (2012). De-hubbing of airports and their recovery patterns. *Journal of Air Transport Management*, 18, 1–4.

Swoboda, F. (2001, April 28). Airline antitrust case dismissed. *The Washington Post*. www.washingtonpost.com/archive/business/2001/04/28/airline-antitrust-case-dismissed/f5861525-eb76-4b9b-889e-93621416830c/

Synder, B. (2017, March 2). *The cranky flier: United's big summer expansion confirms its focus on connecting traffic*. https://crankyflier.com/page/90/?theme=cambay

Unnikrishana, M. (2015, March 12). *United Airlines set to start Chicago O'Hare re-banking*. Air Transport World (online).

US Department of Transportation. (2015, April). *Aviation consumer protection/air travel consumer report*. www.dot.gov/airconsumer

Wang, W. & Wang, X. (2019). Why do airlines prefer multi-hub networks? *Transportation Research. Part E, Logistics and Transportation Review*, 12.

Review Questions

1. Describe the three types of route structures. Define or illustrate each.
2. Why was the linear system widely used in early airline history but is now rare?
3. What are the advantages and disadvantages of a point-to-point route system?
4. What are the advantages and disadvantages of the hub-and-spoke system?
5. Explain why many large airlines operate several hubs.
6. How do aircraft routings and passenger connections differ in a directional versus an omnidirectional hub? Why are directional hubs less expensive to operate?
7. What is a "rolling hub" and what savings does it provide in comparison with a traditional hub-and-spoke operation?
8. Compare focus cities with rolling hubs.
9. What is the difference between a rolling hub and a directional hub?
10. Why does point-to-point service cost less to operate than connecting service through a hub?
11. What are the requirements for a successful hub city?
12. Why are small cities dependent on hub-and-spoke systems for air travel?
13. Explain why low-cost airlines often operate a single fleet type whereas full-service network carriers use several different aircraft types.
14. Airfares for flights originating in an airline's major hubs are often higher than in other city pairs of similar distance and city sizes. What may cause this price disparity?
15. In the same city pair, should an airline be able to charge a different fare for a point-to-point service versus a single connection at its hub?
16. How can airlines that do not operate any of the same routes be competitors? Provide a current example.

4 Product Offering

Upon completion of this chapter, you should be able to:

- Rank airline passenger wants and needs and explain how these influence airline product offerings.
- Compare and contrast the business models of each airline category.
- Provide airline examples for each airline category.
- Explain how the worldwide growth of low-cost carriers has forced full-service network carriers to adapt.
- Assess the ability of long-haul low-cost carriers to compete with full-service network carriers.
- Compare and contrast the business models of the three categories of cargo airlines.

4.1 Introduction

Having considered in Chapter 3 how an airline chooses a route architecture to connect its destinations, this chapter examines the other product features an airline decides to offer its customers. This choice ranges from basic transportation, sometimes labeled *no-frills*, offered by some low-cost carriers such as Europe's Ryanair to exclusively business-class service proffered by a few highly focused airlines. Most large *full-service network carriers* (FSNC), in contrast, offer a variety of products from premium service in the first-class cabin to economy coach class with limited amenities and many restrictions. A critical part of the product for many FSNCs is the regional airline partners that provide service under contract on a network carrier's thinner routes, usually with flights between small cities and the FSNC's hub airports. The chapter concludes with an examination of the business models of cargo airlines. The product an airline chooses to provide depends largely on the passenger segments it wishes to serve. The airline product is one of four variables in the *marketing mix*. The others are price, promotion, and place (or distribution). Price and distribution are addressed in later chapters.

DOI: 10.4324/9781003290308-4

4.2 Strategic Choices

4.2.1 The Marketing Concept

In the early 20th century, firms emphasized efficient production of consumer goods, perhaps best illustrated by Henry Ford's assembly line production of the Model T. Later, as mass production became commonplace and consumers' demand for basic goods was satisfied, marketing philosophy shifted to an emphasis on sales. After World War II, as the variety of consumer goods and services exploded, marketers began to realize that producing and selling a product with little regard to what the consumer wanted was not a sustainable path to profits. Modern marketing theory holds that a company facing intense competition, as do airlines in the US, Europe, Asia, and increasingly throughout the developing world, must determine customers' needs and wants and then offer a product satisfying those desires at a price that yields a profit. This is known as the *marketing concept.* Any company that fails to satisfy customers' needs will lose market share to more-able competitors.

The supersonic airliner is a classic example of a failure to observe the marketing concept. The Anglo-French Concorde was built because it was technologically feasible and a statement of European aeronautical expertise. However, the designers didn't seem to consider whether a sufficient number of passengers were willing and able to pay higher fares necessary to cover the aircraft's operating costs to save two to three hours on transoceanic flights. Only 20 aircraft were built. British Airways and Air France flew the Concorde on limited routes for 27 years beginning in 1976, but only because the British

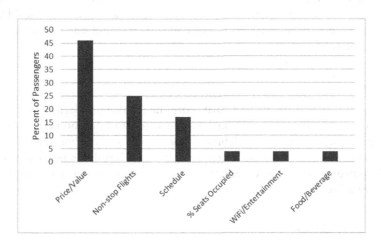

Figure 4.1 Passenger Criteria in Airline Choice.
Note: Percent of passengers listing criterion as the first choice. Data source: Airlines for America, Air Travelers in America, Key Findings of a Survey Conducted by Ipsos. February 2022.

and French governments covered much of the acquisition costs. Taxpayers subsidized the development and sales of an aircraft few would use for travel.

So what do airline passengers want? Lots of things: low price; many destinations; high flight frequency; non-stop service; assigned seating; generous seat room, easy distribution and real-time information on their flight delivered to personal electronic devices; easy airport access; quick check-in; passenger lounges; attentive inflight service; excellent inflight food, inflight internet access and entertainment; and new jet aircraft.

Doganis (2019) lists five key product features that affect a passenger's choice of airlines – price, schedule, comfort, convenience, and image. The majority of airline passengers travel for vacation or to visit friends and family, so the price/value is the most important criterion for choosing an airline (see Figure 4.1). Although leisure passengers are the majority, business travelers often generate a greater proportion of total revenue. For business passengers, schedule convenience will weigh heavily in their choice. Whether an airline emphasizes price or schedule should be determined by the target passenger segment(s) and will vary by route. For business markets, high flight frequency would be an important competitive weapon. In vacation markets, the same airline might choose less frequency but utilize larger aircraft with high-density seating to lower seat-mile costs. Segmentation may be extended to several more dimensions to precisely determine the wants and needs of small passenger segments. With extensive data available from a passenger's travel history and frequent flier information, airlines can tailor a product to the needs of the individual passenger. For example, an airline might offer a golf package to a frequent flier known to be an avid golfer.

Still, airlines struggle to meet all these wants. An airline, for example, cannot offer low prices with an array of expensive amenities and still be profitable. Rather, it must decide what passenger segments to serve and carefully balance price with the product features.

4.2.2 Generic Strategies

Many business strategists believe that corporate strategic decisions must be based on a core concept defining how the firm can best compete in the marketplace. Michael Porter, a celebrated Harvard Business School professor, developed three generic business strategies: cost leadership, differentiation, and focus. By pursuing one core strategy, the firm attempts to obtain a sustainable competitive advantage.

4.2.2.1 Cost Leadership

A firm pursuing cost leadership attempts to maintain the lowest cost production for a given level of quality. This strategy involves determining the core product features customers demand and delivering these with minimum resources, high efficiency, and high volume. Customers, of course, want other product features but many will trade extra features for a lower price. Passengers won't sacrifice safety for a lower price but may accept high-density seating for a cheaper fare. Examples of firms pursuing cost leadership include Walmart, whose original slogan, "Always the low price," captures this generic strategy. Southwest Airlines is probably the best-known airline example, but Europe's low-cost airlines, including Ryanair and Wizz Air, are more dedicated to this generic strategy.

4.2.2.2 Differentiation

A differentiation strategy creates unique products for varied consumer groups. Tailored product features create brand loyalty that translates into the ability to charge a premium price. As with the cost leadership strategy, differentiation can be pursued on a broad, industry-wide scope, or a narrower basis. With a broad scope, a firm provides tailored products for many customer segments and exploits synergies in producing several different, but closely related products. General Motors is a classic, though now less dominant, example of a corporation that produces a product for every large consumer segment ranging from Chevrolet to Cadillac. Toyota is another example from the automobile industry. In the airline industry, most full-service network airlines such as American Airlines, Japan Airlines, or British Airways offer a product for each consumer group from highly restricted economy coach to first class.[1]

4.2.2.3 Focus

The focus strategy attends to the needs of a particular market segment, usually one not well served by firms pursuing cost leadership or differentiation strategies. A niche strategy is another apt term. Focus strategy firms concentrate on a single product and seek unchallenged specialization. Continuing with the automobile industry examples, Rolls Royce comes to mind. Early in the 21st century, several new-entrant airlines such as Eos and MAXjet began all-business-class service between New York and London meeting with some early success. Airline history, however, is replete with attempts to cater exclusively to business travelers, all of which eventually failed. Perhaps not surprisingly, so did Eos and MAXjet. Many small airlines serve niche markets that are not attractive to larger carriers. In the cargo industry, FedEx limits its business to small packages – a focus strategy.

4.2.3 Industry Evaluation: Porter's Five Forces

Continuing to follow Porter, he suggests that the starting point for choosing a strategy is an evaluation of the industry along five dimensions or forces of competition.

Figure 4.2 illustrates how opposing market forces determine the degree of completion among competing firms within an industry. For the airline industry in the US, Europe, and increasingly in Asia, these forces vary from moderate to

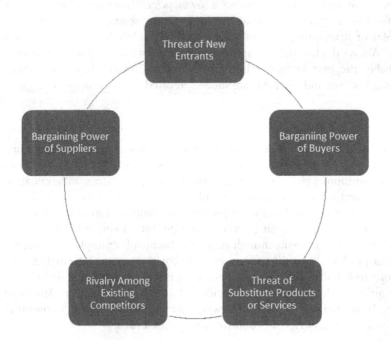

Figure 4.2 Evaluating Industry Competition.

high strength resulting in intense competition. Each force is summarized below (IATA, 2011).

4.2.3.1 Threat of New Entrants – High

In the mature economies of North America and Europe, there are few new entrant airlines, but even with all potentially profitable markets receiving airline service, incumbent carriers must consider the threat of new entry carriers. For example, two new carriers, Breeze and Avelo, entered the US domestic market in 2021. In developing and rapidly growing economies, new entrants, especially low-cost carriers, proliferate.

4.2.3.2 Bargaining Power of Buyers – High

Although there are millions of individual passengers, information on airline products and prices is now easily available, cheap, and transparent because of internet distribution. Switching costs are low and many leisure passengers view the airline coach seat as a commodity.

4.2.3.3 Threat of Substitute Products or Services – Medium and Increasing

Although there are no substitutes for fast, long-distance travel, high-speed rail is a competitor for short-hauls in Europe, Japan, China, and the US northeast corridor. For short distances, the automobile is a strong substitute. Video conferencing grew rapidly during the COVID-19 pandemic and threatens a substantial portion of business travel. Business jets whether wholly owned, time-shared, or chartered compete with airlines' premium products.

4.2.3.4 Bargaining Power of Suppliers – High

Aircraft and engine producers are concentrated oligopolies. Airline labor unions are powerful in most countries. Many airports are local monopolies with only the largest cities offering secondary airports as competition. Global distribution systems are still powerful but direct distribution has provided airlines with more bargaining power. Global market forces determine the price of fuel, the second largest expense for most airlines, leaving airlines little power to bargain for lower prices.

4.2.3.5 Rivalry Among Existing Competitors – High

Even though most markets are oligopolies, airlines are highly competitive. Limited product differentiation and a perishable product promote price competition.

This framework suggests that the airline industry is intensely competitive and, therefore, a challenging environment for sustained airline profitability. Airlines must design and produce a product that yields some competitive advantage to survive and prosper. Porter suggests that airlines should choose one of the three generic strategies; however, the emergence of the hybrid airline challenges this academic recommendation.

4.2.4 Dimensions of Product Choice

An airline has a broad range of product choices with which to meet passengers' wants and needs. The product decision process begins with the airline's strategic vision. This vision, from which it intends to exploit some market opportunity, is the subject of strategic management and beyond the scope of this text. With an overarching vision and a choice from Porter's strategies, the airline should next evaluate passengers' wants and needs and then target some segment(s) of passengers for service. With these strategic decisions in place, Figure 4.3 depicts a series of specific product choices.

4.2.4.1 Geographical Scope

The geographical scope can range from a very limited area to worldwide. Full-service network carriers serve major business and popular vacation destinations around the globe, some through their global alliance partners, ensuring that a customer need not seek another carrier to satisfy travel needs. In contrast, some small airline route structures are very limited. Two airlines based in the Hawaiian Islands provide an interesting contrast. Mokulele Airlines serves just

Airline Product Design

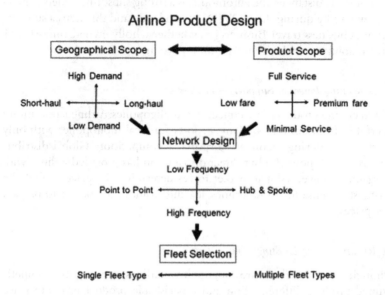

Figure 4.3 Airline Product Design.
Source: Adapted from Holloway, 2002.

four destinations within the Islands whereas Hawaiian Airlines has expanded from its origin as an inter-island carrier to a predominately long-haul airline serving Hawaii with connecting service from Asia, Australia, and New Zealand to the US mainline.

The geographical scope is tied to the choice of cities to be served. Some airlines choose primarily vacation and holiday destinations – the European tour operators are an example – whereas others target business centers. The larger carriers serve both.

Routes vary greatly in passenger demand. Regional airlines serve mostly low-demand routes connecting smaller cities to major hubs, usually under contract with large network carriers. At the other end of the spectrum are the world's densest routes. Korean Air and Asiana Airlines together operate an astonishing 60 daily flights between Seoul and Jeju Island in South Korea. These airlines, incidentally, plan to merge in 2022 potentially creating a monopoly on the world's busiest route. The next two highest density routes are Hanoi-Ho Chi Minh City in Vietnam and Sapporo-Tokyo in Japan.

4.2.4.2 Product Scope

Product scope choices range from no-frills basic air transportation to all business class, or a combined offering with several products designed to match the needs of most passenger segments. Of the airlines offering core transportation, Ryanair is perhaps the best-known example. Aircraft are configured in single-class, high-density seating. Amenities are few and for those Ryanair charges a fee. FSNCs offer three or four cabin classes to provide a product for most passengers' needs Singapore Airlines, for example, features a broad product scope on its Airbus 380 aircraft boasting four cabins, including individual suites, business class, and two levels of economy class on some aircraft. Suite's appeal to the high-end traveler willing to pay very high fares whereas the lowest level economy class is intended for passengers whose priority is a low price.

4.2.4.3 Network Design

Route structure options are examined in Chapter 3 and range from point-to-point to hub-and-spoke with most large carriers incorporating some elements of both generic designs. Allegiant, EasyJet, and Wizz Air are purely point-to-point airlines whereas the FSNCs such as Delta Air Lines, British Airways, and Emirates operate mostly a hub-and-spoke system. Flight frequency also varies by market and carrier. The US low-cost carriers Allegiant, Frontier, and Spirit Airlines serve many US and nearby international markets only a few days a week whereas high-demand markets are served many times per day, perhaps none more frequent than Korean Air's 30-plus daily flights each way between Seoul and the Jeju Island.

4.2.4.4 Fleet Selection

Fleet selection will follow from the targeted passenger segment(s), product scope, and network design. If the network includes cities of greatly differing sizes and/or routes of varying lengths (domestic and international cities), several aircraft types may be required. Low-demand markets can be profitably served only by small-capacity aircraft. Before beginning a decade-long and painful restructuring, American Airlines operated 14 fleet types in 30 configurations tailored to specific markets. It has since simplified or rationalized its fleet of 819 aircraft to just four types, some with several models, in eleven different configurations (American Airlines, 2021).

Determining the appropriate dimensions of product scope is an iterative process as decisions on one dimension will influence others. For example, a decision to serve a market with high frequency will influence the optimal size of the aircraft.

Airlines are characterized by their product. Along a spectrum are full-service network carriers, regional airlines, hybrid carriers, and low-cost airlines. Each is discussed in detail in the following sections.

4.3 Full-Service Network Carriers

At one end of a continuum are the full-service network carriers. Air China, Lufthansa, and United Airlines are examples of full-service network carriers. These airlines offer a broad product scope, including (O'Connell, 2011):

- Service to hundreds of destinations via a global hub-and-spoke route system. Through one of the three global airline alliances, service is extended to many more worldwide destinations. FSNCs serve primary airports favored by business travelers occasionally supplemented with additional service to secondary airports. Several daily frequencies are offered in most city pairs providing passengers with flights that meet their desired travel times.
- Large fleets numbering in the hundreds of aircraft with several different types to meet route requirements that vary greatly in length and demand. Regional airline partners operating smaller aircraft extend the network coverage from hub airports to smaller cities.
- Aircraft are configured with two or more cabin classes with a vast portfolio of in-flight products in business and first-class such as flatbeds, quality food and beverages, advanced in-flight entertainment systems with multiple channels, internet, and mobile phone connectivity. Economy cabin amenities are more limited and are often only for purchase. The coach cabin is increasingly divided into two sections with more seat room and service in premium economy coach and a basic economy section specifically designed to compete with the low-cost carrier product.
- Amenities offered to business and frequent customers include airport lounges, fast-track security, preferential boarding, and occasionally even limousine service.
- Distribution is available through many channels, including travel agents, online travel agents, airline call centers, websites, and social media.
- Loyalty programs include frequent flyer awards and incentives to travel agents.

4.3.1 Product Range

Though FSNCs provide a product for most passenger segments, profitability depends on capturing business travelers who represent only 10% to 15% of long-haul passengers but account for up to half of the revenue for airlines such as Lufthansa and British Airways (Mouawad, 2013). Although first-class cabins have traditionally been offered to capture passengers traveling for business, many airlines have dropped first class for an upgraded business class cabin featuring flat beds on international flights with other amenities that rival what had been first class. The next product innovation was premium economy class targeting business travelers whose employers will not pay for business class but do allow an upgrade in coach. FSNCs continually revise their products, but any competitive advantage is usually fleeting as competitors can copy most product innovations.

Faced with a loss of their price-sensitive passengers to growing LCCs, FNSCs created a new economy product with high-density seating and a host of fees for amenities such as early boarding, seat selection, and beverages and snacks. Upon exiting bankruptcy in 2005, Air Canada was an early adopter of

Table 4.1 Air Canada's Fare Options

Air Canada Montreal to London One-Way Fares								
Cabin	Economy				Premium Economy		Business Class	
	Basic	Standard	Flex	Latitude				
Fare	$263	320	728	1,365	1,192	1,535	2,420	5,260
Refundable	no	no	no	yes	no	yes	fee	yes

Note, fares and conditions obtained from aircanada.com

a pricing innovation that bundled amenities into several fare and cabin classes. For a one-way flight from Montreal to London, there are eight fares in three cabins as shown in Table 4.1. Four fares are available in the economy cabin with higher fares, including seat selection, checked baggage, and flight change. It is, however, the inclusion of a fully-refundable ticket that makes the greatest difference in price in all three cabins.

4.3.2 FSNC Example: Delta Air Lines

Delta Air Lines is a full-service network carrier and the world's second-largest airline in 2021 with 2019 revenues of $47 billion. It flies more than 200 million passengers each year to 300 destinations on six continents. Delta has more than 75,000 employees worldwide and operates a mainline fleet of more than 800 aircraft.

Delta operates a hub-and-spoke route system with domestic hubs in Atlanta, Boston, Detroit, Los Angeles, Minneapolis-St. Paul, New York-LaGuardia and JFK, Salt Lake City, and Seattle-Tacoma. The fleet comprises Airbus A350, 330, 320, 220, and Boeing 767, 757, 737, and 717 aircraft. Aircraft figuration varies with type and route. All aircraft have business class, premium economy, and basic economy seating. Some wide-body aircraft also have a first class featuring lay-flat seats. Onboard amenities also vary by fleet type and route but include Wi-Fi, satellite TV, and video. In addition to its 5,000 daily flights, three regional airlines – wholly owned subsidiary Endeavor Air and independent Republic Airways and SkyWest – provide another 2,500 Delta-branded flights (Delta Air Lines, 2021).

The airline is a founding member of the SkyTeam global alliance with 19 member airlines. Indicative of the complexity of airline alliances and joint ventures, Delta also partners with airlines that are not SkyTeam members and holds equity stakes in Aeromexico, Air France-KLM, China Eastern, Korean Air, LATAM, and Virgin Atlantic. The COVID-19 pandemic, however, forced Aeromexico, LATAM, and Virgin Atlantic into bankruptcy causing a write-down of $2 billion.

4.4 Regional Airlines

Regional airline (RA) is an accepted but poorly defined term referring to passenger airlines flying aircraft smaller than mainline jets and providing service to
cities with insufficient demand to be profitable with FSNC service. Regional
airlines vary from the world's seventh-largest airline by the number of aircraft,
SkyWest, to the smallest airlines such as Australia's King Island Airways which
flies a single route with five aircraft. Some regional airlines often operate only
under contract to FSNCs whereas many others are independent. Regional
airlines operating on behalf of FSNCs are discussed first.

4.4.1 FSNC Affiliated Regional Airlines

The predecessors to today's regional airlines, once known as commuter airlines,
emerged for more than 50 years providing feeder service to larger airlines. The
first such partnership between the US local service carrier Allegheny Airlines,
now part of American Airlines through a series of mergers, and Henson Airlines
began the world's first *codeshare* agreement in 1967. Henson and other commuter airlines operated small capacity turboprop aircraft that passengers often
found undesirable. With the introduction of the regional jet (RJ) in the early
1990s, regional airlines expanded rapidly. As part of restructuring in the early
years of the new millennium, US legacy carriers retired their older, inefficient
mainline jets in favor of RJs operated by their regional partner airlines. While
capacity did not increase, the number of aircraft and frequency was more than
double that of the grounded fleet. Regional airlines now operate 40% of all
US domestic flights. Figure 4.4 charts the growth of US regional airline traffic
measured in revenue passenger miles from 1980 until 2020. Note the explosive
growth between 1995 and 2005 following the introduction of regional jets. Of
course, the regional airlines suffered from the pandemic along with their FSNC
partners.

   The regional airline provides an extension of the full-service network carrier's
product, not a separate offering. Indeed, the industry publication *Aviation Daily*
terms these airlines "network extenders." Yet, regional airlines are distinct from
their larger partners. Regional airlines:

- may be subsidiaries of the network carrier such as Delta Air Lines' Endeavor
  Air or independently owned, of which Skywest Airlines is the largest.
- operate under their own airline operating certificate and government
  regulatory oversight but many elements of the operation, including scheduling, pricing, promotion, and distribution are managed by the network
  carrier.
- Flights share the network airline's brand and flight code. Aircraft are
  painted in the FSNC's livery with the same or similar logos, trademarks,
  and some version of the FSNC's name such as *Delta Connection* and
  Lufthansa Regional.

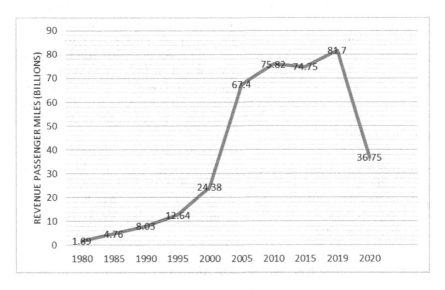

Figure 4.4 Regional Airline Historical Traffic.
Data source: Regional Airline Association Annual Reports.

- Flight schedules are integrated with the network partner to facilitate connections to the FSNC's flights. Check-in and baggage handling are coordinated so that passengers need only to check in their luggage once, at the start of their trip.
- Passengers traveling on the regional airline earn the network airline's frequent flyer points.
- For many markets and routes, the regional airline has a lower cost structure, primarily from lower employee wages and higher productivity, than the network carrier. Indeed, this is the incentive for the FSNC to contract with the RA.
- RA operate a fleet of regional jets typically ranging in capacity from 30 to 90 seats (Gillen et al., 2015).

Traditional contracts between the FSNC and the regional airline paid the regional airline a fixed fee for each departure plus an additional fee per passenger; hence the term *fixed-fee contract*. Incentives for flight completion, on-time, and baggage handling performance were often included. The FSNC usually paid some operational costs, particularly fuel. Under more recent and still less common *revenue-sharing arrangements*, the regional airline receives a percentage of the ticket revenues. All costs are borne by the regional airline which greatly increases its business risk (Gillen et al., 2015).

The network carrier and the regional airline mutually benefit from the outsourcing arrangements. The network carrier is afforded lower-cost access to small and moderate size markets increasing passenger feed to its hubs. By contracting with one or more regional airlines, the network carrier benefits from competition for contracts and the ability to quickly react to market changes and opportunities. The regional airline, on the other hand, benefits from the network carrier's brand recognition and creditability, marketing and distribution coverage, premium pricing, purchasing power, network efficiencies, and loyalty programs. Nonetheless, the regional carrier is subject to the control of its major airline partners with little ability to operate outside of these contractual partnerships.

Regional airlines play a surprisingly large role in the US air transportation system, one belied by the terms regional airline and regional jet. In 2019, the regional airlines operated 10,500 daily flights, 40% of all domestic passenger flights, and flew 20% of all domestic passengers. Of the more than 400 US commercial airports, 63% relied exclusively on regional airline service (Regional Airline Association, 2021).

4.4.2 Regional Airline Example: SkyWest Airlines

SkyWest Airlines flies on behalf of its four FSNC partners: Alaska, American, Delta, and United. Its route system, as seen in Figure 4.5, is not regional but rather spans the Continental United States extending into Canada, Mexico, and the Caribbean. Although few airline passengers would recognize SkyWest as a

Figure 4.5 SkyWest Route System as of August 2022.
Note. Courtesy of SkyWest Airlines.

major airline, it's one world's largest operating a fleet of 486 aircraft, providing service to 246 cities, and employing 14,888 (SkyWest, 2021).

SkyWest serves several roles for its partners. Most flights feed traffic to the FSNC's hubs. Using SkyWest's partnership with United as an example, SkyWest flights support United's domestic hubs in San Francisco, Denver, Houston, and Chicago. Regional carriers provide exclusive service on many routes, but others are flown by the FSNC during peak demand supplemented with its regional airline's flights during low demand. This allows the FSNC to offer high frequency without significantly increasing capacity that would drive lower yields (an increase in supply results in lower prices if all other factors are unchanged). With the introduction of the new generation of regional jets that provide a comparable level of passenger comfort to mainline aircraft, FSNCs have expanded their networks adding long, thin routes such as Minneapolis–Calgary (upper center of route map).

When regional jets began replacing turboprop aircraft, industry analysts anticipated they would be used in point-to-point service bypassing major hub airports. However, an inspection of Figure 4.5 shows that this role has gone mostly unfilled.

4.4.3 Independent Regional Airlines

Although the largest regional airlines usually have some affiliation with a network carrier, more operate independently shouldering all of the management responsibilities and risks of a traditional airline. The Norwegian carrier Wideroe is a quintessential regional airline serving more than 40 destinations, many isolated along Norway's east coast. With a fleet of 40, it is the world's largest operator of Bombardier DHC-8 turboprop aircraft. Surprisingly, it was also the first airline to purchase the E2 upgraded model of the Embraer E-190 regional jets (pictured in Figure 4.6). As of 2021, Wideroe's had three E2s in operation and was holding options on 12 more. Wideroe routes are predominately point-to-point but it codeshares with Air France, KLM, Finnair, and Scandinavian Airlines. Another example of an independent regional airline is South Africa's Airlink. For most of its history, Airlink was the regional airline for financially struggling South African Airways. In 2020, it severed ties and launched its own brand proudly asserting that "Airlink is a privately-owned, premium, full-service regional airline serving a comprehensive network of destinations throughout Southern Africa" (Airlink, 2021). Airlink, now Africa's second largest airline, flies to 45 destinations in 13 African countries with a fleet of 50 Embraer ERJ-135s, 170s, and 190 regional jets plus a few older British Aerospace Jetstream 41s.

4.4.4 Regional Aircraft

The first regional jets introduced in 1993 had a seating capacity of 30 to 50 passengers. Canada's Bombardier Aerospace and Brazil's Embraer were the leading manufacturers. Compared with mainline jet aircraft, the early regional jet models offered lower trip costs, but higher cost per available seat mile (CASM) than the mainline jets they replaced. Even with its higher seat-mile costs, the RJ was attractive in many smaller markets. By offering higher frequency without adding capacity, the RJ suited the needs of the business traveler willing to pay higher fares for convenient, frequency service. Of course, seats that would otherwise go unsold were offered at lower fares.

Commercial aircraft often become economically obsolete before their technical obsolescence. This was true of the first generation of regional jets, but other factors were at play. The small capacity of the earlier regional jets was a result of scope clauses in collective bargaining agreements between FSNCs and their pilot unions. By limiting the capacity and number of aircraft that regional airlines could operate on behalf of their larger airline partners, major airline pilots sought to protect their work. Consequently, the first regional jets were designed to accommodate these labor agreement restrictions but sacrificed the economic efficiencies of larger aircraft, especially fuel economy. For example, a Bombardier 50-seat CRJ-200 uses 19 gallons of fuel to fly each passenger 500 miles whereas a larger current generation mainline jet would use just 7.5 gallons per passenger over the same distance.

FSNCs negotiated less restrictive scope clauses during the consolidation and restructuring in the first decade of the century. Regional jet manufacturers responded with a new generation of larger and more efficient jets. Bombardier and Embraer's new aircraft abandoned the tail-mounted engines for the now-standard under-wing design. With capacities of up to 160 seats, these aircraft have little in common with the earlier regional jets. Rather, the new regional jets encroach on the market for Airbus and Boeing's smallest models blurring the distinction between regional and mainline jets. Bombardier's new C-series was highly acclaimed, but early sales disappointed. Faced with huge development costs and ongoing losses, Bombardier sold the C-series to Airbus which it rebranded as the A-220. Airbus has enjoyed good sales, but mostly to FSNCs rather than regional airlines. Embraer's new offerings were the similarly designed E-170, 190, and 195. Development of regional aircraft in Japan and Russia foundered having been indefinitely delayed or abandoned. The Comac C919 may find a significant domestic market in China, but global sales are uncertain. This leaves Embraer to dominate the regional jet market for years. Yet, despite its near-monopoly, Embraer has also struggled in recent years as airlines opted for larger aircraft (Flottau, 2021, 2022). The early and latest Embraer aircraft models are pictured in Figure 4.6. The first-generation Embraer ERJ 145 operated by American Airlines' wholly owned subsidiary Envoy Air is configured with 47 main cabin seats and three seats with additional legroom. The Embraer 190-E2 in Wideroe's livery has business and economy cabins for a total of 110 seats.

Turboprop aircraft, the backbone of early regional airline fleets, have enjoyed a modest resurgence. For Winderoe, the turboprop is well suited to its many small and remote destinations. The turboprop also has better fuel efficiency and operating economics over segments under about 500 miles. Alaska Airlines'

Figure 4.6 Regional Aircraft.

Source: Images from Wikimedia Commons. Left image from John Davies. Licenses https://creativecommons.org/licenses/by-sa/4.0/deed.en and https://en.wikipedia.org/wiki/GNU_Free_Documentation_License.

wholly owned regional subsidiary Horizon Air replaced its early regional jets in favor of the Bombardier Q400 turboprop. Similarly, the Canadian carrier WestJet launched its wholly owned regional subsidiary Encore with the Q400. ATR, the leading turboprop manufacturer, is planning upgrades to its product while Embraer has proposed a clean-sheet turboprop aircraft. Bombardier, which sold its regional jet business to Japan's Mitsubishi Heavy Industries in 2019, will likely end production of the Q400.

4.4.5 Regional Airlines Worldwide

Regional airlines first developed a substantial presence in the US but now occupy an important place in the global airline industry. The European Regions Airline Association (2021), the representative of Europe's regional airlines, lists 55 airline members of which seven are FSNC affiliated and more than 20 are independent regional carriers. There are structural differences between the regional airlines in the US and those in Europe and Asia. US FSNCs partner with several wholly owned and independent regional airlines whereas elsewhere regional airlines typically feed traffic only to their parent company. KLM Cityhopper, for example, operates solely to support KLM and Australia's Sunstate Airlines serves only Qantas. The regional airlines listing in Wikipedia, incomplete but the best available, includes another 60 regional airlines operating outside the US, Canada, and Europe ("List of Regional Airlines," 2021).

4.5 Low-Cost Carriers

The most important driver reshaping the global airline industry over the past 40 years is the emergence and rapid growth of low-cost airlines (LCC). From a mix of plan and circumstance, Southwest developed the LCC business model in the mid-1970s. By 2000, only the US and Europe had a sizable LCC presence; by the end of the decade, LCCs were operating in South America, the Middle East, India, Asia, and Australia. Today, new LCCs are sprouting quickly, especially in Asia, but the attrition rate is also high. Of the 155 LCCs listed in Wikipedia, Asia-Pacific was home to most LCCs with 70 followed by Europe with 42. ("List of low-cost airlines," n.d.)[2]

Low fares are the bedrock of LCC marketing, but several other factors have aided the rapid LCC growth. The emergence of internet distribution channels and airline websites greatly improved the ability of passengers to find and compare airline schedules and fares (fare transparency). Especially since the recession of 2001, a portion of business travelers has defected to LCCs that now provide an alternative to FSNCs in nearly all major domestic markets in North and South America, Europe, and Asia.

LCCs have captured market share from FSNCs with typically low prices, but low prices also generate additional passenger demand allowing for continued LCC growth. LCC's worldwide market share more than doubled in the 15

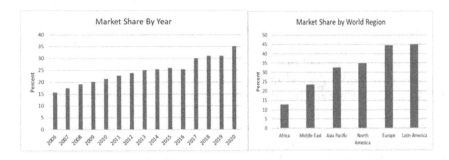

Figure 4.7 Low-cost Carrier Market Share (percent).
Data source: Statista, 2021.

years from 2006 to 2020 (Figure 4.7 left panel). Market share is highest in
Europe and Latin America (Figure 4.7 right panel).

4.5.1 Business Model

The LCC business model stands at the other end of the continuum from the
full-service network carrier. As developed and perfected by Southwest and
copied by many other airlines, the archetypal model offers a simple product
supported by relatively simple operations. The model includes the following
features (Budd & Ison, 2017; Doganis, 2019):

- Point-to-point route structure with relatively short segments of less than
  400 nautical miles, often to secondary and uncongested airports. Flight
  frequencies are high in dense markets but the model can also support low
  frequency.
- A single fleet type, either the Boeing 737 or Airbus 320 series aircraft,
  reduces training and maintenance costs. Aircraft are configured in single-
  class with high-density seating.
- Aircraft and other asset utilization are high with aircraft block hours
  exceeding 11 hours per day. The flight schedule extends beyond peak
  demand hours with some flights operating at inconvenient times. Airport
  turn-times are short, around 25 minutes.
- Fares are lower than traditional airlines and usually one-way, non-refundable,
  transparent, simple, and unbundled. Only basic transportation is included
  in the published fare with fees for all other services and amenities. Prices
  increase with the approaching departure date; usually, only a single fare is
  available at any given time. Ticketing is point-to-point without connections
  to other flights.
- No interlining, code sharing, or alliances with other airlines.
- Loyalty programs, if any, are simple and not shared with other carriers.

- Distribution is direct via the airline's website avoiding travel agents and global distribution systems.
- Seating is open; passengers board and take any available seat.
- Minimal in-flight service consists of snacks and drinks available for purchase.
- Personnel may be well paid, enjoy profit sharing and bonuses, but have few restrictive work rules and are highly productive.
- Ground handling and heavy aircraft maintenance are outsourced.

A common definition of an LCC is an "air carrier that has a relatively low-cost structure in comparison with other comparable carriers and offers low fares and rates" (ICAO, n.d.) However, many airlines typically labeled LCCs deviate substantially from several elements of the model. More recently, carriers closely adhering to the pioneering LCC business model and focusing on maintaining the lowest costs have been labeled ultra-low-cost carriers (ULCC), a term coined by Spirit Airlines. ULCCs, especially in Europe, can achieve costs approaching half those of the full-service network carriers. In the US, ULCCs' cost per available seat mile has averaged about 70% of the FSNCs, in part, because US FSNCs have lower costs than their European counterparts (MIT, n.d.). Bachwick & Wittman (2017) find that ULCC competition in a market lowers the average fare by 20.5% compared to a reduction of 7.7% with the presence of an LCC.

4.5.2 Ancillary Revenues

LCCs pioneered la carte pricing or unbundling, a practice that quickly spread to most airlines. Fees first imposed for checked baggage spread to nearly every other imaginable product feature. Led by the low-cost carriers, total ancillary revenue for the world's airlines has grown rapidly for years until the pandemic laid waste to the industry (Figure 4.8). The pandemic notwithstanding, ancillary revenues continued to grow as a percentage of total revenues. In 2020, the European LCC Wizz Air was the most aggressive and successful airline in pursuing ancillary revenues obtaining 55.9% of its total revenue from ancillary fees just topping Spirit Airlines at 55.8% (IdeaWorks, 2021). In contrast, the highest percentage of ancillary fees among the FSNCs was American Airlines at 21.8%, and much of this revenue is related to sales of frequent flier points to third parties (see Chapter 7 for more discussion of ancillary revenues).

Table 4.2 shows just a few of Wizz Air's fees, some of which vary by season or the method of payment. The bewildering range and variety of fees can easily overwhelm and mislead passengers.

Ancillary revenues are a critical part of the LCC business plan and make substantial contributions to profits. Spirit Airlines argues that passengers should only have to pay for product features they use and value; thus, ancillary fees provide passengers with options and variety they value. Less obvious is the effect of these fees in reducing costs. If beverages, for example, are provided at no additional cost, most passengers will gladly accept the offer. When faced

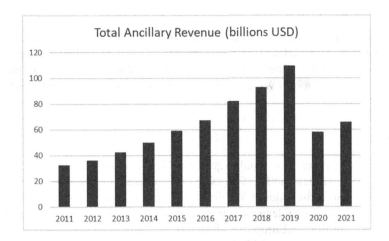

Figure 4.8 World Airline Total Ancillary Revenue.
Source: Statista, 2021.

Table 4.2 Wizz Air Ancillary Fees

Product Feature	Fee (converted from euros to USD)
Checked baggage per flight, per passenger per bag	Bag weight up to 10 kg., low season $9.60 to $67 Bag weight up to 32 kg., high season, $36.72 to $135
Excess bag weight per kilo, per item, per flight	$12.43
Booking fee per flight, per passenger	$9 to $14.70
Cancelation fee	$68
Airport check-in	$39.55
Assigned seat per flight, per passenger	$4.50 to $36.16
Sitting together	$6.80
Food and beverage	$2.00 and up per item

Source: Wizz Air, 2021.

with a charge, however, most decline. The airline reaps substantial savings. Similarly, fees for checked baggage reduce the number of bags passengers bring on their trips. Spirit's fee for carry-on baggage resulted from an unintended consequence of its checked baggage fees. To avoid checking bags, passengers resorted to carrying more bags aboard. With high-density seating, overhead storage space for carry-ons was inadequate so carry-on bags had to be moved to the belly cargo space near the scheduled departure time. Departure delays increased. Spirit solved this problem by charging more for carry-ons than for checked bags.

4.5.3 LCC Example: Wizz Air

Despite the rapid growth and financial success of LCCs, only a few of the airlines typically classified as LCCs incorporate all or most of the elements of the business model. One analysis of the top 20 European LCCs found that only two, Ryanair and Wizz Air, closely adhered to the model (Klophaus et al, 2012).

4.5.3.1 Wizz Air

This young, aggressive, and rapidly growing ULCC closely adheres to the archetypal LCC business model. "Our obsession and compulsion are to operate at the lowest cost in the industry. We drive efficiencies across every part of our operations to continue to decrease our unit cost and deliver on our mission of retaining our position as Europe's undisputed price leader airline" (Wizz Air, 2021b). From its first flight in 2004, Hungarian-based Wizz Air has grown to serve 190 destinations in 51 countries with a fleet of 153 A320 series aircraft (as of mid-2022). Wizz plans a fleet of 268 aircraft by 2027. All new aircraft are the larger A321 model increasing the average number of seats per aircraft from 205 to 230 (Wizz Air, 2021a). Wizz Air benefits from the financial strength of its main investor, Indigo Partners whose portfolio also includes Frontier Airlines (United States), Volaris (Mexico), and JetSMART (Chile, Argentina). With a huge order for an additional 255 Airbus A320 aircraft announced at the 2021 Dubai Air Show, Indigo has an order book of 1,145 A320 aircraft dedicated to its family of LCCs.

Wizz operates a point-to-point route system from 44 aircraft/crew bases. Passenger connections to other flights are not available. Aircraft are configured in single class, high-density seating with just 28-inch seat pitch (not much legroom). Aircraft daily utilization exceeds 12 hours. Before the effects of the pandemic, Wizz reported a very low cost per available kilometer of 3.86 Euro cents (4.44¢ per mile) (Wizz Air, 2021b). Booking is only through Wizz Air's website. Fares include only core transportation or, as Wizz Air states, "pay only for what you add." As noted previously, ancillary revenues are more than 50% of total revenues.

Wizz Air also illustrates an important difference between European and US LCCs. European LCCs, including EasyJet, Ryanair, and Wizz Air, maintain many bases with dedicated aircraft and crews. Crews and aircraft crew originate each morning from their base and return at the end of the operating day. This system greatly simplifies flight scheduling and eliminates crew overnight expenses. In the US, Allegiant and new entrants Avelo and Breeze follow the European model. However, Spirit and Frontier, the largest US ULCCs, operate more complex route systems with many crews overnighting in hotels. Crew bases are focus cities some of which are the hubs of limited hub-and-spoke networks.

4.5.4 Long-haul LCCs

With the rapid growth and financial success of many LCCs, extending the model to long-haul markets seemed a logical opportunity for growth by incumbent LCCs and new entrants. Indeed, there have been many attempts dating back decades. Laker Airways – an early, well-known example – began operations as a European charter airline in 1966. In the late 1970s, Laker's Skytrain branded service introduced low-fare flights between London and New York undercutting existing prices by more than 50%. Laker offered point-to-point service with high density, single-class seating on its DC10 aircraft. Enjoying initial success, Skytrain embarked on a rapid and somewhat chaotic expansion on several Atlantic routes but did not survive a retaliatory price war with incumbent US and UK airlines and the recession of 1981–1982. Many other failures followed, including Iceland-based WOW Air in 2019 and Norwegian in 2021. So why haven't long-haul LCCs (LHLCs) enjoyed the same degree level of success as traditional LCCs?

The LCC model was built on short flights. Several of the features of the LCC business model that yield cost savings on shorter flights are less effective on long hauls. Passengers are less accepting of high-density seating and less willing to sacrifice in-flight services and amenities as flight time increases thus depriving LHLC of a significant cost advantage over FSNC. High aircraft utilization is easier to achieve on long segments as the aircraft are airborne for several hours between stops, but the larger aircraft typically used on long routes require longer airport turn-times. Many operational costs where LCCs enjoy an advantage over FSNCs, including crew, ground service, and distribution are a smaller portion of total operating costs on longer flights. In contrast, fuel cost, for which LCCs have no advantage, becomes proportionately greater. FSNCs have been able to unbundle fares closely matching the LHLC product in their basic economy cabin.

Many long-haul markets are highly seasonal. Whereas FSNCs can deploy their fleets seasonally, LHLC often cannot. Finally, the demand on point-to-point long-haul routes is usually insufficient, so the LHLC must generate connecting traffic from short-haul flights either through a hub-and-spoke network or through partnerships with other airlines. Secondary airports may not provide the needed feed traffic or have the facilities to support long-haul flights. In either case, the airline incurs the substantial expense of connecting passengers.

These factors bode poorly for long-haul LCCs. Among the several failures, there are few successes, including Qantas' subsidiary Jetstar. The long-haul subsidiary of the AirAsia Group, AirAsia X, seemed a successful LHLC, but government restrictions on international flights during the pandemic forced it into bankruptcy. In June 2022, the airline announced a resumption of operations after a two-year hiatus proclaiming a "new era for long-haul, low-cost air travel" with seven new routes from its Kuala Lumpur hub ("AirAsia X is back!" 2022).

4.5.4.1 Example: ZIPAIR

Japan Airlines launched this wholly owned subsidiary amid the pandemic in 2020. By the summer of 2022, it operated five point-to-point routes from Tokyo to Singapore, Seoul, Bangkok, Honolulu, and Los Angeles with less than daily service. ZIPAIR fashions itself as a "New Basic Airline" that defines a new standard that is neither a full-service carrier nor a low-cost carrier. New Basic is not just about controlling costs but incorporating the aesthetic details typified by Japanese airlines and creating a new airline that shortens the sense of time" (ZIPAIR, 2021). The Boeing 787-8 aircraft are configured with high-density seating plus a small business class with full lay-flat seats. Service is point-to-point without connections. Fees apply to most services and amenities, including checked baggage.

Not long after JAL introduced ZIPAIR, ANA, Japan's other FSNC, said it too would launch a new subsidiary operating Boeing 787s on medium-length international routes starting in 2023. Air Japan (not to be confused with Japan Airlines) will provide a "completely new kind of air travel experience that is neither a full-service nor low-cost carrier (LCC), combining the best of both worlds while also featuring Japanese-style ideas and quality" (ANA, 2020).

Air Japan, Jetstar, and ZIPAIR are subsidiaries of airline groups headed by a legacy FSNC. The next section explores the long history of FSNCs attempting to compete with LCCs by creating a wholly owned, low-cost subsidiary, including other LHLC examples.

4.5.5 LCC within Full-Service Network Carriers

With LCCs grabbing market share and depressing ticket prices in the US, Europe, Asia, and increasingly elsewhere in the world, it's not surprising that FSNCs have attempted to compete by establishing LCC subsidiaries. The idea has been tried many times in the US but has never succeeded. The last attempt was United's Ted – the last letters of United, or, as critics often claimed, the end of United. Ted began flying out of United's Denver hub in 2004 to compete with the LCC Frontier Airlines which is also headquartered in Denver. Ted was fully integrated into United's domestic hub-and-spoke network; consequently, the point-to-point route system that is the foundation of the LCC business model was not available for cost savings. Ted was quietly withdrawn from service in 2009. This was United's second attempt at an airline within an airline following Shuttle-by-United that operated from 1994 until 2001. Similarly, Delta tried twice with Delta Express and Song; both failed as did Continental Lite and US Airways' Metrojet.

In contrast to the US FSNCs that tried and then abandoned subsidiary LCC strategy, the practice has been widely adopted elsewhere. Air Canada tried Zip and Tango in an attempt to compete with WestJet. Both failed. Perhaps believing in the third-time charm, Air Canada established a subsidiary named Rouge in 2013 that has been more successful. It was shuttered during the pandemic but resumed service in late 2021. In Europe, International Airlines Group, the holding company for British Airways and Iberia, established Iberia Express and purchased Vueling to compete in Europe where Iberia's high costs are uncompetitive. Vueling's cost structure is some 40% below Iberia's. In 2017, IAG created yet another LCC brand dubbed Level but operated by Iberia. After several pandemic disruptions, Level operates long-haul routes from Barcelona to the US and South America with Airbus A330 aircraft in a two-class configuration. Similarly, Lufthansa has a constantly evolving stable of subsidiary airlines. China's big-3 FSNCs each host several LCC subsidiaries. Other Asian carriers with LCC offshoots include Garuda's Citilink; Philippine Airlines' Airphil Express; Korean Air's Jin Air. In Japan, both of the country's FSNCs, All Nippon Airways and Japan Airlines, have LCC subsidiaries. Singapore Airlines is a particularly interesting example and the case study at the end of this chapter.

An LCC subsidiary is set up to circumvent the high labor costs and restrictive work rules that burden the parent; nevertheless, the business models of FSNC and LCCs are so different that an LCC subsidiary under central management may not maintain the focus on cost control essential for LCC success. If subsidiaries are not fully separated, the FSNC parent risks confusing its brand. Labor unions, fearing loss of jobs and wage competition, often fiercely resist establishing or expanding an LCC subsidiary. In 2021, the British Airways' pilot union rejected plans by its parent IAG to establish yet another LCC based at London Gatwick.

4.6 Hybrid Airlines

The distinction between full-service network airlines at one end of a spectrum and low-cost carriers at the other is useful for elucidation but simplistic. The dearth of airlines fully embracing the LCC business model suggests that many airlines are positioned elsewhere along a continuum of product offerings. Faced with intense competition and low-profit margins, airlines frequently modify their product in an attempt to exploit perceived gaps in the market and secure sustainable profits. Consequently, the sharp line that once divided network carriers from low-cost carriers has blurred as FSNCs shed amenities and embraced unbundled products and a la carte pricing while LCCs have increased connectivity and upgraded in-flight service. Many carriers not readily classified as either FSNC or LCC have been termed *hybrid* carriers, a name that is unimaginative but descriptive.

Hybrids occupy a large range on the spectrum of product offerings and, therefore, are difficult to define with a list of business model characteristics. Hybrids do not have the extensive international route networks of full-service network carriers but do typically have marketing arrangements with other airlines, including code sharing. The in-flight product is of higher quality than the bare bones of LCCs, including features such as greater legroom, in-flight entertainment, higher quality meal and beverage service, and, in some instances, a business or first class cabin. Among the many hybrid airlines, two very similar US airlines are examples.

4.6.1 Examples: Alaska and JetBlue

In 2013, JetBlue's then-CEO Jeff Barger explained the hybrid model by stating that JetBlue has no desire to mimic either the ultra-low-cost model of Spirit Airlines or that of network carriers. During a 2013 earnings call, Barger assured analysts, "There is room for more than two models on the industry landscape and we are proving that" (JetBlue, 2013).

JetBlue, having begun operations in 1999, is a relative newcomer whereas Alaska Airlines has more than an 80-year history, but the two carriers share many similarities. Their route structures are nearly mirrored images. Alaska's routes extend from Alaska down the US West Coast and into Mexico via hubs in Seattle, Portland, and Los Angeles. It also serves Hawaii, some cities in Canada, and the US Midwest and East Coast. JetBlue's route structure is concentrated on the East Coast with extensions to major western cities, into the Caribbean, the northern reaches of South America, and the start of European service to London from New York. Both carriers have established close partnerships with American Airlines and varying degrees of cooperation with many other, mostly foreign, airlines. As part of its American Airlines partnership, Alaska joined the Oneworld Alliance in 2021. American and JetBlue formed the "Northeast Alliance" in 2021 aimed at closely coordinating flights in many markets. Alaska and JetBlue target higher-end leisure and cost-conscious business travelers (Centre for Aviation, 2013) and serve a mix of destinations attractive to each segment. Both use traditional global distribution systems as well as direct distribution through call centers and websites. Each offers a frequent flier loyalty program. Both have a quality product that rivals the best of the US FSNCs. JetBlue provides free Direct TV in all seats, a product innovation it introduced first.

Alaska has operated many aircraft types in its long history but has moved to standardize its fleet with Boeing 737s, a project that was complicated by its 2016 acquisition of Virgin America, an Airbus 320 operator. Alaska will gradually replace the A320s with the Boeing 737 MAX 9. All aircraft are configured in two cabins with first and economy classes. JetBlue's fleet, on the other hand, consists of two fleet types: the Embraer 190 and Airbus 320 and 321 models. Aircraft were all single-class until a premium business class was introduced in 2014 on a sub-fleet of A-321s dedicated to transcontinental service. The seat room is better than LCCs where high seat density confers a cost advantage but decreases passenger legroom. Spirit Airlines, for example, configures its A-320 aircraft with 178 non-reclining seats versus 150 seats for the same model aircraft at JetBlue.

Though both airlines are competitive on price, service quality is equally, if not more, important. Both have won many passenger service awards. Alaska and JetBlue are of similar size with 2019 annual revenues between eight and nine billion dollars and similar profit history.

Airline analysts have long suggested that Alaska and JetBlue merge. Instead, after a protracted battle, Spirit Airlines agreed in 2022 to be acquired by JetBlue.

As these examples illustrate, hybrid airlines have no unique business model and are instead better categorized by what they are not — full-service network airlines or low-cost carriers.

4.7 Focus Carriers and Tailored Products

Focus is the third of Porter's generic corporate strategies. The airline industry offers several examples of carriers whose product is tailored to meet the needs of a small segment of passengers, a niche market.

4.7.1 All-Inclusive Charter Airlines

Charter airlines offered low fares long before the emergence of today's LCCs. In the US-regulated era, supplemental carriers competed with scheduled airlines for leisure travel, primarily to Las Vegas and vacation destinations in Florida. The supplemental carriers did not survive under deregulation. In Europe, all-inclusive tour companies dominated holiday travel between Northern Europe and the Mediterranean with later expansion to other sun destinations in the Caribbean.

European charter airlines prospered in the 1960s and 1970s from their ability to avoid restrictive Air Service Agreements (also known as bilateral agreements) between countries that limited capacity and set fares for scheduled carriers. European tour operators contracted with charter airlines to provide the air portion of all-inclusive holiday packages. Charter fares were 40% to 70% less than competing services on scheduled airlines. With this price advantage, tour operators often provided the only air service between the UK and Germany to Greek Island and Turkish resorts.

As Europe's LCCs grew, however, vacationers gradually chose the LCCs' frequent flights and low fares over inclusive tour packages. With the emergence of Internet distribution, vacationers were able to arrange their travel, including air, hotels, and rental cars that the tour companies bundled (Doganis, 2019). Germany's TUI Group is the largest old-line tour operator. In addition to travel agencies, hotels, and cruise ships, TUI owns five airlines with some 150 aircraft. "This integrated offering enables us to provide our 27 million customers with an unmatched holiday experience in 180 regions." (TUI, n.d.). The failure of the legendary UK travel company Thomas Cook is the case study in Chapter 10.

4.7.2 All Business Class Service

An all-business-class airline seems an obvious niche to be profitably exploited. The concept is compelling: offer a high-quality business product in major business markets at a fraction of the fare charged by the FSNC incumbents; attempt to capture or create enough traffic to fill a low-density cabin (Boyd,

2007). Similar to the early US experiments with an LCC within an existing carrier, several all-business-class airlines first emerged in the US – all failed. A few domestic examples include Air1 and Air Atlanta, both of which failed in 1987; MAXjet Airways operated from 2003 until it failed in 2007. International routes had even more allure, especially between New York and Western Europe. Eos Airlines, L'Avion, and Silverjet tried and failed. British Airways operated between London City Airport and New York Kennedy for more than a decade with Airbus 318s – the smallest of the A320 series – configured with just 32 seats. With the pandemic-caused collapse of demand, BA discontinued the service and retired its last A318. As of 2022, only one example remains – La Compagnie offers daily service between Paris Orly and Newark Liberty with Airbus A321neo configured with 72 seats. Of course, the airline plans to expand.

The all-business-class model suffers from several weaknesses. First, business travel fluctuates by day of the week. To fill otherwise empty seats on weekends and mid-week, the airline resorts to discount fares that are not profitable with low-density seating. Second, there are only a few markets with sufficient business demand to support all-business-class service, and airports in these markets are frequently slot restricted. London's Heathrow is the prime example. With only a few potential markets, the all-business-class airline cannot expand to reach an economic fleet size. Then, many business travelers prefer to travel on one or two major airlines to maximize frequent flier benefits. The all-business-class airlines serve too few markets for the frequent flier programs to be attractive for most business travelers. Third, the success of full-service network carriers depends on attracting high-fare business passengers. If the all-business-class airline is successful in capturing any significant portion of this segment, larger rivals can retaliate with increased service and lower prices, exactly what American Airlines did when faced with competition from Eos in the New York–London market.

4.8 Cargo Airlines

Most cargo airlines also follow a focus strategy; however, the air cargo business is markedly different from its passenger counterpart. Air shipment of cargo is several times more expensive than via ground or sea transport, so there must be some compelling reason for a shipper to choose air transport. The common types of freight most often shipped by air are: (a) emergency shipments such as food and medicine required for natural disasters; (b) high value to weight and volume; (c) high risk of theft or other loss; and (d) perishables such as vegetables or other time-sensitive freight. Transport of goods from Asia to the US via ship takes several weeks versus 24 hours or so via air. Perishable goods such as vegetables and flowers can only be shipped by air to avoid spoilage en route. Routine air shipment can be economical for high-value items by weight. Computer parts are often air shipped to low-wage countries for assembly with finished goods air-shipped back to the US. Business contracts and other legal documents are time-sensitive and of high value. Consumer goods ordered online may be shipped by air to meet customer expectations of fast delivery.

Inadequate roads and railroads in undeveloped countries make air transport an expensive but viable alternative to ground shipping. Theft of cargo is another problem that can be mitigated by air shipment (Doganis, 2019).

The needs of air cargo shippers are not those of airline passengers. Airfreight is often time-sensitive, but shippers are indifferent to routing if the delivery is on time. Therefore, aircraft routing between a cargo origin and destination may be changed frequently with intermediate stops added or deleted to meet demand. In contrast to passengers who usually travel round trip, freight transport is one way resulting in uneven directional demand. Demand from Asia, especially China, to the US and Europe is much higher than in the opposite direction. This imbalance leads to low prices and loads on aircraft returning to Asia. Finally, airfreight varies greatly in size, weight, and handling requirements. Animals are often shipped by air and require careful handling as do hazardous goods. Hazardous materials shipment is restricted to special all-cargo aircraft and must be labeled "cargo aircraft only" or CAO.

Three distinct types of air cargo carriers meet shippers' differing needs: (a) combination; (b) integrated; and (c) all-cargo carriers.

4.8.1 Combination Carriers

Most FSNCs offer freight services with cargo carried as belly hold, hence the name combination carrier.[3] Before the pandemic, combination carriers supplied about half of all air cargo capacity. To supplement cargo capacity on passenger aircraft, some European and Asian carriers also operate dedicated freighter aircraft on both scheduled and charter flights; however, the substantial belly-hold capacity of wide-body aircraft flown on many international routes has reduced the need for a separate fleet of cargo aircraft. FSNCs provide frequent flights with substantial cargo capacity, but schedules meet passenger demand that may not be ideal for cargo sales. The combination carrier market is dominated by middle-eastern and Asian airlines with the Gulf carriers Qatar Airways and Emirates generating the greatest tonne-miles in 2020 (IATA, 2021b). The grounding of many international FSNC flights during the pandemic created a shortage of airfreight supply contributing to the global supply chain distributions.

4.8.2 Integrated Carriers

Integrated cargo carriers (ICC), also called express carriers, operate worldwide, door-to-door networks, mostly shipping small packages across multiple hub-and-spoke systems. Aircraft range from small turboprops to wide-body jets. For door-to-door delivery, a huge fleet of ground delivery vehicles is also required. Delivered time is guaranteed. Sophisticated tracking systems allow the carrier and its customers to see the location of any package in the system in near real-time. The industry giants are FedEx and United Parcel Service (UPS) with DHL having a strong international presence. FedEx and UPS were the two largest carriers of international air cargo in 2020 (IATA, 2021b). A more recent entrant is Amazon Air (formerly Amazon Prime Air), the brand name for several all-cargo airlines operating under contract with Amazon.

4.8.3 All-Cargo Airlines

All-cargo airlines operate only freighter aircraft, mostly converted former passenger planes. These airlines are relatively small compared with FSNCs with a substantial component of their business derived from long-term contracts with FSNCs, mainly Asian carriers. This model, similar in some ways to the role of regional airlines, is known by the acronym ACMI that stands for *aircraft, crew, maintenance, and insurance.* The FSNC provides all other functions, including marketing, scheduling, ground services, and, critically, fuel. FSNCs find that all-cargo airlines offer a more flexible and cost-effective cargo service than providing the service with their aircraft and crew. All-cargo airlines also fly scheduled service under contract to the largest freight forwarders and on-demand charter flights.

All-cargo airlines can transport freight that combination and integrated carriers do not accept. Hazardous materials (also called dangerous goods) include explosives, compressed gases, flammable and combustible goods, corrosives, poisons, infectious substances, and radioactive materials that regulations restrict to all-cargo airlines. These airlines can also transport outsized cargo and other freight that requires handling only all-cargo airlines can provide.

The largest all-cargo airline is Luxembourg-based Cargolux although the combined tonne-miles of US-based Atlas and Polar Airlines, owned by the same holding company, are greater.

4.8.3.1 Example: Cargolux

Cargolux Airlines International, based at Luxembourg airport, operates all-cargo scheduled and charter flights to 75 worldwide destinations with a fleet of 30 Boeing 747 aircraft. It was the launch customer for the B-747-8F, the latest and last of the iconic Boeing 747s. As only an all-cargo airline can do, it transports live animals, pharmaceuticals, hazardous goods, perishables, artwork and other valuables, and difficult loads ranging from outsized and heavy freight such as flight simulators and shopping center escalators (Cargolux, 2021). Transporting this varied cargo requires specialized expertise not needed in airline passenger service.

4.8.4 Distribution

The air cargo industry has developed a unique distribution model. Millions of airline passengers arrange their travel through various distribution channels (see Chapter 10) and find their way to and from their origin and destination airports. The shipment of airfreight is more complex requiring specialized packaging, delivery to the airport, and, for international cargo, detailed documentation. These are tasks that most shippers either cannot or do not wish to handle. For express packages, the integrated carriers provide all aspects of door-to-door transportation. In contrast, middlemen known as *freight forwarders* control 85% of non-express air cargo carried on combination and all-cargo airlines. For a typical shipment, the freight forwarder:

- Contracts with airlines to fly consigned cargo.
- Picks up the freight at the shipper's location and trucks it to the forwarder's facility near the airport.
- Packages the freight to airline requirements, usually combined with other shipments bound for the same geographical area.
- Prepares the documentation required by airlines and governments.
- Trucks the packaged freight to the selected airline at the departure airport.
- Clears the freight through outbound customs.
- Picks up the cargo at the destination airport and clears inbound customs.
- Trucks the cargo to its destination facility where it is unpacked.
- Trucks the cargo to its receiver.

The world's largest freight forwarders control much of world air cargo and ocean shipping. With their size and scope, they exercise substantial market power. The world's largest forwarder is DHL Supply Chain & Global Forwarding, itself part of the Deutsche Post DHL Group that includes DHL airlines. Switzerland's Kuehne+Nagel, Germany's DB Schenker, Danish DSV, and China-based Sinotrans Limited round out the top five.

With the growth of e-commerce, freight forwarders and cargo airlines have expanded their services to include inventory management and warehousing. A retailer, for example, may hold some of its inventory at a UPS warehouse adjacent to one of UPS's air freight hubs. A customer's order is filled from the warehouse and placed on a UPS flight in the evening for delivery to the customer's door the next morning. Companies providing fully integrated logistics are termed *global logistics suppliers*.

4.9 Summary

In the new millennium, full-service network carriers across the world find their yields depressed and market share under attack from aggressive low-cost carriers. Faced with high costs and lower prices, FSNCs have reevaluated and modified their product offerings. Short-haul routes have been transferred to regional airlines and LCC subsidiaries created within the FSNCs. Following the LCC practice of product unbundling, FSNCs have imposed separate fees for baggage, limited food service, and other product features amenities previously included in the ticket price. Passengers have grudgingly accepted this bewildering assortment of additional fees.

The largest low-cost carriers also face a rapidly changing environment. In the US and Europe, large underserved and overpriced markets that once presented expansion opportunities have been filled. There's little low-hanging fruit. Expansion is often in markets already served by other airlines competing on low fares. Smaller markets that still command high yields are too thin to serve with point-to-point routes, so connecting traffic is essential to enter these markets. Finally, LCCs, envious of business travelers willing to pay high fares for quality service, have upgraded their products.

Faced with these market realities, both FSNCs and LCCs have adopted many of the product features of the other. Table 4.3 summarizes the evolving products. The result is a fading of the distinction between the full-service network carrier and the low-cost carrier with hybrid carriers occupying the middle space. This continuum of product offerings serving a broad range of consumer wants and needs is typical of mature, competitive industries. In the automobile industry, for example, there's a car to satisfy nearly any driver's desire.

A look at the world's largest airlines illustrates the geographic and business model scope of the airline industry. Table 4.4 lists the largest 25 airlines ranked by aircraft fleet size. Other measures such as revenue passenger kilometers,

Table 4.3 Fading Distinction between FSNC and LCC

Low-Cost Carrier Attribute	LCC Product Evolution	FSNC Product Evolution
Unit Cost Advantage	The foundation of the LCC business model	Restructuring has narrowed the cost gap
Point-to-Point Routes	Increasing connectivity	Maintains connecting traffic advantage
Limited geographical coverage	Establishing code-shares, establishing affiliate airlines	Members of global alliances
Service to secondary airports	Recent expansion has emphasized primary airports, e.g., Ryanair and Spirit Airlines	
High Aircraft Utilization		Depeaking and rolling hubs increase utilization
Direct Distribution	Adding GDS distribution	Pushing website booking and sales
No Frills Service	Adding in-flight entertainment, seat room, business class	Unbundling, eliminating food and other in-flight amenities in coach class, dropping first class
Low, simple fares	Applying revenue management	Eliminating restrictions
Single aircraft type	Adding 2nd fleet type	Reducing fleet types

Source: Adapted from Meehan, D. (2006). Aviation industry outlook for 2006.

Table 4.4 World's Top 25 Airlines

#	Airline	Model	Aircraft	#	Airline	Model	Aircraft
1	American Airlines	FSNC	1119	14	UPS	ICC	237
2	Delta Air Lines	FSNC	853	15	Lufthansa	FSNC	207
3	Southwest Airlines	LCC	622	16	All Nippon Airways	FSNC	188
4	United Airlines	FSNC	584	17	Qatar	FSNC	187
5	China Southern Airlines	FSNC	560	18	Air France	FSNC	186
6	China Eastern Airlines	FSNC	559	19	Hainan Airlines	FSNC	184
7	SkyWest Airlines	RA	457	20	LATAM	FSNC	182
8	Air China	FSNC	428	21	Shenzhen Airlines	FSNC	182
9	FedEx	ICC	313	22	Japan Airlines	FSNC	175
10	Ryanair	LCC	253	23	Aeroflot	FSNC	164
11	Turkish Airlines	FSNC	251	24	Sichuan Airlines	FSNC	163
12	JetBlue Airways	Hybrid	248	25	Qantas	FSNC	158
13	Alaska Airlines	Hybrid	245				

Notes: Business models: full-service network carrier (FSNC), low-cost carrier (LCC), regional airlines (RA), hybrid, integrated cargo carrier (ICC).

Source: Air Transport World, July–August 2021.

enplaned passengers, or total revenue would alter the rankings, but most of the airlines would remain. The top three are unchanged when measured by operating revenue, revenue passenger kilometers, or enplaned passengers. Two of the airlines may be a surprise. Southwest is the world's third-largest airline, second if measured by passenger enplanements, and the regional carrier SkyWest Airlines is seventh.

The world's three largest airline markets – the US, China, and Europe – are represented by nine, six, and four airlines, respectively, in the top 25. All of the airline business models discussed in this chapter are also represented. Full-service network carriers predominate with 14. There are two low-cost carriers, one in the US and one in Europe; two hybrid airlines; two integrated cargo carriers; and one regional airline.

Airline jargon is confusing, exacerbated by inconsistency across airlines. This chapter offers a typology of airline business models, but the terms employed are not universally accepted. Full-service network carrier is descriptive, but other terms, including full-service airline, full-fare airline, major airline, and legacy carrier are common. Low-cost carrier is widely used but budget airline, low-fare airline, discounter, no-frills carrier, and the latest, ultra-low-cost-carrier, will also be encountered. Regional airline is common but not descriptive. Small jet provider and network extender are more descriptive but not in wide use. Hybrid is a recent term that, fortunately, seems almost universal.

The remnants of the old US Civil Aeronautics Board classification system create another potential confusion. The CAB classified airlines by total revenue in three categories: major, national and regional. Regional and major are still in some use, but no longer refer to total revenue.

As we have seen in the previous chapter, an airline must make a strategic choice of network architecture to connect the destinations it chooses to serve. Similarly, it has a wide choice of the product it provides ranging from no-frills basic air transportation to an augmented product with lavish inflight and ground amenities. Like most businesses, airlines cannot profitably serve all passenger wants and needs; therefore, an airline identifies those passenger segments it plans to serve and tailors a product for those needs. Porter's popular generic strategies help conceptualize an airline's choice of competitive scope. There are many examples of the striking variety of airline services, for example, Ryanair's low-priced, core air transportation with some extra services available, but only at an additional charge, to the unsparing service provided by many international carriers such as Singapore Airlines. Recently, many airlines, following the lead of Air Canada, have packaged their services so that passengers can choose a level of service desired with corresponding prices.

This and the previous chapter address the airline product. The next chapter covers two related topics – developing a flight schedule and managing daily flight operations.

4.10 Case Study: Singapore Airlines

Singapore is a city-state whose population of 5.7 million is crowded into less than 300 square miles, about one-third the size of Luxembourg, yet it is home to one of the world's largest international airlines. Founded just after World War II in 1947, Singapore Airlines (SIA) is renowned for its superior service. It operates a hub-and-spoke route system flying to 78 destinations in 32 countries. The fleet of 148 aircraft comprises Airbus A380, A350–900, Boeing 777–300ER, 787–10, 737–8 and 800, plus seven all-cargo Boeing 747–400F. Passenger aircraft are variously configured with 15 different interiors. As a full-service airline, Singapore offers a product for most passenger segments best illustrated by the four cabins on the A380 ranging from first-class to basic economy. While Singapore promotes its business class product, nearly three-quarters of seats on the A380 are basic economy. To supplement the belly-hold capacity of its passenger fleet, Singapore Cargo Airlines operates all-cargo Boeing 747s. Singapore is a member of the Star Alliance maintaining codeshare agreements with more than 30 airlines, most of whom are alliance members.

The Singapore Group has at times included several subsidiary airlines and investments in unrelated airlines. To protect its image, when Singapore wished to expand into other than a premium market, it created a new brand.

4.10.1 SilkAir

SilkAir started in the late 1980s as a holiday/charter airline in the mold of Europe's all-inclusive charter airlines. By 1992 when it adopted its present name, SilkAir had morphed into a short and medium-haul subsidiary with a fleet of narrow-body aircraft serving destinations in Southeast Asia and feeding Singapore's hub. This model is akin to a regional airline except for the sharp boundary between narrow-body aircraft and Singapore's widebodies used for long-haul international flights. It's as if all of United Airlines' US domestic flights were operated by its Ted subsidiary with the United Airlines name appearing only on over-the-ocean widebody flights.

4.10.2 Tiger Airways

In the early 2000s, Singapore joined other FSNCs in developing a low-cost carrier subsidiary. Tiger Airways grew rapidly with its fleet of Airbus A320 serving regional destinations in Southeast Asia overlapping some SilkAir routes. Other Tigerair branded subsidiaries sprouted in Taiwan, the Philippines, Mandala, and Australia.

4.10.3 Scoot

In the next decade, Singapore decided to expand with a long-haul, low-cost airline, so it created Scoot with cast-off 777s from the parent company. Scoot pushed into Australia first, then to China, India, the Middle East, and Honolulu with a flight via Osaka/Kansai.

4.10.4 NokScoot

Wanting to expand Scoot into other countries but constrained by foreign country ownership restrictions, NokScoot was created in 2013 as a joint venture with Thailand's Nok Air and Singapore's wholly owned subsidiary Scoot. Like Scoot proper, it also operated Boeing 777–200 aircraft on routes from Bangkok to China, India, Japan, and Taiwan.

4.10.5 Others

Singapore Airlines bought a 49% stake in Virgin Atlantic in 2000 for $966 million. By 2013, it had acquired a 20% stake in Virgin Australia. Vistara, a joint venture between Singapore and the Indian conglomerate Tata Sons inaugurated Indian domestic service in 2015.

4.10.6 Too Ambitious, Too Complex

Perhaps not surprisingly, this vision of a conglomerate airline structure didn't fare well. Singapore sold its underperforming stake in Virgin Atlantic to Delta Air Lines in 2013 for $360 million, a loss on its original investment of about $600 million. NokScoot didn't survive the pandemic and was liquidated in 2020. Virgin Australia fared only slightly better being acquired in bankruptcy by the private-equity fund Bain Capital. It faces an uncertain future.

Tiger and Scoot were both created as low-cost airlines although with vastly different fleets. In 2016, Singapore decided there was no reason to keep two brands. It retired the Tiger name and put everything under Scoot. As of 2022, it serves 60 destinations spread across Asia, Australia, India, and Europe with a mixed fleet of 39 Airbus A-320 series for short and medium routes and 20 Boeing 787s for long-haul service.

How about SilkAir? The problem here was that SilkAir operated within Singapore Airlines' hub-and-spoke system. Connecting passengers may fly a segment of their itinerary on SIA and another on SilkAir encountering different standards of service. A first-class passenger on a Singapore flight would suffer a service downgrade service when connecting to a SilkAir flight. To solve this problem, Singapore decided to upgrade the SilkAir product and then fully integrate SilkAir into Singapore Airlines, a process completed in 2021. In its 2019–2020 Annual Report, Singapore promised the merger with SIA would "help strengthen and enlarge the SIA Group's short-haul offerings, enhancing benefits to our customers."

Singapore Airlines Group emerged from the restructuring with two distinct airline brands: a low-cost medium and long-haul carrier in Scoot and Singapore Airlines as a full-service network carrier. Lesson learned? Perhaps, but the Annual Report casts doubt on SIA's commitment to simplicity and conservative growth.

> The SIA Group is at the heart of Singapore's aviation ecosystem and anchors Singapore's air hub status. … Beyond Singapore, the SIA Group's multi-hub strategy complements its efforts to strengthen and grow the Singapore aviation hub. By investing in airlines outside Singapore, the Group is able to set up new hubs in growing markets, extending its reach and tapping into new traffic flows. For example, the investment in Vistara enables the Group to have a presence in India's fast-growing airline market. In 2019, Vistara launched its first international flights and expanded its codeshare agreement with SIA and SilkAir on international flights.

In the interim, the Tata Group, Singapore's partner in Vistara, acquired Air India that will directly compete with Vistara.

Concepts: full-service network carriers, regional airlines, low-cost carriers, long-haul low-cost carriers, LCCs within FSNC, inclusive charter airlines, all-cargo airlines, global alliances, multiple fleet types, restructuring and simplification

Sources: Cranky Flier, 2018a, 2018b; SeatGuru; Singapore Airlines 2019, 2021; and several Wikipedia entries.

Notes

1 This explanation of differentiation as applied to the airline industry differs somewhat from Porter's original concept. The strategy as described here is also known as "full-line generalist."
2 The total number and distribution of LCCs by region varies substantially by source. The Wikipedia list is representative and likely the most authoritative.
3 The term is traditionally applied to airlines that operate both passenger and all-cargo aircraft. We adopt the broader definition following Doganis (2019) and Lange (2019).

References

AirAsia X Is back! Rejuvenated airline spreads its wings to London, Istanbul and Dubai. (2022, June 15). *Aviator.*

Airlink. (2021). *About us.*

ANA. (2022, March 8). *ANA Holdings unveils AirJapan, a new international airline brand taking off in the second half of fiscal 2023* [Press release]. www.anahd.co.jp/group/en/pr/202203/20220308.html

Bachwick, A. R., & Wittman, M. D. (2017). The emergence and effects of Ultra-Low cost Carrier (ULCC) business model in the U.S. airline industry. *Journal of Air Transport Management, 62,* 155–164.

Boyd, M. (2007, January 29). *Hot flash.* The Boyd Group International.

Budd, L., & Ison, S. (2017). Air transport management: An international perspective. Routledge.

Cargolux. (2021). *About us.* www.cargolux.com/about-us/Introducing-Cargolux

Centre for Aviation. (2013, June 26). North American hybrid airlines offer a range of possibilities as consolidation takes hold: Part 2.

Delta Air Lines. (2021). Corporate information. Retrieved from www.delta.com/cont ent/www/en_US/about-delta/corporate-information.html

Doganis, R. (2019). *Flying off course: Airline economics and marketing* (5th ed.). Routledge.

European Regions Airline Association. (2021). Our members. www.eraa.org/members hip/our-members

Flottau, J. (2021, April 12). What will happen in the regional jet market? *Aviation Week Network.*

Flottau, J. (2022, June 9). Embraer builds its future as it works through five-year recovery plan. *Aviation Week Network.*

ICAO. (n.d.) *Glossary.*

IdeaWorks. (2021). *The 2021 CarTrawler yearbook of ancillary revenue.* https://ideaworks company.com/wp-content/uploads/2021/09/2021-Ancillary-Revenue-Yearb ook.pdf

JetBlue. (2013). Q2 2013 JetBlue Airways earning conference call.

Klophaus, R., Conrady, R. & Fichert F. (2012). Low cost carriers going hybrid: Evidence from Europe.

Largest airlines in the world. (2021, December 1). In *Wikipedia.* https://en.wikipedia. org/wiki/Largest_airlines_in_the_world

List of low-cost airlines. (2021, December 2). In Wikipedia.

List of regional airlines. (2021, November 23). In Wikipedia.

Mouawad, J. (2013, August 3). The race to build a better business class. New York Times. Retrieved from www.nytimes.com/2013/08/04/business/the-race-to-build-a-bet ter-business-class.html?pagewanted=1&_r=0

O'Connell, J. F. (2011). Airlines: An inherently turbulent industry. In J. F. O'Connell & G. Williams (Eds.), *Air transport in the 21st century.* London: Ashgate Publishing Group.

Singapore Airlines. (2013). On board experience. www.singaporeair.com/en_UK/fly ing-with-us/suites/

Singapore Airlines. (2019, 2021). Annual reports.

SkyWest (2021, September 30). *About.* www.skywest.com/about-skywest-airlines/facts

TUI. (n.d.) About TUI Group. www.tuigroup.com/en-en/about-us/about-tui-group

Wizz Air. (2021a, December 2). Information and services. https://wizzair.com/en-gb/ information-and-services/investor-relations/investors/information-dashboard

Wizz Air. (2021b). Wizz Air Holdings Plc Annual report and accounts 2021.

ZIPAIR. (2021). Company.

Review Questions

1. What is the "Marketing Concept" and why do marketers consider it essential in a competitive economy?
2. List some things (product attributes) airline passengers want from their flight?
3. Why can't an airline offer a passenger everything she or he wants?
4. List and describe the business models of each airline category.
5. Use the Airline Product Design model to compare two airlines from different categories.
6. Why are FSNCs dependent on passengers traveling for business for profitability?

7. How do LCCs, and especially ULCCs, keep costs low compared with other categories of passenger airlines?
8. Why have many FSNCs split their economy cabin into premium and basic economy sections?
9. How do hybrid carriers fill the gap between LCCs and FSNCs?
10. How do the needs of air freight shippers differ from airline passengers for air transportation?
11. List and describe the business models of the three types of air cargo airlines.

5 Flight Schedule Development and Control

Upon completion of this module, you should be able to:

- Recall the economic objectives of the flight schedule.
- Recount various compromises needed to construct a viable schedule of services.
- Explain the pros and cons of slack resources to improve schedule reliability.
- Explain the different crewmember scheduling processes and the pros and cons of each.
- Describe the operations control functions located in a typical Airline Operations Control Center.

5.1 Introduction

An airline's product – the destinations served, the route structure connecting those destinations, and the features included with the core air transportation service – is frequently reviewed and refined. Occasionally airlines undertake a significant overhaul of their product, as the US full-service network carriers have done in the wake of the financial crises of the first decade of the new millennium and following the COVID-19 pandemic. The airline's flight schedule, however, is more frequently revised with major changes typically published twice a year (for the winter and summer seasons). Following an overview of the airline planning process, the first section of this chapter addresses the complex task of developing a flight schedule. Once the flight schedule plan is completed, it falls to the operations managers to operate the schedule daily. As any frequent air traveler knows, flight schedules are subject to disruption for many reasons. The methods available for tactical flight schedule management are the subject of the second half of this chapter.

5.2 Airline Planning Process

Airline planning encompasses strategic forecasts developed by a small staff to short-range tactical planning involving several departments and many managers.

DOI: 10.4324/9781003290308-5

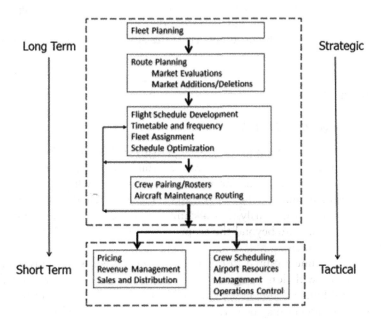

Figure 5.1 Airline Planning Process.

Figure 5.1 is a flow chart of the airline planning process. Strategic plans extend over five or more years. Because of the long lead times to acquire new aircraft, the development of the fleet plan and new aircraft orders follow directly from the long-range plan. Product planning, including partnerships with other airlines ranging from code-sharing to joint ventures, emerges from the airline's strategic vision and perceived growth opportunities. These strategic decisions are made by the airline's executive management and, at large airlines, supported by dedicated planning staff.

At large airlines, preliminary market evaluation begins some years in advance of implementation with decisions to enter or exit a specific market usually finalized one to two years out.

Flight schedule development builds on the existing schedule and begins about one year in advance of operations. The final product is the airline's timetable: a listing of all *city pairs* served, flight numbers, departure and arrival times, and aircraft type. Printed timetables, whose origins date to the early days of ocean shipping and later to railroads, were once ubiquitous but are now rare. The electronic versions are still available on some airlines' websites.

Once the flight schedule is finalized, the operating departments are charged with its execution. A centralized group of managers and staff located in the *Airline Operations Control Center* (AOCC) coordinate the daily activities of pilots and flight attendants, mechanics, and station personnel during routine flight operations and implement revised plans when the flight schedule is disrupted by weather, mechanical failures, and a host of other unplanned events.

The following sections discuss each step in the airline planning and control process.

5.3 Strategic Planning

5.3.1 Long-Range Plan/Fleet Selection

Long-range planning, extending over five to as many as 20 years, begins with a corporate vision and mission statement. Some airlines target aggressive annual growth of 15% or more while mature airlines subordinate growth to emphasize return on investment and shareholder wealth. A SWOT (strengths, weaknesses, opportunities, and threats) analysis is a common framework for long-range planning. These topics are beyond the scope of this text and are left to books on strategic management.

Because of the time required to obtain new aircraft, fleet planning is often contemporaneous with long-range planning. For large airlines, new aircraft orders of several hundred aircraft are common with deliveries spread over five to 10 years. New aircraft may be slated for capacity growth, as replacements to upgrade the existing fleet or, more commonly, for both. Economic recessions and other external shocks disrupt long-range plans usually necessitating a subsequent major revision. Following the Great Recession of 2008, fast-growing Emirates Airlines illustrated the close connection between the strategic plan and aircraft acquisition.

> In 2010, in line with the airline's strategic growth plan, Emirates significantly increased its order for new aircraft. Underscoring its incredible growth, the airline is currently the world's largest operator of both the Airbus A380 and Boeing 777. Emirates' current order book stands at more than 230 aircraft, with a total value of approximately $84 billion as of November 2011. In combination with what is already the youngest and one of the most modern fleets in worldwide commercial aviation, this commitment to the future reflects our goal to develop Dubai into a comprehensive, global, long-haul aviation hub. (Emirates, n.d.-a).

Emirates successfully implemented this growth plan until the COVID-19 pandemic struck at its heart when in March 2020 the UAE government suspended all scheduled passenger flights as part of the country's pandemic response (Emirates, n.d.-b). Two years later, Emirates had placed its entire Boeing 777 fleet and over half of our A380s back in active service. With the worst of the pandemic passed by 2021, United Airlines announced an order for 270 single-aisle Boeing and Airbus planes, the biggest aircraft purchase from a US airline in a decade and the largest order in United's history (Chokshi, 2021).

5.3.2 Product Planning

The next step in the planning process identifies markets for potential expansion and, less frequently, for deletion. US low-cost carrier (LCC) Spirit Airlines, which had 100 new aircraft on order as of mid-2015, claimed to have identified 500 new routes that met its criteria for (a) large markets with more than 200 passengers per day each way; (b) high average fares; and (c) potential to achieve a 14% operating margin (Spirit Airlines, 2015). While targeting expansion, Spirit has not been shy to drop markets that fail to meet its financial expectations. Southwest Airlines offers a somewhat different perspective. Speaking to investors in late 2012, CEO Gary Kelly explained that planned and potential expansion of its service to Mexico, Canada, and the northern tier of South America could support substantial fleet growth. "If you added up all of the opportunities that are represented by that route map (of potential international routes), on a rough base of 700 airplanes there are 200 or 300 airplanes' worth of growth opportunities, all else being equal" (Compart, 2012, p.1). Unlike Emirates and Spirit, however, Southwest is a mature carrier that has shifted its focus from growth to return on investment. Kelly went on to explain that further expansion was subject to meeting a 15% return on invested capital (Compart, 2012).

Facilities to support expansion plans also require long lead times. Much of this responsibility falls to airport operators. To support its expanding airline industry, China plans to build an additional 215 airports by 2035 (Lee, 2020). In contrast to much of the rest of the world, US airlines play a major role in airport terminal design, construction, and finance. For example, Southwest Airlines fully funded the construction of a new five-gate international terminal at Houston's Hobby Airport that opened in 2015 ("Houston Airport System," n.d.).

The widespread restructuring of US airlines in the first decade of the 21st century and later in European carriers should have been part of strategic planning, but circumstances often compel these decisions with much less lead-time. Ironically, airlines don't consider bankruptcy as part of a strategic plan, yet it has been the driving force in all US legacy airline restructuring.

As the time horizon shortens to between one and three years, market evaluations become more detailed. Decisions include product upgrades and pricing policies, code-sharing agreements and alliance participation, and predicting competitors' behavior. Detailed evaluations of potential new destinations and routes result in the selection of new services between 12 and 18 months in advance of the first flight, although, for competitive reasons, public disclosure is usually withheld until much closer to the start of the new service. Some existing routes may also be dropped on a somewhat shorter timeline. Chapter 4 covers product choice, while partnerships and alliances are the subjects of Chapter 11.

5.4 Flight Schedule Development

The flight schedule is the airline's core product designed to solve the customer's time-space problem. Recalling the *marketing concept*, the flight schedule is designed to meet the customer's need to travel to some distant place at a certain time. The flight schedule, sometimes known as the *schedule of services*, lists the destinations or routes operated, the flight frequency and times, and the type of aircraft assigned to each flight. The schedule development task falls broadly under the marketing discipline, but many airlines have a specialized schedule planning or airline planning department.

Except for new-entrant airlines, each new flight schedule is a revision of the previous schedule. Route structure architecture is a long-term commitment, while service in some markets will evolve. A hub-and-spoke carrier, for example, may add point-to-point service in some city pairs as traffic grows or in response to competitive pressure. A change from tightly timed connecting complexes to a rolling hub is a more extensive and complex schedule revision. Passengers, however, appreciate schedule stability, so airlines operate many flights at the same times and with the same flight number for years. Because most airlines accept reservations up to one year before the flight, work on flight schedule revisions commence from more than a year out and continue to a few months before the flight. Booked passengers must be notified of schedule changes implemented after reservations have been made. Before the pandemic, close-in schedule changes were rare. The pandemic, however, upended airlines' ability to accurately forecast demand, so they often were forced to react to unexpected bookings, sometimes canceling flights just days before departure.

Printed hard copies of the flight schedule or timetables were once the primary means of providing flight information to potential passengers, but the internet has rendered the printed timetable obsolete. Now, electronic timetables are available on most airline websites. Figure 5.2 is an excerpt from a classic printed 1978 Alitalia timetable showing flights to and from Singapore, Stockholm, Stuttgart, Sydney, Tananarive (Madagascar), Teheran, Tel Aviv, and Tokyo. The timetable shows the origin and destination, days of the week on which the flight operates, times and flight number, aircraft type, class of service, and whether non-stop or requiring a connection. The timetable is the final product of the flight schedule development process.

5.4.1 Objectives

Development of the flight schedule is a complex task, not only because of the vast number of variables and possibilities to be considered, but also because of the required tradeoffs among revenues, costs, reliability, and constraints. Figure 5.3 depicts four often conflicting objectives the schedule planner attempts to balance. Each objective is considered next.

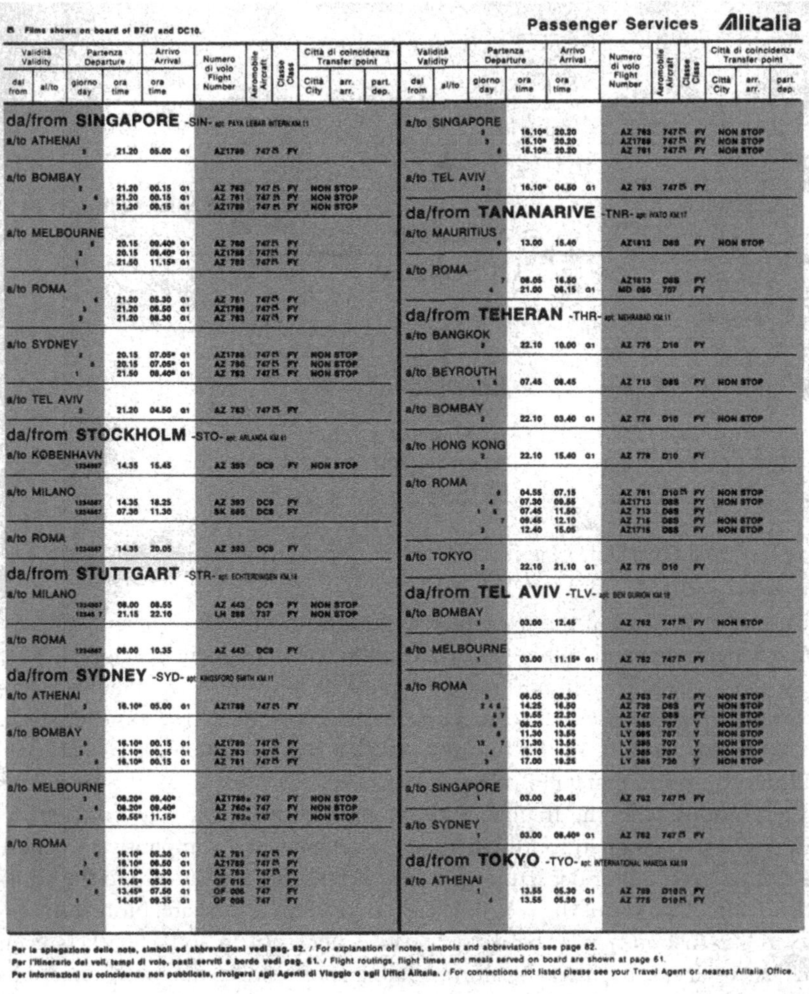

Figure 5.2 Alitalia Timetable.
Source: From Wikimedia Commons.

5.4.1.1 Revenue

The flight schedule seeks to maximize network revenue by matching flights and capacity with passenger demand. For leisure passengers, price is usually more important than schedule when choosing an airline; however, the high-yield business passengers generally consider an airline's schedule more important than price (Baseler, 2002). An airline targeting the business passenger

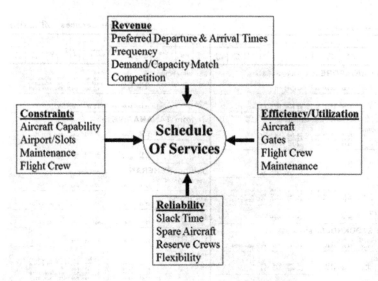

Figure 5.3 Flight Schedule Objectives.

must offer flights when the passenger wishes to travel with sufficient capacity to meet peak demand. Business travelers also favor frequent service because travel plans change. In business markets, frequent morning and late afternoon/ evening flights are essential to meeting passenger desires, but some off-peak service is also needed. In the business traveler-rich Houston–Dallas city-pair, Southwest Airlines generally offers flights every 30 minutes in the mornings and afternoons, with the frequency dropping to hourly mid-day. For markets with aggressive competition, high flight frequency is a competitive weapon. Sometimes even minor departure and arrival time changes can add to the competitive attractiveness of the schedule and increase market share. Non-stop service (point-to-point) may be a competitive necessity in some markets even when such service bypasses the airline's hub airports thus reducing connections in other markets. For routes with less competition, such as those to smaller cities, high frequency is less important, but morning and late afternoon/evening service is still desirable.

Revenue potential increases with the number of city pairs served, so hub-and-spoke carriers must maximize connections through hub airports with the most profitable markets having the most convenient connections. Passengers prefer non-stop flights, but if an intermediate stop at the hub is required, a continuing flight on the same aircraft (called *direct flights* in airline jargon) is preferable to a change of aircraft. The planner must consider the aircraft flow across the hub to maximize same-aircraft connections and hence schedule convenience.

The schedule is a compromise between flight capacity and frequency. Is it more profitable to operate one 400-seat aircraft twice a day or a 100-seat aircraft eight times per day? One solution is to offer a higher capacity mainline aircraft during peak demand supplemented with regional jet service during off-peak times.

Optimal departure times for international flights will be determined by the length of the flight and time zone changes. Most passengers do not favor late evening arrivals. From the US, late afternoon departures allow for morning arrivals in Europe and lower South America. Gulf airlines operate early morning connecting complexes for their long-haul international flights to provide mostly daylight departure and arrival times in Asia, Europe, and the Americas. Connection times from international to domestic flights must be long enough to include the time necessary for passengers to process through customs and immigration.

5.4.1.2 *Efficiency and Utilization*

The schedule planner lowers the airline's production cost, known as *Cost per Available Seat Mile* (CASM or *unit cost*), with high utilization of aircraft, crew, and other assets. High utilization spreads fixed costs over more available seat miles (ASM), thus lowering the CASM. Aircraft capital costs, for example, are fixed regardless of the hours flown. If the aircraft is leased, the lease payment is made monthly whether the aircraft is flown many hours per month or only a few. Similar reasoning applies to other fixed costs like maintenance facilities, airport gates, and terminal space — the more they are used, the lower the cost per use.

Higher aircraft utilization not only lowers unit costs but also increases revenue potential as revenue is generated only when aircraft are flying. Historically, LCCs enjoyed a substantial advantage over their network competitors in utilization, but in recent years that advantage has diminished. In 2013, LCCs utilized their aircraft over two hours per day more than network carriers (12.49 versus 10.17); however, by 2019, utilization was essentially the same for both categories of carriers, dropping for the LCCs to 10.37, while the network carriers increased to 10.79 (Massachusetts Institute of Technology [MIT], 2021). This reflects a trend over the past 10–15 years where the carriers once considered LCCs have slowly shifted their operational strategies making them more like network carriers. Canada's WestJet is a prime example. The newer *ultra-low-cost carriers* (ULCCs), Allegiant, Frontier, and Spirit, have taken over the lead for utilization, replacing the original LCCs. Although the average utilization for the ULCCs was only 10.81 in 2019, if Allegiant[1] is factored out of the average, the two other ULCCs average close to 12 hours a day (MIT, 2021). Southwest's decrease in utilization is mostly a result of extended aircraft ground turn times needed to accommodate more passengers and baggage on larger aircraft. For the newer international flights, additional time is required for aircraft and crews to process through customs and immigration.

High pilot and flight attendant utilization per day lowers crewmember cost per segment and requires fewer total crewmembers to operate the flight schedule. Before legacy carrier restructuring, network carriers averaged about 50 flight hours per month per pilot, whereas currently Southwest Airlines pilots fly an average of nearly 70 flying hours per month.

5.4.1.3 Reliability

The flight schedule is subject to disruptions, particularly for weather and aircraft mechanical problems. The schedule must incorporate sufficient slack resources to absorb delays and provide competitive on-time service. Without some slack, a schedule that looks good on paper may be disastrous in actual operation and end up driving passengers to competitors. Slack can be built into a schedule in several ways. To protect operations against aircraft mechanical problems, some aircraft are held as spares and out of the schedule to be substituted for aircraft that run late or suffer mechanical problems. Likewise, reserve flight crews (pilots and flight attendants) fill in for other crewmembers in the event of illness, illegality, or off-schedule operations. The schedule design should allow for aircraft substitutions. High flight frequency and the regular rotation of aircraft and crews through the network carrier's hub airports provide opportunities to swap or switch aircraft and crewmembers, flexibility not enjoyed by carriers utilizing a point-to-point system. Finally, slack time between arrivals and departures allows the airline to recover from late arrivals. The US Department of Transportation ranks airlines by on-time arrivals (within 15 minutes of scheduled arrival time) and publishes the data monthly. Partly in response, US airlines have increased scheduled block times to pad arrival times a little so that unforeseen delays (like higher winds en route) do not result in a late arrival (Yimga & Gorjidooz, 2018). On-time arrivals statistics are improved but at the cost of lower utilization.

5.4.1.4 Constraints

The flight schedule must also be feasible given numerous constraints. For example, a schedule that requires more aircraft than the airline operates violates a constraint and is not feasible.

Aircraft have operational capabilities not suited to all routes. Different aircraft types have greatly varied ranges and load-carrying capabilities. Airbus and Boeing's newest long-range, twin-jet aircraft, the A-350 and B-787, have enabled airlines to serve long-haul routes that were not previously economically and operationally feasible. Aircraft takeoff weight is restricted by altitude and temperature, which can limit fuel, payload, and range. Flights operating from high elevation airports such as Bogota in Columbia (8,361 feet or 2,548 meters) suffer from takeoff weight restrictions. High summer temperatures in the Gulf region impose similar restrictions. Again, the adverse impact varies greatly by type of aircraft. If aircraft are scheduled to operate at near maximum range, a day with unusually strong headwinds may require an en route fuel stop or a payload limit. Cargo and/or baggage may be left at the origin so that additional fuel may be uploaded to enable a non-stop flight. Of course, passengers will be displeased to find their luggage was not loaded on their flight.

Airport capacity is often limited by available runways, taxiways, gates, and counter positions. A few US and many European and Chinese airports are slot-limited. Night curfews impose a similar limitation on the schedule planner.

Aircraft must receive regular maintenance that requires periodic removal from the flight schedule. Additionally, crewmembers are subject to maximum flight time and rest requirements arising from regulations and contractual provisions, although these constraints are often not considered in building the flight schedule. Instead, the flight operations department is left to manage these limitations as crews are assigned to the previously developed flight schedule.

5.4.2 Fleet Assignment

The initial flight schedule may be developed without fully assigning specific aircraft types to each flight. British Airways, for example, operates six aircraft types (including those of its regional subsidiary) ranging in capacity from 469 seats on the Airbus 380 to 98 seats on the Embraer 190 (British Airways, n.d.). Even within a single fleet type, seating capacity may vary. Wizz Air operates only Airbus A320 series aircraft but seating capacity varies from 180 to 239 for different models (Wizz Air, n.d.). The first consideration in fleet assignment is the aircraft capability for the route. Long-haul international routes have been the domain of wide-body jets whereas regional jets are restricted to shorter routes. Within these performance limitations, the airline still has the flexibility to assign aircraft with varying seat capacity and interior configuration. The objective is to match capacity with demand. Higher capacity aircraft should be assigned to the flights with the highest demand. The Airbus A321XLR slated to begin service in early 2024 will provide airlines with a narrow-body aircraft option to open new long-haul routes and replace wide-body jets on others.

Changes to the fleet assignment can be made well after the timetable is finalized to meet evolving demand. A few airlines are effectively utilizing *dynamic scheduling* which changes fleet assignments within a few days of operation. Dynamic scheduling is addressed in more detail later in this chapter.

5.4.3 Trade-offs

Flight schedule development involves innumerable trade-offs because of the conflicting objectives of maximizing revenue, minimizing costs, and enhancing reliability. High aircraft utilization maximizes revenue and minimizes costs but also reduces slack which jeopardizes reliability. Spare aircraft generate no revenue while incurring high fixed costs but assigning all available aircraft to flights substantially compromises the airline's ability to recover from disruptions. The operation of red-eye flights greatly increases the utilization of aircraft, but ticket prices must be lowered to attract passengers to undesirable departure and arrival times. Business travelers value high flight frequency, but frequent departures with larger aircraft can result in too much capacity which will lead to lower average prices. Regional jets can be used to increase flight frequency without adding excess capacity, but the regional jet CASM may be twice that of larger mainline aircraft. The list of potential trade-offs is long and the final flight schedule is the result of many compromises.

5.4.4 Optimization

With each major schedule revision, a draft is circulated to operating departments for suggestions and approval. Departments may identify constraints that hadn't been considered. A station manager might anticipate gate conflicts due to local airport construction. The flight operations department may point out that the assigned aircraft type will have payload restrictions on a particular route in high summer temperatures. The introduction of a new aircraft type brings with it added complexity, for example, pilot training may have a longer lead-time than schedule planners had considered.

The schedule development is an iterative process as revisions are made in response to inputs from the operating departments. Ultimately, the schedule is a compromise that imposes burdens on some operating departments or forces the schedule planners to accept a schedule that doesn't meet their objectives for efficiency and profitability. Aggressive scheduling often leads to messy recoveries from irregular operations, reducing revenues, and damaging passenger relations and loyalty. During the chaotic summer of 2022, many airlines painfully learned that their staffing was inadequate to reliably operate the summer flight schedule (see the Case Study in this chapter).

With so many conflicting goals and required compromises, producing a profit-maximizing flight schedule is a daunting, seemingly impossible task. The cost of operations can be estimated with reasonable certainty, with volatile fuel prices the greatest unknown, but revenues are subject to many more variables and uncertainties. Software applications are available to identify optimal departure times, maximize passenger flows through the network, and model costs. Conceptually, a single software application should be able to produce an optimal schedule of services; it is just a large optimization problem to maximize profits subject to constraints. However, developing a single solution is beyond current technical capabilities because of the huge number of variables and complexity of the profitability function. Instead, separate applications focus on optimizing a single aspect of the schedule. A large network carrier would employ many applications. This functional listing of software applications provides an appreciation for the scope and complexity of the flight schedule development process and serves as a review of the schedule development process.

- Market Size – Estimates the total market demand in terms of passengers traveling in each city pair. An international network airline might evaluate 30,000 origination and destination (O&D) markets. Historical data, trends, seasonality, aggregate pricing, and other macroeconomic data are combined to create individual city-pair demand forecasts.

- Market Share – Estimates of an airline's market share begin with the Market Size forecast. Then considering the relative quality of service versus competitors, the software estimates the share of the market the airline can expect to capture.
- Fleet Assignment – Specific aircraft fleet types are assigned to the schedule of services so that capacity meets estimated demand subject to constraints. Choices might involve a wide-body versus narrow-body aircraft on a particular route segment.
- Passenger Spill – Estimates the number of passengers who will not find an available seat given the proposed schedule. This is an estimate only because there is no way of knowing exactly how many potential passengers looked at the airline's schedule and didn't find what they were looking for.
- Through Assignment – Even if a stop at a hub is required to reach their destination, passengers prefer not to change aircraft. These *direct flights* provide some marketing advantage over a competitor requiring a change of aircraft. Because many aircraft routings are possible through the hub, optimizing through flights will affect passenger choice and potential revenue.
- Dependability Prediction – In practice, the schedule will be disrupted by weather, mechanical failures, and a host of other problems. This application estimates the reliability of the proposed schedule in actual operation. (Sabre Holdings, n.d.)

As each application is run, the schedule is revised for a better solution resulting in an integrated, iterative process of schedule development, profitability forecasting, and fleet optimization.

5.4.5 The Passenger Service System

The flight schedule is stored electronically in the airline's *Passenger Service System* (PSS), its central information technology (IT) repository. The PSS is developed and customized by each airline; individual systems vary in sophistication and capability. Functions that typically reside within the PSS, in addition to the flight schedule, are passenger reservations (called PNRs for passenger name records), frequent flier awards, flight information system, check-in system, and many more (Arciuolo, 2014). Passengers are familiar with the PSS from flight check-in when the passenger service agent accesses the system with a series of rapid-fire keystrokes, or, more recently, from direct interaction through an airport check-in kiosk. Unseen are the countless other airline employees who continuously access the PSS to perform their duties. The reservations process is covered in Chapter 10 – Distribution.

The "back end" of the PSS contains interfaces between the PSS and other internal IT systems that use PSS information for daily airline management.

For example, some airlines extract data from the check-in or departure control system to compute the aircraft weight and balance for each takeoff. Another is the airline's revenue management system that allocates the seat inventory to price classes. Revenue management is the subject of Chapter 9.

5.5 Asset Assignment

Once the flight schedule is finalized, operating departments begin assigning personnel and other physical assets needed to operate the schedule. Staff may need to be hired and trained, or additional airport gates and support equipment obtained. On the other hand, if flight frequency is reduced in some markets or transferred to a regional airline, personnel may be furloughed or transferred to other locations. Required lead times vary greatly. Hiring and training new pilots may require six or more months of lead time, while adding airport staff may require only a few weeks. Airport facilities for a new destination may be readily available or difficult to obtain. Many smaller and mid-sized airports often have counter and gate space available and actively seek expanded airline service. On the other extreme, at busy major airports, needed space and gates may be difficult to procure.

Before the flight schedule is ready for implementation, two important asset assignment tasks remain: first, assignment of specific aircraft (tail numbers) to the flight schedule; and, second, assignment of pilots and flight attendants to every scheduled flight. These tasks are completed concurrently between four to six weeks before the actual flight operation.

5.5.1 Aircraft Assignment

Assignment of individual aircraft to the flight schedule is the less complex of the two tasks. Maintenance requirements are the principal driver of aircraft assignments. Aircraft must undergo relatively simple inspections every few days and more complex routine maintenance every few weeks. The time interval for heavy maintenance in which major components are inspected, serviced, or replaced varies by the airline with most completing the required tasks in phases. The assignment of aircraft to the flight schedule must ensure that aircraft are routed to maintenance facilities when inspections and work are due. Routine inspections are performed at many stations, usually overnight, whereas a limited number of airline facilities are capable of heavier maintenance. For network carriers, heavy maintenance facilities are often at hubs.

Figure 5.4 (left panel) shows a typical routing for one aircraft in one day. This aircraft originates in Atlanta and flies to the airline's Denver hub for a connecting complex. Of course, aircraft from other spoke cities arrive in DEN nearly simultaneously to make up the complex. Once passengers and baggage are transferred from other inbound flights, this aircraft continues to Seattle. Next, the aircraft returns to the Denver hub and then flies to Boston. Finally, the aircraft returns to the hub for a third connecting complex of the day before

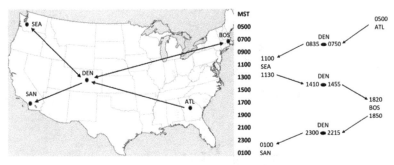

| Single Aircraft Flight Assignment | Time-Space Diagram for Aircraft Assignment |

Figure 5.4 Aircraft Assignment.

continuing to San Diego. Figure 5.4 (right panel) displays the aircraft routing in a station time-space diagram. Denver local time (MST) is displayed on the vertical scale for illustration (airlines would use Coordinated Universal Time). The dots at Denver represent complexes. Note that this aircraft transits Denver for three of the four daily complexes. A larger airline such as United Airlines, which also has Denver as one of its hubs, would operate several more complexes.

On completion of the day's flights, the aircraft remains overnight in San Diego where it will originate the next day's flying. For this small network carrier, the aircraft would regularly transit the Denver hub but would likely fly to different cities on each subsequent day's flights. At some point, the aircraft would terminate a day's flying in Denver for scheduled maintenance.

5.5.2 Aircraft Flow Chart

For operational use, the flight schedule is displayed as an aircraft flow chart, similar to a Gantt chart. Figure 5.5 is a typical aircraft flow display. This flow chart is based on one of the many airline software applications available from Lufthansa Systems and shows a portion of the flight schedule for Lufthansa Group member Swiss International Airlines. Each row displays the sequential flights for one aircraft over two-and-a-half days from Monday through part of Wednesday. A graphical rectangle, called a puck, represents a single flight. The puck can be configured to display data of interest, including departure and arrival times, actual bookings, or special comments.

Looking closely at Figure 5.5, time blocks are shown below the days of the week. These would normally be coordinated universal time, but local times could also be displayed. The origination and destination (O&D) cities are displayed on each side of the puck with the flight number and departure and arrival times within the puck. The first row shows the flight schedule for aircraft #1. It departs Boston (BOS) on Monday at 01:40 am as flight number 053 and

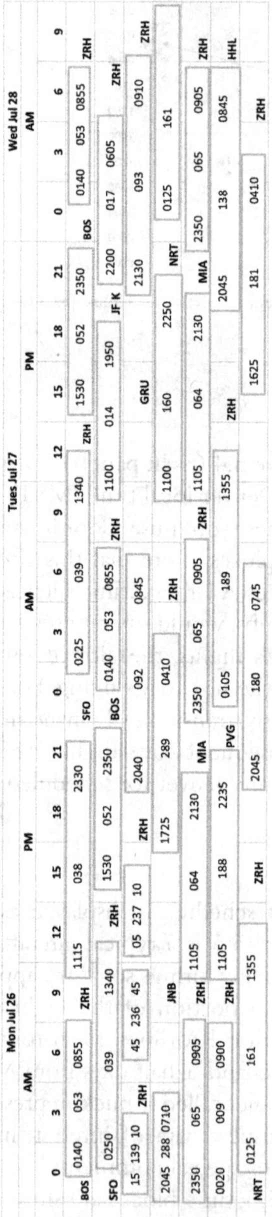

Figure 5.5 Aircraft Flow Chart.
Data source: Lufthansa Systems.

arrives in Zurich (ZRH) at 08:55 am The scheduled block time is seven hours 15 minutes. The ground or turn time in ZRH is two hours 20 minutes, typical for a long-haul international flight. The aircraft then continues from Zurich to San Francisco (SFO) as flight 038 and then returns to ZRH as flight 039 on Tuesday. Next, the aircraft operates another round trip to BOS with the return to ZRH (on the far right of the figure). Similarly, aircraft #2 is scheduled to operate flights from SFO to ZRH, then to BOS, and return to ZRH followed by a round trip to New York (JFK).

The flights for just seven aircraft are visible on this screen, but a large airline would have hundreds of aircraft of various types assigned to its flight schedule so that managers can scroll down to view other aircraft, scroll right to see the later days of the week, or otherwise customize the view of the flow chart as needed.

5.5.3 Crew Scheduling

Assigning pilots and flight attendants to staff every flight in the timetable is an immensely more complex task than the assignment of specific aircraft to the schedule. A large airline might have 800 aircraft but more than 8,000 pilots and more than twice that many flight attendants. Second, aircraft assignment is subject to few restrictions whereas the scheduling of pilots and flight attendants is subject to legal and contractual requirements that are usually different for pilots and flight attendants. Depending on the fluctuating cost of fuel, crew costs may represent the single largest cost for airlines, so efficient crew scheduling can significantly reduce operating costs. With many variables to consider, assigning crews to flights becomes a very large optimization problem that often needs significant attention from management.

Crew scheduling can be broken down into two primary steps – pairing development and pairing assignment.

5.5.3.1 Paring Development

The first step in crew scheduling involves building *pairings*. Pairings are work schedules consisting of a sequence of flights that begin and end at the crewmembers' assigned base (also known as *domicile*). The length of the flight sequences ranges from one to several days. Multiple-day pairings will include intervening rest periods, usually overnight away from the domicile. Airlines flying longer legs are more apt to require multiple day parings to facilitate the long legs and required rest periods. Airlines flying shorter legs can often avoid the extra costs involved in overnight stays by developing pairings that start and end in the crewmember's home base. There are innumerable variations in the design of pairings with differing degrees of effectiveness, efficiency, and cost. As with the flight schedule itself, compromises are required to accommodate conflicting objectives of minimizing cost while ensuring reliability. For example, daily pilot utilization might be increased by having pilots frequently change aircraft at a network carrier's hub airports, but if the crew's inbound flight is delayed, so too will be the next flight the pilots are scheduled to operate even if the aircraft for the outbound flight arrived at the hub on time. Thus, a pairing that requires pilots to change aircraft often can cause operational problems that snowball throughout the day. Pairings will differ for each crew domicile. For a large airline, thousands of different pairings will be required to cover the entire flight schedule.

In a simple example, Figure 5.6 shows how pairings are developed from the aircraft flow chart. The daily aircraft flow chart used as an illustration shows just four aircraft of the same type connecting six cities over a single day. For simplicity, we assume the airline operates the same schedule every day of the week in contrast to Figure 5.5 where the aircraft flow must extend over several days because of the long international flights. In practice, however, even most domestic airlines operate a different schedule midweek and on weekends, which introduces another level of complexity. Five pairings (P101, P201, P202, P203, and P301) are shown for Panama City, Panama (PTY) based pilots with pairing shaded differently to correspond to individual flights in the flow chart. The simplest pairing is P101, a one-day work schedule consisting of two flights on aircraft #1, PTY to LIM, and return. Pairings 201, 202, and 203 are each two days, showing the crews remaining overnight (RON) at JFK, MIA, and JFK, respectively. Pairing 301 is a three-day trip, including a two-night stay in Mexico City (MEX) that is required because the crew arrives in MEX too late at the end of day 1 to operate the originating flight from MEX the next morning. Note that all five pairings begin and end in the crew base (PTY).

Pairings determine crew utilization and efficiency. A pairing that has crews spending too much time on the ground between flights or with too many aircraft changes can result in longer, less productive crew days. Advanced computer algorithms help airlines develop the most efficient pairings to maximize productivity.

5.5.3.2 Pairing Assignment

Once pairings are developed, there are generally three ways to assign the parings to crews – line bidding, preferential bidding, and fair assignment (Fennell, 2017). The method used depends on the carrier's overall operations philosophy and its labor contract provisions.

5.5.3.3 Bid Line System

The bid line scheduling system was the original process used by airlines for decades. Using a *bid line* process, schedulers assemble pairings into nameless bid lines that typically cover a calendar month. Pairings are usually separated by three or four days off to ensure crewmembers meet required rest and duty limitations. Each crew base will have roughly the number of bid lines as there are crewmembers in that position. For example, if a base has 400 captains,

Aircraft Flow Chart

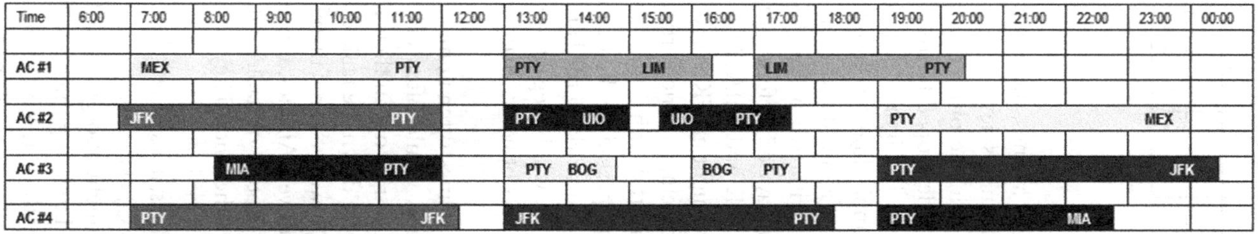

Time	6:00	7:00	8:00	9:00	10:00	11:00	12:00	13:00	14:00	15:00	16:00	17:00	18:00	19:00	20:00	21:00	22:00	23:00	00:00
AC #1		MEX				PTY		PTY		LIM		LIM			PTY				
AC #2		JFK				PTY		PTY	UIO		UIO	PTY		PTY				MEX	
AC #3			MIA			PTY		PTY	BOG		BOG	PTY		PTY			MIA		JFK
AC #4		PTY				JFK		JFK				PTY		PTY			MIA		

Pairings

P101
PTY	LIM	13:00	16:20	3+20
LIM	PTY	17:00	20:20	3+20
				6+40

P203
PTY	JFK	19:00	00:25	5+25
	RON at JFK			
JFK	PTY	13:00	18:25	5+25
				10+50

P201
PTY	JFK	7:00	12:25	5+25
	RON at JFK			
JFK	PTY	6:35	12:00	5+25
				10+50

P301
PTY	BOG	13:00	14:40	1+40
BOG	PTY	16:00	17:40	1+40
PTY	MEX	19:00	00:00	5+00
	RON at MEX			
	RON at MEX			
MEX	PTY	7:00	12:00	5+00
				13+20

P202
PTY	UIO	13:00	15:00	2+00
UIO	PTY	15:30	17:30	2+00
PTY	MIA	19:00	22:40	3+40
	RON at MIA			
MIA	PTY	8:20	12:00	3+40
				11+20

BOG	Bogota, Columbia
LIM	Lima, Peru
MEX	Mexico City, Mexico
MIA	Miami, USA
PTY	Panama City, Panama
JFK	New York Kennedy, USA
UIO	Quito, Ecuador

Figure 5.6 Aircraft Flow Chart and Crew Pairings.

	1	2	3	4	5	6	7	8	9	10	11	12	13	14	15	16	17	18
Line #1		101	301	MEX	MEX					201	JFK	202	MIA				203	JFK
Line #2				203	JFK	301	MEX	MEX							101	301	MEX	MEX
Line #3			101	202	MIA				301	MEX	MEX	101					201	JFK
Line #4	201	JFK	202	MIA					101	203	JFK				201	JFK	202	MIA

Figure 5.7 Bid Lines.

they will develop about 400 captain bid lines. At larger airlines, there will be hundreds of bid lines per position per base. Crewmembers review the bid lines and submit a rank-ordered list of the lines (a bid) in the order that suits them. Figure 5.7 shows four bid lines (Lines #1 – #4) developed for the first through the 18 of the month using the pairings shown in Figure 5.6 (actual bid lines would include the entire month). As an example, line #1 shows the crewmember off on the first day of the month, working a one-day trip on the second followed by a three-day trip starting the third with two overnights in MEX. Following that trip, the crewmember has four days off before starting two back-to-back two-day trips with overnights in JFK and MIA. Another three-days off is followed by a two-day trip with an overnight in JFK. It's worth pointing out that the lines shown in Figure 5.7 are different, with no particular pattern. One line (#2) has nine workdays through the 18th, while the others have more. Also, note that line #4 starts a four-day work period on the first day of the month. If the crewmember that gets that line is also working the last few days of the prior month, the chances are good that there will be legality issues that will have to be resolved by removing some of the days of work. That is one of the negatives of the bid line system.

Before the widespread use of personal computers, bid lines, often numbering in the hundreds, were printed on paper. Crewmembers would often cut the pages into strips, a thin strip for each line, then lay the strips out in their rank order. Once they were satisfied with the order, they wrote down the ranked line numbers and submitted the list to crew scheduling. Nowadays, crewmembers at airlines that still use the bid line system, like Southwest Airlines, can usually download the bid lines electronically to their laptop or tablet where sophisticated sorting applications can help them sort the lines. They submit their sorted bids electronically.

After the crew scheduling department receives the bids, it assigns the lines in seniority order. For example, the senior first officer (F/O) at a crew base would get his or her first choice, next, the #2 first officer might also get their first choice (assuming it was different than the #1 F/O). Line assignments would continue through the end of the F/O list. Theoretically, the last, most junior, F/O doesn't even need to bid because the only line remaining unassigned is the one nobody else wanted.

The most positive aspect of this system is that crewmembers can see specific flights, overnights, and workdays and bid accordingly. They can also manually trace the schedule award if there is any concern that the award was not correctly completed.

On the negative side, crewmembers do not know their future schedules beyond the current bid cycle. The next month could bring something totally different. Also, the bid line system, while great for high seniority crewmembers who can get desirable schedules every month, is not so good for junior crewmembers who rarely get a desirable schedule. Another negative is that the bid lines are built without regard for variances in individuals' availability. Crewmembers with planned vacation or planned training will be assigned a line, then pairings in that line that overlap the planned vacation or training periods will have to be removed. These pairings are then assigned to another crewmember, possibly by regenerating new lines and accomplishing another smaller bid cycle. Similarly, if a crewmember's awarded line includes flights on the first days of the month, and the previous month's schedule had flights on the last days of the month, one of the pairings will have to be modified or removed.

5.5.3.4 Preferential Bidding Systems

Advances in computer processors and software over the decades now allow for more sophisticated scheduling (AIMS International, n.d.). Instead of developing anonymous monthly lines, many airlines now use a preferential bidding system (PBS). Using PBS, crewmembers submit their preferences for a variety of options, including workdays, overnight cities, start and end times, and so forth. Crewmembers add weights to the parameters to signify their relative importance. Software programs then allocate the pairings into a personalized monthly schedule on a seniority basis to best-fit crewmembers' desires. For example, if the senior pilot in the base wants to work weekdays, with showtimes after 7:00 am, release times before 4:00 pm, and overnights in certain cities, a schedule can be built with pairings to accomplish exactly that. Further down the seniority list, pilots' schedules will comply with fewer preferences. The PBS generates lines that account for the crewmember's previous month's schedule, vacations, and training eliminating the need to rework bid lines after award. PBS has become the favored process among many airlines as it is more efficient than bid lines while still considering seniority.

Preference bidding systems provide substantial efficiency improvements but have yet to be widely embraced by crewmembers. Some do not like PBS because they cannot see the exact schedule they might get (as compared to bid lines). Additionally, it is impossible to trace the schedule award – they must trust the computer got it right.

5.5.3.5 Fair Assignment

A crew assignment process called *fair assignment*, or sometimes *rostering*, is a process being used by a few airlines (mostly LCCs). It is a no-bid, equitable distribution of workload. This process attempts to equalize block hours, days off, and various other aspects of the workload to maximize both crewmember utilization and crewmember satisfaction. Pairings are distributed according to established protocols. Fair assignment includes a fixed workday schedule (discussed below). Although probably the best process for crew utilization and efficiency, fair assignment mostly ignores crewmember seniority, making it a hard sell to unionized, seniority-driven work groups.

5.5.3.6 Fixed Scheduling v. Variable Scheduling

The assignment of pairings into monthly work schedules whether based on bid lines, PBS, or fair assignment, can be based on either a fixed or variable schedule.

Fixed scheduling continuously follows a repeating on/off schedule, allowing crewmembers to know their workdays and days off well in advance. Often, the knowledge of days off in advance is one of the more important aspects of the schedule for the crewmember, with weekends off usually a high priority. With fixed schedules, weekends off will cycle through everyone's schedule, giving even junior crewmembers an occasional weekend off. Following the sequence, crewmembers know far in advance the days they are working and their days off. Note in Figure 5.8 that each pilot follows a five-on/four-off fixed schedule. A few days or weeks before the work period starts, the pilot is notified of the specific flight legs for the work period. Fixed scheduling is often used in conjunction with fair assignment.

Variable scheduling is a more commonly used process. Using variable scheduling, the days crewmembers work vary month to month, with no particular sequence. With variable scheduling, a crewmember gets their new month's schedule a few weeks before the end of the current month. Variable schedules are often coupled with the bid line system or PBS.

European LCC Ryanair uses a fair assignment process to produce a fixed schedule. Pilots work five days on followed by four days off, and cabin crews (flight attendants) work five days on followed by three days off (Ryanair Group, n.d.). Ryanair also has over 80 crew bases across Europe and North Africa (Ryanair Group, n.d.), six to 10 times as many as the largest carriers in the United States. With Ryanair's generally short-haul flights, having this many crew bases allow it to schedule all pairings as single-day trips (also called "turns") – in other words, every day the crew members start and end at their home base (Airline Prep, n.d.). This simplified process reduces costs for the airline in several ways. First, fair assignment scheduling is simple and cost-effective, saving time for scheduling personnel. Additionally, the one-day trips save the airline hotel and per diem costs, which can be a major saving. As an example of the scope of the

	1	2	3	4	5	6	7	8	9	10	11	12	13	14	15	16	17	18	19	20	21	22
Pilot A	X	X	X	X	X					X	X	X	X	X					X	X	X	X
Pilot B		X	X	X	X	X					X	X	X	X	X					X	X	X
Pilot C	X	X					X	X	X	X	X					X	X	X	X			
Pilot D					X	X	X	X	X					X	X	X	X	X				X

Figure 5.8 Fixed Scheduling.

potential savings on rooms, Southwest Airlines, which uses variable scheduling and 11 crew domiciles, typically schedules crews for three-day to four-day trips, usually with two or three overnights in hotels in non-domicile cities. Southwest pays for 4,000 to 5,000 hotel rooms a night.

With fixed scheduling, Ryanair crewmembers also enjoy long-range schedule planning. They know exactly which days they will be working far into the future. With a nine-day cycle (five-on/four-off), Ryanair pilots cycle through weekend and holiday duties throughout the year. Pilots' schedules are published monthly, and changes are given with a seven-day notice period. Although pilots do not have a choice of flights, they can trade with other crewmembers once schedules are published (Airline Prep, n.d.).

5.6 Tactical Management

When the assignment of aircraft and crewmembers to the flight schedule is complete, responsibility for the execution of the schedule transfers to tactical or operations managers (refer to Figure 5.1). The carefully crafted schedule will be subject to real-world stresses, including weather, aircraft mechanical failure, passenger and crew issues, as well as a variety of other tribulations. Several departments – principally flight operations, maintenance and engineering, and airport services – work together to operate the airline's flights. Working in parallel, but mostly unseen, are revenue management personnel who continually adjust prices and control the allocation of seat inventory in response to competitors' actions and ever-changing demand. Others work to distribute the airline product through the airline's website, reservations offices, and third-party traditional and online travel agents. These functions are addressed in later chapters.

5.6.1 Airline Operations Control Center

The pilots, mechanics, and station personnel who operate the airline's flights are scattered across the airline's route system. Historically, each of these employees reported to different managers who were frequently in different geographical locations resulting in "silos" with conflicting priorities and poor communication among the functional groups. Over the past few decades, airlines have moved daily operations managers and staff to a single operations center to improve coordination, communication, and tactical decision-making. The development of the *Airline Operations Control Center* (AOCC) – the name varies across airlines – is one of the major airline management evolutions ranking alongside network development and revenue management. Figure 5.9 is a diagram of the functions located within a typical AOCC along with a photo of American Airlines' AOCC. Large AOCCs have a staff of several hundred people, typically operating in three shifts and always open. United Airlines also sports a new AOCC that United calls a NOC (Network Operations Center). It has 341 desks with about 300 people on each shift (Sloan, 2013). One of the

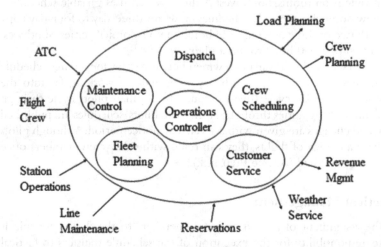

Figure 5.9 Airline Operations Control Center.
Source: Photo from American Airlines.

newer positions found within the customer service node in many AOCCs is a social media expert.

At the center of the AOCC are one or more operations controllers supported by personnel from several functional areas.

- *Dispatchers* are licensed professionals responsible for flight planning, issuing flight plans to captains, and following each flight's progress. The dispatcher and captain are legally jointly responsible for the safe operation of the flight. Often working in conjunction with the dispatcher are load planners who track the passenger and baggage loading on each flight and issue instructions for the distribution of baggage and cargo to ramp agents and total weight and center of gravity computations to the captain.

- *Crew Schedulers* track individual crewmembers as they move through the airline's route network, maintaining up-to-date status, and calling in reserve crewmembers or readjusting crewmember schedules as necessary when schedule disruptions occur.
- *Maintenance Controllers* coordinate with line mechanics for aircraft maintenance, especially when malfunctions occur, ensuring that required parts are available to meet aircraft and mechanics have access to the appropriate aircraft maintenance program procedures to troubleshoot and correct malfunctions.
- The specific aircraft on the flight schedule must occasionally be reassigned because of disruptions to the schedule or unforeseen maintenance requirements. *Fleet planners* track individual aircraft to ensure that the aircraft schedule allows for required maintenance.
- *Customer Service* staff ensure that changes to the flight schedule are communicated to station personnel and passengers and arrangements are made to accommodate disrupted passenger itineraries. American Airlines Robert W. Baker Integrated Operations Center includes a social media desk within the customer service function to provide passengers with immediate responses to flight problems via the various social media networks. Social media specialists have access to the most current operations information and are often the first to know of operational issues affecting passengers (Hegeman, 2016).
- The AOCC also houses or is directly adjacent to a *crisis center* that is activated for an incident, accident, or another major event.

As an aside, passengers are occasionally frustrated during flight delays when the airport passenger service agents do not provide timely information about revised plans. Operations managers in the AOCC are developing required schedule revisions, but this process takes some time. In the case of an aircraft mechanical problem, mechanics are not able to estimate the repair time until the troubleshooting procedures are complete, and these procedures take time. The AOCC manager cannot make a decision without an estimated repair time. In the interim, passengers wait impatiently for information.

The challenges faced by AOCC range from routine to crisis. Executing the flight schedule during normal operations requires extensive communication to ensure that a myriad of flight support tasks are closely coordinated. Aircraft must be maintained and certified as airworthy and crews must be in the proper location for their next scheduled flight. Aircraft must be fueled, cleaned, catered, and baggage loaded. Station personnel ensure that passengers are checked-in and boarded in time for the scheduled departure. These and other routine activities are orchestrated and completed simultaneously within a tight time window. Each day, minor modifications to the plan are required because of many small disruptions such as ill passengers, mechanical problems, and security processing. At other times, however, flight operations are subject to major disruptions, often weather-related, requiring large-scale modifications to the flight schedule to return the airline to published service.

5.6.2 Flight Schedule Disruptions

Many factors, adverse weather being the most common, disrupt the flight schedule. For the 10 years from 2012 through 2021, about 20% of US domestic flights arrived more than 15 minutes behind schedule indicating some type of schedule disruption (Bureau of Transportation Statistics, 2021). Without sufficient slack in the flight schedule, one delay can cascade through the system. Hub-and-spoke systems are especially vulnerable to weather at a hub airport that can disrupt entire complexes that then ripple across the airline's network.

Aircraft mechanical problems and air traffic control restrictions, the latter often weather-related, are well-known to most passengers. Less obvious causes for delays include pilot and flight attendant absence, sickness, or legality restrictions. At airports, ground support equipment may be unavailable due to mechanical problems. Airport construction can cause delays, especially when a runway is closed for routine repair or major upgrade. In recent years, delays for security have been more prevalent.

On 8 August 2016, Delta Air Lines suffered a major power outage in its Atlanta Operations Control Center (OCC) caused by a fire in a power control module at the center. Although the fire was quickly extinguished, standard power was out for an extended time. Delta's computers failed to switch to backup power sources, and as a result, were down for about five hours. The 1,000 flights canceled that day rippled through to the next two days leading to another 1,000 cancelations. Customer service agents resorted to handwritten boarding passes for Delta flights. Normal operations were restored after about a week, costing Delta $150M (Zhang, 2017; Owusu, 2016).

In response to this incident, Delta took steps to form an irregular operations (IROPS) team tasked to coordinate and respond to disruptive events around the system. They also changed the placement of crews and aircraft in their system to better handle disruptions. These actions seemed to have helped – another power failure in Atlanta about 16 months later was longer (11 hours) but better managed. Although about 1,400 flights were canceled, the system was mostly back to normal in two days (Zhang, 2017).

Although difficult to calculate, the Federal Aviation Administration (2020) estimated the cost of delays for 2016 through 2019. Table 5.1 lists their results. These estimates included losses to airlines and passengers in addition to the lost demand and indirect losses.

Table 5.1 Cost of Delay Estimates (in the US)

	2016	*2017*	*2018*	*2019*
Airlines	$5.6B	$6.4B	$7.7B	$8.3B
Passengers	$13.3B	$14.8B	$16.4B	$18.1B
Lost Demand	$1.8B	$2.0B	$2.2B	$2.4B
Indirect Losses	$3.0B	$3.4B	$3.9B	$4.2B
Total	$23.7B	$26.6B	$30.2B	$33.0B

Source: From the Federal Aviation Administration.

5.6.3 Managing Irregular Operations

When major disruptions to the flight schedule occur, a recovery plan must be developed and implemented to restore the published flight schedule. Methods vary depending on circumstances and individual airline culture and philosophy. American Airlines, for example, will sacrifice today's schedule to operate normally tomorrow. Differing and conflicting sub-goals, such as retaining revenue or minimizing passenger inconvenience, are difficult to balance and currently beyond the capabilities of software to solve or optimize. Like the schedule development process itself, compromises are necessary.

Operations managers have many recovery options to employ singly or in combination depending on the severity of the disruption.

- *Reserve Crews.* Standby or reserve crews, a form of slack resource, may be assigned to replace crew members who are sick, become illegal for continued flight duty because of regulatory or contractual limitations, miss connections to their next flight, or to staff rerouted or additional flights. Some airlines maintain up to 30% of all crewmembers on standby or reserve status to preserve the flight schedule reliability.
- *Aircraft Swap.* Aircraft may be swapped to a different flight or route. This usually happens at a hub. A passenger announcement of a gate change is often evidence of an aircraft swap.
- *Spare Aircraft.* Just as the airline maintains reserve crews, it will also have a limited number of unassigned spare aircraft to substitute for late or broken aircraft or to insert into the schedule for additional flights. This form of slack resource is, of course, expensive.
- *Delay.* Flights may be delayed but operate on the originally scheduled routing.
- *Cancel.* Flights can be canceled. Massive pro-active cancelations in anticipation of severe weather have become more common in recent years. In the US, this is in part a reaction to Department of Transportation Tarmac Delay regulations imposing severe fines on airlines for flights that incur extensive delays while waiting for takeoff.

- *Rerouting.* Finally, aircraft can be rerouted with additional stops en route or two or more flights can be combined, and perhaps operated with larger capacity aircraft.

The possible alternatives are many and complex. A network airline with high flight frequency through its hubs has more options to address irregular operations than does a low-frequency, predominately point-to-point carrier. The judgment and experience of the operations controller are critical to a successful recovery. The operations controller must take into account several, often conflicting, objectives. For a network airline, 50%–60% of all passengers connect at a hub to reach their destination. The controller must consider whether connecting flights will be available so that passengers are not stranded at the hub. Pilots and flight attendants must be in place with at least legally required rest to operate the following day. Consequently, late flight arrivals can result in delays the following day that continue to propagate through the network. Similarly, aircraft must also be in place for the following day's departures. Because pilots are usually only qualified to fly one type of aircraft, staffing for the following day must be considered in addition to the current day's operation. Maintenance requirements also restrict aircraft assignment options. Specific aircraft must be in place for required scheduled maintenance.

Several options are also available to get disrupted passengers to their destinations. Re-accommodation on a later flight, or, for airlines with multiple hubs, routing through another hub may be possible. Passengers can sometimes be accommodated on another airline (known as off-line accommodation). If passengers cannot be flown to their destination in a reasonable time, they may remain overnight awaiting an available flight the next day, or in extreme situations, some days later. Airlines maintain a database of their most valued passengers and will accommodate these passengers first. Much of the information in the database comes from the airline's frequent flyer program.

Most of these options are costly to the airline. The European Union requires airlines to pay for passenger expenses even in circumstances fully outside of the airline's control. In the US, however, passengers are often surprised that their airline declines to take responsibility for additional expenses incurred to complete their travel. Each airline outlines its policies and responsibilities in the Contract of Carriage, a legal document that, unfortunately, few passengers read.

Making optimal decisions for passenger accommodation and revenue projection involves a vast decision space; any schedule change generally affects all functional areas. Controllers charged with developing and implementing the recovery plan must process an immense amount of information. This requires experience and aptitude. Increasingly capable computer-based decision support systems (DSS) assist with this complex decision-making.

Though there are many examples of irregular operations and the difficulty in returning to the published flight schedule, JetBlue's operational meltdown that began with the Valentine's Day storm of 2007 is perhaps the most horrific. At the time, JetBlue's operational strategy was to operate all flights, no matter how late, reasoning that passengers would rather arrive late than have their flight canceled. With this strategy, JetBlue, unlike most other carriers, did not cancel flights in anticipation of winter snowstorms. During the previous winter, this philosophy served it well when a projected major storm was less intense than forecast and JetBlue operated flights when other carriers did not.

During the Valentine's Day snow and ice storm of 2007 in the Northeast US, JetBlue again planned to operate all scheduled flights. The problem started when aircraft could not depart New York's Kennedy Airport in freezing precipitation. Arrivals, however, could still land, and soon, gates filled with arriving aircraft. Aircraft waiting for departures began stacking up, eventually, JetBlue started to cancel flights, but all gates were occupied. As a result, aircraft ended up sitting with passengers for up to 10 hours on the ramp. They experienced "toilets overflowing, nothing to eat but snacks and, ironically, cabins so hot and stuffy that doors had to be opened to let fresh air in" (Carey & Aalund, 2007). At JFK, airport buses eventually moved stranded passengers from their aircraft back to the terminal.

Severe operational problems continued for a full week with JetBlue canceling about 25% of its flights each day as it attempted to restore scheduled operations. Flight crews exceeded regulatory duty times and could not fly until completing required rest intervals. Both planes and crewmembers were out of position to operate needed flights. The airline's 20 crew schedulers were overwhelmed, unable to put together a feasible recovery plan, and often unable to contact crewmembers needed for rescheduling. JetBlue had neither sufficient specialized personnel nor adequate crew scheduling and flight control software to plan and execute a successful recovery plan.

The problems and extensive negative publicity caused JetBlue to publish a Passengers' Bill of Rights. Further, most of the existing operational management and even the airline's founder, David Neeleman, were subsequently replaced.

This example notwithstanding, several factors favor lengthy delays awaiting takeoff rather than a flight cancelation. Both pilots and passengers frequently prefer to wait for departure when flights are delayed after gate departure. Returning to the gate usually means sacrificing a place in line for takeoff. Finally, with high load factors, few seats are available for rebooking, a fact many passengers realize.

Major disruptions aren't confined to newer and smaller airlines. Chaos sometimes strikes at even the largest and most experienced airlines as British Airways learned with its ill-fated move to its new Terminal 5 at London Heathrow. Within hours of opening on March 27, 2008, the $8.5 billion complex where British Airways is the sole occupant quickly became a nightmare. The terminal's high-tech baggage system failed, resulting in thousands of bags being misplaced, more than 400 flights being canceled or delayed, and hundreds of irate passengers left stranded in its first 10 days of operation. The opening was a major public relations disaster. British Airways had touted the modern terminal as a major step in ending the so-called "Heathrow hassle" of long lines, overcrowded conditions, and lost bags that has long plagued Europe's busiest airport.

5.6.4 Dynamic Scheduling

Individual aircraft are assigned to the flight schedule a few weeks before operation, but frequent changes to the specific aircraft assignments are common to accommodate various disruptions to the flight schedule. With the introduction of several aircraft models of the same type, especially the Airbus 320 and the Boeing 737 "families" of aircraft, a new opportunity to increase revenues and profits arose by swapping aircraft models of different capacities to accommodate changing demand. Alaska Airlines, for example, operates a Boeing 737 family (-700, -800, -900, and MAX 9) as well as an Airbus 320 family (320 and 321neo) series aircraft (Alaska Airlines, n.d.). Seating capacities range from 124 to 178 with the 737 family, and 150 to 190 with the Airbus family (Alaska Airlines, n.d.). As the date of departure approaches, the airline can estimate passenger demand for each route with ever greater precision. This presents an opportunity to swap aircraft to better match capacity with the projected demand and thereby increase revenue. A flight initially scheduled for the smallest Boeing 737–700 model but experiencing higher than projected passenger bookings might be rescheduled with the larger -800, -900, or MAX 9 models. The same can work with a flight scheduled for the Airbus. Similarly, if a 321 was scheduled, a 320 might be swapped into that flight if the load is less than forecast, allowing the larger 321 to be used elsewhere.

While this process of *dynamic scheduling* may increase revenues, it also presents several problems. First, the aircraft maintenance requirements must be respected. Each aircraft swapped to a new flight may not return to its original schedule until at least two and probably several more flights have been completed. Demand on these subsequent flights may not be well matched to the aircraft's capacity; thus, the decision to swap aircraft must consider the demand on a sequence of several flights that will likely dilute the potential revenue gain. In Alaska's case, aircraft models also differ in first-class seating capacity. If a smaller model than originally scheduled is swapped to a flight, some reserved first-class passengers may be denied first-class accommodations. Because these passengers are among the airline's most valued, the negative consequences probably outweigh any potential for increased revenue. Finally, pilots are usually qualified to fly all models of the aircraft type, so swapping within the family generally does not affect the assignment of pilots; however, the larger models may require one or more additional flight attendants introducing another complication. Although the potential revenue enhancement from dynamic scheduling has been recognized for some years (Clark, 2000), the operational complications have limited the practice.

5.7 Continuous Improvement

The JetBlue Valentine's Day meltdown (highlighted earlier) illustrates that airlines, like all other businesses, must strive for continual improvement or competitors will eventually take over their market share. Customers want a cheap price, but they also expect and demand quality. One classic example of the use of quality as a competitive weapon is the Japanese assault on the US domestic auto industry in the second half of the 20th century. Higher quality Japanese automobiles gradually won market share. Despite impressive gains in recent years, the perception of poor manufacturing quality continues to haunt US producers today.

Quality must be defined by the customer, not the producer. Thus, research is necessary to determine what customers value and will pay for. This is part of the marketing concept discussed in Chapter 4.

Quality function deployment is a concept of designing quality into a product rather than something obtained after production by inspection and correction of defects. The basis for this concept is the conviction that high-quality products are ultimately cheaper to produce and sell. Consider the cost to JetBlue of its operational nightmare that other airlines were able to avoid.

Another critical component of both the *total quality management* and quality function deployment movements is employee involvement. Employees often know where to seek improvement and can offer solutions.

5.7.1 Goals

The heart of continuous improvement is an intuitively appealing, four-step process diagrammed in Figure 5.10. In the first step, specific performance goals are set that can be objectively measured. While this may be an obvious management task, many companies do not set clear objectives and/or develop measures to determine if objectives are achieved. This can lead to reactive management – putting out fires. In one study, less than half of the companies knew the profitability of each product. One-third had no regular review. Half failed to compare their prices with competitors, analyze warehousing and distribution costs, analyze causes for returned merchandise, conduct formal evaluations of advertising effectiveness, or review salesforce call reports (Baker, 2001).

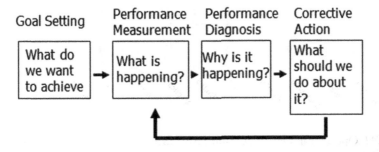

Figure 5.10 Control Process.

Establishing effective goals requires careful consideration captured in the acronym SMART – goals should be specific, measurable, attainable, relevant, and time-bound. An airline, for example, might establish a goal that 90% of flights will arrive within 15 minutes of the flight schedule, considered on time within the industry. This goal is specific, measurable, and relevant. Whether it is attainable, or even desirable, depends on the airline's unique circumstances. Harkening back again to the marketing concept, objectives should be developed from the passengers' perspective and attainment should add passenger value. An on-time objective that is appropriate for one airline may not be so for another. Passengers' wants, competition, and costs must be balanced in setting objectives.

5.7.2 Measurement

The second step in continuous improvement measures actual performance. Airlines track hundreds of performance statistics; Figures 5.11 and 5.12 show two examples. The US Department of Transportation in the Air Travel Consumer Report publishes these and several other metrics monthly to allow passengers to evaluate airlines' operational performance. Figure 5.11 graphs on-time performance (OTP) for the largest US airlines for 2021. Hawaiian Airlines has led the pack for on-time performance for years; however, it's worth noting that Hawaiian enjoys typically better flying weather and less congested airspace than any other listed airline, hence their better on-time performance. The full-service network carriers performed better than the carriers usually categorized as low-cost (the authors categorized JetBlue and Southwest as hybrid carriers). The low-cost carriers' high aircraft utilization (discussed earlier in this chapter), can lead to on-time performance degradation.

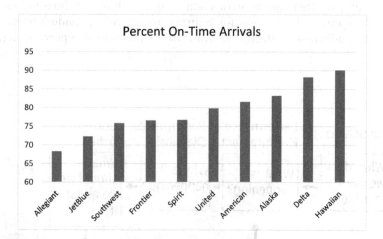

Figure 5.11 Operational Metrics – On-time Arrivals.

Note: Percentage of flights arriving on time for 2021. From US Department of Transportation, Air Travel Consumer Reports.

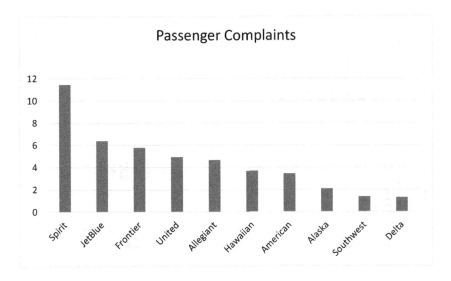

Figure 5.12 Operational Metrics – Passenger Complaints.

Note: Passenger complaints per 100,000 enplanements for 2021. From US Department of Transportation, Air Travel Consumer Reports.

Timeliness is an important factor in consumers' choice of an airline. Poor on-time performance can lead to reduced market share. Passengers may also be entitled to compensation for late flights that varies by world region.

Figure 5.12 shows the number of passenger complaints (per 100,000 enplanements) filed with the Department of Transportation, Bureau of Transportation Statistics for the same period (2021). Southwest had led in this category for years, but Delta took first place with Alaska a close third. Often, passengers booking the lowest-fare flights do so without fully considering the customer amenities typically not included on these ULCC carriers. As a result, passengers are disappointed with the lower-quality product offered by ULCCs, all of which ranked in the bottom half.

While executives should continually analyze many different measures to determine whether their performance is satisfactory, these two metrics suggest that poor on-time performance leads to customer dissatisfaction.

A poor on-time record is one instance of actual performance that doesn't meet established objectives. All departments should have clearly established performance goals. A goal of the reservations department, for example, might be to answer calls within 15 seconds. Customer complaints, although often unpleasant for managers, are another valuable source of information about corporate performance from the customer's perspective.

Every airline department should maintain detailed statistics to track and diagnose performance problems. Table 5.2 is an example of operational performance goals, actual performance, and deviations from objectives. The

Table 5.2 Operational Performance Summary

OPERATIONAL PERFORMANCE SUMMARY

	Daily Statistics			Month To Date		
	Day	Forecast	Deviation	MTD	Forecast	Deviation
OPERATIONS	Actual %	Goal %	Dev. %	Actual %	Goal %	Dev. %
Kickoff flights on time	89.5	85	4.5	71.1	85	-13.9
Kickoff flights < 16	100	90	10	83.5	90	-65
Departures/delay < 16	77.1	80	-2.9	71	80	-9
Arrival/delay < 16	70.8	85	-14.2	64.6	85	-20.4
Arrival/delay < 60	91.7	95	-3.3	84.7	95	-10.3
Arrival/delay < 2 hr	94.8	99	-4.2	92.1	99	-6.9
DEPARTURES COMPLETED						
Actual	96			1981		
Scheduled	96			1984		
Cancelations	0			3		
DELAY PERCENTAGE	Over 5 min					
Stations	45.2	14		22	167	
Maintenance	22.6	7		24.4	185	
Crew	3.2	1		13.9	106	
ATC/Weather	19.4	6		30.5	232	
Gate	9.6	3		4.1	31	
Operations	0	0		4.7	36	
AC Damage	0	0		0.4	3	
FAA Ramp	0	0		0	0	
Unknown/Unresolved	0	0		0	0	
Total	100	31		100	760	

month-to-date statistics reveal serious deficiencies as none of the listed goals were met. Diagnosis might begin by recognizing that morning originating flights (kickoff flights in Table 5.2) are not departing on time so subsequent flights are delayed throughout the day. Of the many potential causes analysts might consider, any or all of the following may apply:

- Are aircraft that underwent overnight maintenance or relocation not being positioned in the morning at the gates as soon as needed?
- Are security lines especially long in the morning delaying passengers or flight crew arrival at the gate?
- Is the fueling process too slow?
- Is baggage processing and loading too slow?
- Are flight crews routinely arriving late?
- Is the catering service holding up the departure?

There is a myriad of possibilities, and more than one cause is likely. In addition to data analysis, front-line personnel should be asked for their observations and diagnosis. The solution could be as simple as fixing a broken ice machine.

5.7.3 Corrective Action

Implementing corrective action is the final step in the control process. The appropriate corrective action, of course, depends on the diagnosis. If the morning originating flights are on time but delays increase throughout the day, then slack resources or longer turn-times may be added to the flight schedule. Alternatively, delays may occur at hubs due to passenger and baggage transfers. More personnel or equipment might alleviate the problem. If flight delays are primarily incurred at one airport, potential corrective actions could range from adding staff or facilities, more training, and additional equipment, to addressing poor labor relations.

American Airlines provides an example. American's on-time performance has historically ranked in the middle of US airlines; its fourth ranking for 2021 (Figure 5.11) is typical. In the spring and summer of 2016, however, it fell to ninth place, below all of its major competitors. If sustained, this poor performance would have led to a loss of market share to its better-performing competitors. American identified the root cause as deteriorating operations at its Los Angeles hub (LAX) after adding more than 20 new daily flights. Delays at LAX rippled across its system. It took several corrective actions (Cranky Flier, 2016), including:

- To ensure on-time departures for the morning originating flights, maintenance personnel met every inbound crew on LAX overnighting aircraft to understand maintenance write-ups and communicate them to Maintenance Operations Control so that repairs were completed overnight.
- Minimized aircraft swaps that might affect LAX originating flights.

- Enhanced dispatcher and pilot coordination to "speed up" flight plans to reduce delays caused by crew duty time limitations.
- Suspended some aircraft cosmetic work at its heavy maintenance facility in Oklahoma to provide more support for daily operations.

By the fourth quarter of the year, American returned to its middle-of-the-pack on-time performance where it has since remained.

The control process is continuous and iterative. Once corrective action is implemented, then data must be analyzed to determine if the correction is effective, and then the process begins anew. A consistent failure to meet a goal may be the result of an unattainable or inappropriate goal rather than performance.

5.8 Summary

Developing a flight schedule, or schedule of services is an extremely complex task subject to many constraints such as aircraft availability and capability, airport and gate limitations, maintenance requirements, and many others. Trade-offs must be made between revenue and cost in an attempt to maximize potential profits. Many software applications assist with the task, but none can replace skilled airline planners. In daily operation, the flight schedule is subject to disruptions, the weather being chief among the causes of irregular operations. Most airlines now place various tactical operations managers in an Airline Operations Control Center. Specialists in dispatch, crew scheduling, aircraft maintenance, and passenger service coordinate their decisions through an overall operations manager. In the case of severe disruption, the operations manager has several tools available to restore the schedule, including aircraft and crew swaps, slack resources, and combining and/or canceling flights. Collection and analysis of operational data are essential to the continuous improvement of the airline's flight operation.

5.9 Case Study: Summer Chaos

5.9.1 The Chaos

Airlines welcomed the resurgent air travel demand in the summer of 2022 and the prospect of renewed profits. Instead, chaos ensued. Travelers were plagued with chronic delays, flight cancelations, lost baggage, and hours-long waits to clear passport control.

In early 2022, governments began lifting pandemic travel restrictions. Travelers responded by flocking to book summer travel. With demand rebounding and starved for revenue, airlines implemented big capacity increases for spring and summer. As Delta Air Lines CEO Ed Bastian explained, "We stretched ourselves to try and grab and secure as much of the revenue pie as we could. We did get caught off guard, as I think everyone in the industry did to some degree." Unfortunately, airlines struggled to operate the expanded flight schedules.

An unprecedented number of flights departed late at major European airports. Among the hardest hit in June and July were Frankfurt, Charles de Gaulle, London Heathrow, and Schipol with flight delays of 46.4%, 42.8%, 40.2%, and 36%, respectively. Overall, 25% of European flights took off late. Many flights never departed. During the last week of June, 400 flights were canceled in just the UK, an increase of 158% from the same seven days in 2019. The US and Canada were not immune. Over the July 4 holiday weekend, more than 12,000 US flights were delayed and hundreds canceled. In Canada, a majority of international flights – 52.9% – departed late from Toronto Pearson. An untold number of vacations were ruined.

Operational problems often began at the airports straining to manage the surge in passengers, no more so than at London Heathrow. Lines for check-in at times stretched outside the terminals. Travelers described scenes of "absolute chaos" after finding long queues at passport control. As 10 or so flights arrived at close to the same time one evening at Terminal 2, passengers queued for six hours to get through passport control with no food or water. Passengers claimed that only two of 10 desks were open. London, however, was not unique. In Sydney, passengers waited for hours to check-in. India experienced similar disarray.

As delays and cancelations rippled across airline networks, mishandled baggage piled up in terminals occasionally overflowing outside. SITA, a Swiss baggage-tracking software provider to airlines worldwide, reported the number of mishandled bags between April and June was five times higher than the previous year. Mechanical baggage systems failed in Toronto and Amsterdam. In what is termed a "creative solution" to return luggage mishandled at Heathrow, Delta Air Lines flew an Airbus A330–200 from Heathrow to Detroit packed with 1,000 pieces of luggage, but no passengers. From Detroit, the bags were flown on to irritated passengers across the US.

5.9.2 Labor Shortage

Airline industry players from airlines to airports failed to appreciate the difficulty in bringing back the staff laid off during the pandemic and in recruiting new workers. Airlines accepted billions of dollars and euros in pandemic subsidies intended in part to retain staff so that flights could be quickly restored when demand returned. Nonetheless, airlines offered early retirements and extended leaves of absence to hundreds of thousands of employees from pilots to cabin crew and ground-handling staff. Many workers, however, did not return when needed. Tim Clark, president of Emirates, summed up the industry exasperation by stating, "The question on everybody's lips is, Where have they all gone? There are hundreds of millions of people that have disappeared from the labor market."

Swissport International Ltd., an aviation services and baggage handler at many airports, tried to fill 17,000 jobs, offering sign-on bonuses of $5,000 at some US airports. Other ground-handling companies faced similar shortages. At Heathrow, airline-handling teams began the summer at 70% staffing. Sydney Airport staged a job fair looking for 5,000 new hires for work ranging from ground support for Qantas Airways to serving hamburgers at McDonald's. Amsterdam's Schiphol airport awarded a summer bonus of €5.25 per hour – a 50% increase for those on minimum wage – to 15,000 workers in security, baggage handling, transportation, and cleaning.

Delays in obtaining airport security clearances often stymied hiring. The International Air Transport Association (IATA) blamed the gridlock besetting UK airports on clearance delays for new staff. Willie Walsh, the director general of the IATA, complained that it took as long as three months to get security badges for new employees, compared with three to four weeks previously. Many potential workers sought jobs elsewhere.

Airlines also struggled with staffing. British Airways had some 3,000 potential recruits stuck in background checks. Similarly, at EasyJet, 140 qualified crew were idled awaiting security clearances. The limited number of flight simulators and instructors delayed training for returning and new-hire pilots. At one point, more than 10% of Delta's pilots were in training and not available to fly.

US airlines blamed poor operations on inadequate air traffic controller staffing, which carriers said led to flight disruptions even during good weather. The Federal Aviation Administration responded, "After receiving $54 billion in pandemic relief to help save the airlines from mass layoffs and bankruptcy, the American people deserve to have their expectations met."

Compounding the misery for airline passengers, workers struck European airlines and airports demanding higher wages to compensate for inflation. Airport staff went on strike in Paris and Hamburg. Air traffic controllers struck in Italy. Pilots walked out at SAS. The Scandinavian airline warned it would cancel half of its scheduled flights, affecting 30,000 passengers per day.

5.9.3 Corrective Actions

The airline industry understood it had to repair its dismal performance. Increasing staffing is the long-term solution; however, in the short run, airlines throttled back overly ambitious schedules. US airlines flew about 15% less in the summer than they had planned to at the outset of the year. At Newark Liberty International Airport, United Airlines further cut 12% of its flights in mid-summer. Lufthansa shelved about 2,000 flights because of staff shortages and strikes. The global flight airline analysis firm Citrium reported that airlines dropped 25,378 flights from their August schedules, of which 15,788 were in Europe. Turkish Airlines lead the cancelations in Europe with 4,408 flights followed by British Airways with 3,600, EasyJet with 2,045, Lufthansa with 1,888, and Wizz Air with 1,256.

Airlines took other, sometimes usual, steps to address staffing shortages. In addition to reducing its flight schedule, EasyJet removed a row of seats from 58 of its smallest Airbus A319 aircraft so it can fly those jets with one less flight attendant. Its chief operating officer, Peter Bellew, also resigned after schedule disruptions. The carrier said it is "absolutely focused on our daily operation" and that it has "taken pre-emptive action to build further resilience for the summer due to the current operating environment." Australia's Qantas has asked office staff at its headquarters to help on the ground at its airports in Sydney, Melbourne, and Brisbane. Lufthansa and Air France-KLM took a different tact by restricting the sale of cheaper tickets to reduce demand, a move likely to be widely adopted by others.

US regional airlines have been severely constrained by a pilot shortage exacerbated by agreements that provide for their pilots to move on to their major airline partners – akin to eating your own. American and United each grounded 100 regional jets that they wanted to fly for lack of pilots. In a surprising, perhaps shocking, step to address the pilot shortage, American Airlines raised pilot pay at its two wholly owned regional carriers by 50%. It then offered triple pay – as much as $600 per hour for senior captains – to pilots who accepted extra flights. The higher pay is set to expire at the end of the summer, but substantially higher permanent pay is likely across the regional airlines.

Finally, some airports, including Heathrow, Gatwick, and Schiphol, made a rare decision to cap daily passengers. Heathrow limited departing passengers to 100,000 a day through September 11, 4,000 passengers below what had been scheduled. It asked airlines to stop selling tickets for summer flights. Several airlines complained; Emirates Airlines initially refused to cut its summer schedule but soon reached an agreement. Frankfurt Airport followed suit in late July reducing its hourly takeoffs and landings.

By the end of July 2022, airline operations improved but periodic delays and cancelations still plagued summer travelers.

Concepts: supply and demand, demand estimation, flight schedule development, labor shortages, irregular operations and remediation, regional airlines.

Sources: The chaotic summer was widely reported in the financial and popular press. This case study draws on Katz & Sider (2022) and many other articles.

Note

1 Allegiant's business model, although successful, results in low aircraft utilization – usually between five and seven hours a day (MIT, 2021).

References

AIMS International. (n.d.). Crew management: Automatic crew assignment. www.aims.
aero/crew-management.html

Airline prep. (n.d.). Flying for Ryanair as a pilot. www.airlineprep.co.uk/career-guides/ryanair-pilot-interview-and-assessment/

Alaska Airlines (n.d.). Our aircraft. www.alaskaair.com/content/travel-info/our-aircraft

Arciuolo, F. (2014, June 16). Frankly speaking. *PlaneBusiness Banter*, 18 (22), 3.

Baker, J. B. (2001). *Marketing: Critical perspectives on business and management*. New York: Routledge.

Baseler, R. (2002). Airline fleet revenue management – Design and implementation. In D. Jenkins (Ed.), Handbook of airline economics (2nd ed., pp 77–106). The McGraw-Hill Companies, Inc.

British Airways. (n.d.) Fleet facts. www.britishairways.com/en-us/information/about-ba/fleet-facts

Bureau of Transportation Statistics. (2021). On-time performance – Reporting operating carrier flight delays at a glance. www.transtats.bts.gov/HomeDrillChart.asp

Carey, S., & Aalund, D. (2007, February 20). JetBlue plans overhaul as snafus irk customers. *Wall Street Journal*, p. A11.

Chokshi, N. (2021, September 21). United Airlines plans a record fleet expansion as travel rebounds. *New York Times*, www.nytimes.com/2021/06/29/business/united-airlines-expansion.html

Clark, P. (2000). Dynamic fleet management. In Gail Butler, & Martin Keller (Eds.) *Handbook of airline operations*. New York: Aviation Week.

Compart, A. (2012, December 17). Southwest says international expansion could support hundreds more planes. *Aviation Daily*, 101.

Cranky Flier (2016, July 14). American tries to fix its operational issues in Los Angeles… and elsewhere. https://crankyflier.com/2016/07/14/american-tries-to-fix-its-operational-issues-in-los-angeles-and-elsewhere/

Emirates Airlines. (n.d.-a) The Emirates Story.

Emirates Airlines. (n.d.-b.) The Emirates Group Annual Report 2019–2020.

Federal Aviation Administration. (2020). Cost of delay estimates. www.faa.gov/data_research/aviation_data_statistics/media/cost_delay_estimates.pdf

Fennell, P. (2017). Crew planning. In P. Bruce, Y. Gao, & J. King (Eds.), *Airline operations: A practical guide* (pp. 125–139). Taylor & Francis Group.

Hegeman, H. (2016, April 1). *PlaneBusiness Banter*, 20(11), p. 3.

Houston airport system (n.d.). Hobby International – The plan. www.fly2houston.com/HOUPlan

Katz, B., & Sider, A. (2022, July 19). Air travel Is breaking down at all levels this summer – Each problem creates ripples, leading to delays, cancellations, lost bags. *Wall Street Journal*.

Lee, A. (2020, March 10). China forges ahead with airport construction binge, despite signs of slowing air traffic growth. *South China Morning Post (Online)*.

Massachusetts Institute of Technology. (2021). Airline data project: Average daily block hour utilization of total operating fleet. https://web.mit.edu/airlinedata/www/2020%2012%20Month%20Documents/Aircraft%20and%20Related/Total%20Fleet/Average%20Daily%20Block%20Hour%20Utilization%20of%20Total%20Operating%20Fleet.htm

Owusu, T. (2016). Why didn't Delta prevent its power outage? *The Street*. www.thestreet.com/investing/why-didn-t-delta-prevent-its-power-outage-13673278

Ryanair Group (n.d.). Annual report 2021. https://investor.ryanair.com/wp-content/uploads/2021/08/FINAL_Ryanair-Holdings-plc-Annual-Report-FY21.pdf

Sabre Holdings (n.d.) Product profiles. www.sabreairlinesolutions.com/home/resour
ces/product_profiles/

Sloan, C. (2013, November 25). An inside look: United Airlines' mission control center.
Forbes. www.forbes.com/sites/airchive/2013/11/25/inside-united-airlines-network-
operations-control/

Spirit Airlines. (2015, June). Investor Presentation. http://ir.spirit.com/events.cfm

US Air, Comair scramble to get back to normal. (2004, December 27). *Wall Street
Journal*, p. A3.

Wizz Air. (n.d.). Meet our fleet.

Yimga, J. & Gorjidooz, J. (2018). Airline schedule padding and consumer choice
behavior. *Journal of Air Transport Management, 78 (2019)*, 71–79.

Zhang, B. (2017). Delta has recovered from the nightmare power outage that shut down
its home base. *Business Insider*. www.businessinsider.com/delta-recovered-from-atla
nta-airports-power-outage-2017-12

Review Questions

1. The schedule of services represents the foundation or basis of the airline product (product attributes such as cabin configuration are also important). What is the economic objective of the schedule? Why does schedule construction consist of a "thousand compromises?" Give an example of one such compromise between revenue maximization and cost minimization.
2. Explain why a mixed fleet, particularly airplanes of different seating capacity, can increase revenue potential and, therefore, profitability? If so, why do Southwest and Ryanair only operate Boeing 737s?
3. How are unit costs (cost per available seat mile, CASM) lowered with higher asset utilization (for example, how is the CASM of an aircraft lowered by flying more hours per day)?
4. How can slack resources (such as a spare aircraft) improve schedule reliability? (Note that this is true for any production process.) If schedule reliability increases revenue through improved passenger satisfaction and repeat business, why don't airlines build more slack resources into their schedule of services?
5. Why is the pairing development process critical to an airline's cost control?
6. What are the pros and cons of the three different pairing assignment methods used by airlines.
7. How might a constraint such as available gates at an airport reduce the schedule of services revenue potential?
8. What are the typical causes of irregular operations? Which is most common?
9. Why is a hub-and-spoke route structure more vulnerable to large-scale weather disruption than a point-to-point or linear system?
10. Why has the AOCC evolved?
11. What functions are typically located within AOCC? Describe the responsibility of each.
12. Why is continuous improvement critical to the success of any business operating in a competitive environment?
13. Describe the Control Process. How can it lead to Continuous Improvement?
14. What are some of the metrics (measures) that an airline tracks to control and improve its operations?

6   Labor Relations

Upon completion of this module, you should be able to:

*   Understand the importance of labor in the airline industry.
*   Explain the characteristics of airline labor.
*   Describe why labor laws vary from country to country.
*   Compare methods to control airline labor costs.
*   Describe the types of unions.
*   Analyze the pros and cons of unions in aviation.
*   Describe the labor protective provisions (LPP).

6.1 Introduction

Herb Kelleher, the co-founder of Southwest Airlines and long-time CEO, once said, "Your employees come first. And if you treat your employees right, guess what? Your customers come back, and that makes your shareholders happy. Start with employees and the rest follows from that" (as quoted in Enquoted, n.d.). Good advice for airline managers; however, as this chapter explains, maintaining positive airline labor relations is challenging.

Commercial aviation has long been a labor-intense industry. In fact, in North America and Europe, labor is traditionally the highest cost category for airlines, typically accounting for 30%–40% of their overall costs.[1] It follows that airline managers must understand the nature of labor in the industry so they can avoid pitfalls leading to managerial nightmares. Unique aspects of the airline product combined with the high unionization of employees further complicate the issue.

This chapter begins with a description of labor in the air carrier industry, including the characteristics of labor specific to airlines. Because airline labor is highly unionized, the next section describes various issues involved in union-management relations, including a discussion of labor costs and merger and seniority concerns. The chapter concludes with a case study.

DOI: 10.4324/9781003290308-6

6.2 Airline Direct Employment

Aviation is critical for the global economy. According to the Air Transport Action Group (ATAG), in 2019 (before the pandemic), there were 1,478 commercial airlines worldwide that carried 4.5 billion passengers and 61 million tons of freight (ATAG, 2020). The industry operates over 33,000 aircraft and directly employs approximately 11.3 million people in four different sectors as shown in Figure 6.1.

6.2.1 Global

In addition to these direct employees, there are about another 76.4 million indirect, induced, and aviation-enabled[2] jobs totaling 87.7 million worldwide.

Those 87.7 million employees produce a total of $3.5 trillion in economic goods and services (direct, indirect, induced, and aviation-enabled), or about 4.1% of global gross domestic product (GDP) (ATAG, 2020). Although 4.1% of global GDP doesn't sound like much, if aviation were a country, it would rank 17th by the size of GDP.

The aftereffects of the terrorist attacks in 2001 and the financial crisis in 2008 were very detrimental to the airline industry, but in comparison, the COVID-19 pandemic has been devastating. Jobs fell by 52.5% from 87.7 million to 41.7 million during the pandemic, with most losses from the passenger airlines. The reductions began about March 2020, just after the start of the pandemic. Rebuilding came slowly, but by mid-2021, most carriers were on a path to building back capacity and employment.

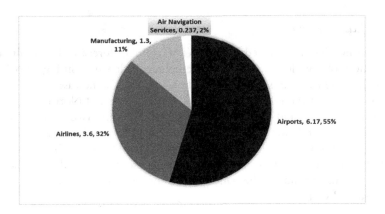

Figure 6.1 Air Transport Direct Employee Breakdown (in millions).
Source: Data from the Air Transport Action Group (2020).

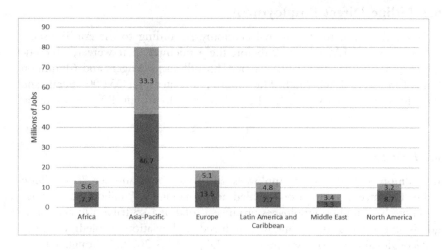

Figure 6.2 Regional Jobs.

Source: data from ATAG (2020). Figures represent estimated aviation jobs in millions. Darker shades show 2018 jobs and lighter shades show projected growth for 2038.

6.2.2 Regional

Figure 6.2 shows the location of jobs by world region. The dark shaded bars indicate the 87.7 million jobs in 2018, of which more than half are in the Asia-Pacific region. Aviation jobs have grown apace with the rapid growth in air travel in Asia-Pacific, especially in China. This trend should continue as the growing population and middle class will stimulate the region's travel demand.

6.2.3 Projected Growth

Oxford Economics, as reported by the ATAG (2020), forecasts 143 million more aviation jobs worldwide by 2038 (the lighter shaded areas in Figure 6.2). Asia-Pacific countries will support 80 million jobs, a 71% increase over 2018 while job growth in the more mature markets of Europe and North America expect a modest 37% and 35%, respectively, over the same two decades. Job losses during the pandemic were comparable across the world regions. Africa, with a higher proportion of international flights, suffered a 58% job reduction compared to 49% in North America. To replace those who retired and others who left the industry during the pandemic, airlines will have to hire at record levels for several years.

6.3 Characteristics of Airline Labor

Airlines are a service industry, and, as such, airline managers face challenging labor issues not encountered in manufacturing and other non-service industries. The distinctive characteristics of airline labor are outlined below.

6.3.1 Highly Skilled and Trained

Airlines typically employ a highly skilled workforce where most jobs are complex and come with a high degree of responsibility. Many employees – such as pilots, flight attendants, dispatchers, and mechanics – must be certified by their governments and require significant initial training, recurring training, and currency requirements. The training pipeline for these employees is often lengthy and long-term planning by management is necessary to ensure adequate staffing levels. Other highly skilled employees – including technicians, engineers, logistics experts, and computer technologists – though not government certified, also require significant training.

In 2022, as the industry tried to return to pre-COVID-19 pandemic service levels, many airlines found they were severely understaffed resulting in many flight disruptions and cancelations. This situation was caused by two main factors which compounded. Anticipating a slow return for demand, airlines reduced their employment through early retirements and/or extended time off. At Southwest Airlines between 2020 and 2021, for example, approximately 16,000 employees elected to take early retirement or an extended time off. Although some 11,400 of those employees returned in late 2021, SWA had a much smaller workforce at the end of 2021 and planned to hire 8,000 employees in 2022 (Southwest Airlines, 2022a). SWA exceeded that plan, hiring 10,000 in the first seven months of the year to return to pre-pandemic employment levels (Arnold, 2022). Additionally, pilot retirements accelerated as the baby boomers reached the maximum retirement age set by most countries. The number surged to about 4,800 pilot retirements in the US alone in 2021 with more than 5,000 retirements a year projected by the end of 2023 (Becker & Cunningham, 2017). Worse, the US Federal Aviation Administration reported a slowdown in the number of pilots qualifying for new airline transport certificates due to the pandemic.

Most airlines report they have been able to find pilots to fill their training classes (ALPA, 2022), but a shortage of qualified pilots caused reduced or dropped service to some destinations. Several possible solutions for the pilot shortage have been proposed.

The US Congress is considering raising the mandatory retirement age from 65 to 67 (Gerhardt, 2022). Although it seems this would provide immediate relief by reducing the number of mandatory retirements, it faces opposition from the Air Line Pilots Association (ALPA), the European Union Aviation Safety Agency (EASA), and President Biden's administration (Air Line Pilots Association [ALPA], 2017), and prospects for passage are dim. ALPA's recommendations to meet future pilot demand include increased awareness of job opportunities in aviation and government funding to help students pay for expensive flight training (ALPA, 2022b).

Some airlines are beginning "ab initio" programs that take prospective pilots with very little or no flight experience and create competent, professional airline pilots through extensive, focused training. Lufthansa has used this method for years, but United Airlines' "United Aviate Academy" was the first mainline US carrier to open its own flight school in 2020. Alaska Airlines also launched its "Ascend Pilot Academy" in 2022 (Goldstein, 2022). Southwest Airlines' Destination 225° offers four different paths for aspiring Southwest pilots (Southwest Airlines, 2022b).

Allowing a single pilot for commercial flights is a long-term potential solution. The European Union Aviation Safety Agency (EASA) is assessing the possibility of single-pilot airline operations as early as 2025. Studies on cockpit functionality and workload alleviation are key to making this a reality (Dubois, 2022).

6.3.2 Highly Paid

High skill levels drive higher average wages for the airline workforce compared to most other industries. Figure 6.3 compares the average wages earned by employees of private industry in the United States and the average wages of employees for US passenger airlines[3] for the period from 2000 to 2020. The sharp decline in 2020 indicates a 10.6% loss in wages due to the COVID-19 pandemic.

The average (pre-pandemic) airline wage in 2018 was about $91,260, much higher than the national private-sector wage of $63,300 (Airlines for America, 2021b; BEA, 2021). However, the total compensation (including wages, taxes, and benefits) for 2018 was almost $30,000 higher at about $120,000 (Airlines for America, 2021a).

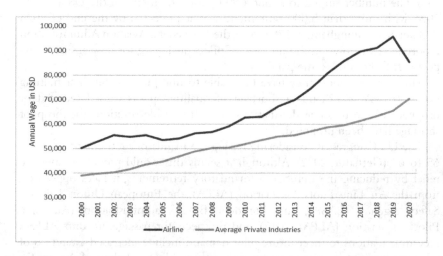

Figure 6.3 Average Private Industry Wages in the US Compared with Airline Wages per Full-Time Equivalent (FTE) Employee.

Source: Data from the United States Bureau of Economic Analysis [BEA], 2021. Wages and salaries per full-time equivalent employee by industry (total wages), and Airlines for America, 2021b.

Along with higher pay, airline employees are highly productive. The ATAG (2020) reports that "aviation jobs are, on average, 4.3 times more productive than other jobs" (p. 11). One measure of employee productivity is the quantity of product produced divided by the number of employees. As an example, a shoe factory's 100 employees might make 100,000 shoes per year, making each employee's productivity 1,000 shoes per year. Airline employee productivity is similarly measured by dividing the available seat miles (ASMs) by the number of full-time equivalent employees.

Although productivity has generally climbed over the two-decade period shown in Figure 6.4, again, we see major productivity losses in post-9/11 (2001 recession), 2008 (recession), and 2020 (pandemic). These drops in product-ivity occurred because airlines can reduce their production faster than they can reduce their employees. Airlines are reluctant to immediately furlough employees to match demand due to the costs associated with re-hiring and retraining. In hard times, the production of available seat miles can be reduced overnight by cutting flights and grounding aircraft. But airlines will often avoid furloughing employees while attempting cost-cutting in other areas. Eventually, if employee reductions must take place, they might be through paid time off, unpaid time-off (furloughs), or retirements (early or on-time), as mentioned earlier. Employee reductions via these programs will lag the available seat-mile losses and result in a short-term productivity loss.

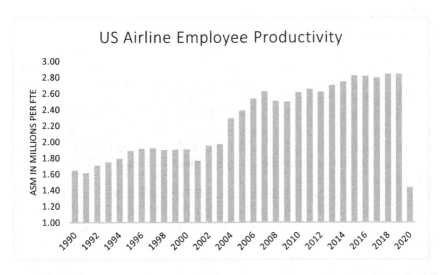

Figure 6.4 US Airlines Employee Productivity.

Source: Available seat-mile data from: From the US Department of Transportation Bureau of Transportation Statistics (n.d.-b). U.S Air Carrier Traffic Statistics through June 2021: System Passenger – Available Seat Miles (January 2000 – June 2021). FTE employee data from Bureau of Transportation Statistics, Employees at US Scheduled Passenger Airlines Month of January 1990–2021 (n.d.-c).

The overall increase in productivity indicated in Figure 6.4 is the result of several management initiatives. Automation, such as passenger check-in kiosks, and automated baggage processing, replaced some employees. Some work was outsourced to third parties. Additionally, less restrictive work rules allowed managers to increase the hours worked while reducing vacations and other time off work. The number of employees per task was reduced wherever possible.

6.3.3 Highly Regulated

Although economic deregulation has spread across much of the world, airline operations and safety remain highly regulated. The Airline Deregulation Act of 1978 freed US airlines of economic regulation, but safety remains highly regulated by Title 49 of the United States Code (USC) and Title 14 of the Code of Federal Regulations (CFR). These laws and regulations provide for licensing and restrict work assignment times for various airline employees, including both flight crews and mechanics. Not only do federal laws impact airlines, but in some cases, state and local laws apply as well. In a 2022 case with Alaska Airlines, California asserted that its labor laws govern work and break rules for flight attendants and pilots based in California. Alaska Airlines is weighing options, but one option is to close or reduce its California crew bases ("Alaska Airlines," 2022).

6.3.4 Highly Unionized

Although overall union density in high-income economies has fallen over the past two decades, airlines remain highly unionized. Most full-service network carriers worldwide have at least some unionized employees. The newer entrant low-cost carriers (LCC), especially those in Europe, have been some of the last to accept employee union representation. Ryanair, founded in 1984 and one of Europe's most successful LCCs, avoided unions for 31 years until accepting a pilot union in 2017. The case study at the end of the chapter explores labor issues at Ryanair in detail. One of the last holdouts, Hungary's Wizz Air, has caught the attention of many labor organizations in Europe and the US that claim unlawful labor practices.

Iceland may have the highest overall workforce trade union density at over 90% (International Labor Organization [ILO], n.d.-b), but Icelandair's unionization is even higher with 97% of its 2,400 employees unionized (Icelandair, 2021). In the US, only about 11% of the overall workforce is unionized, but of the four largest carriers, three (American Airlines, United Airlines, and Southwest Airlines) have between 80% and 85% unionization. Delta has relatively low unionization at only about 20% – but that is still almost twice the overall US average (US Department of Labor Bureau of Labor Statistics [BLS], 2022; Reed, 2020; Delta Air Lines, 2022). The ULCCs Frontier, Spirit, and Allegiant have surprisingly high unionization. Figure 6.5 lists nine selected US airlines and the percentages of unionized workers in 2021.

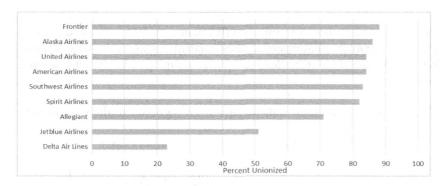

Figure 6.5 Airline Union Representation.
Source: Data from Forms 10K.

6.3.5 Essential to Production

The airline product, the available seat miles, is perishable and cannot be stockpiled. As a result, service industries, including airlines, require continuous work. Strikes, even short-term, have immediate and devastating effects on an airline's operations. If pilots, flight attendants, dispatchers, ramp workers, or baggage handlers decide to walk off the job, flight operations cease. In April 2022, 150 KLM aircraft loading personnel at Amsterdam's Schiphol Airport unexpectedly went on strike. KLM departures stopped and the airport began to fill up with passengers. Airport authorities went so far as to call passengers telling them not to come to the airport. After a few hours, KLM announced it had held discussions with the employees and they agreed to resume work (Garbuno, 2022). Manufacturing industries, on the other hand, can stockpile their product and absorb a short labor work stoppage by selling-off stockpiles while working out their labor issues. This might allow days, weeks, or even months of uninterrupted selling even though production has slowed or stopped.

6.4 International Union Labor

As discussed in Chapter 1, as airline economic liberalization progressed around the world, state-owned airlines often lost the monopoly positions they once enjoyed as new carriers entered the market. Faced with intense competition, all carriers were pressured to cut costs. With labor such a large portion of an airline's costs, labor often took the brunt of cost-cutting measures. This was particularly true if the employees were not represented by unions and airlines could unilaterally impose tougher work rules or reduced pay scales (O'Sullivan & Gunnigle, 2009).

6.4.1 International Labor Laws

The International Labour Organization (ILO), an agency of the United Nations since 1919, establishes international labor standards for its 187 member nations (ILO, 2003). The ILO establishes four core areas endorsed by the international community:

- Freedom of association and the effective recognition of the right to collective bargaining.
- The elimination of all forms of forced or compulsory labor.
- The effective abolition of child labor.
- The elimination of discrimination in respect of employment and occupation. (ILO, 2003, p.7)

The first core area is further developed in the ILO's Fundamental Conventions which state that:

- Workers' and employers' organizations shall have the right to draw up their constitutions and rules, elect their representatives in full freedom, organize their administration and activities and formulate their programs.
- The public authorities shall refrain from any interference which would restrict this right or impede the lawful exercise thereof. (ILO, 2003, p. 12)

These standards ensure that workers are allowed to form unions or associations without interference from their governments and they can best determine how to promote their interests – including the right to strike or lockout if negotiations are unsuccessful. However, the ILO also understands that governments should be able to prohibit certain categories of workers from taking actions like strikes if public safety or continuity of public services are threatened. As an illustration, the ILO specifically mentions several essential services that include among others, police officers, hospital workers, public utility employees, and air traffic controllers (ILO, 2003).

An example of a violation of the freedom to collectively bargain happened in early 2021 in Iceland. Bluebird Nordic, a small Boeing 737 cargo operator, fired all of its unionized pilots and replaced them with "self-employed" pilots (Icelandic Airline Pilots' Association [FIA], 2021a). The pilots union, the Icelandic Airline Pilots' Association (FIA), sued the carrier and the court found for the union, citing that the carrier had violated the employees' freedom of association (FIA, 2021b).

Based on ILO guidelines, individual countries enact their own labor laws to ensure fair work conditions for employees. Although they may follow ILO guidelines, labor practices vary greatly from country to country, especially concerning collective bargaining and labor actions like strikes. In most countries, airline workers are not governed separately from other labor groups. In a few countries, however, airline workers are considered essential, limiting industrial actions. The ILO provides a reference for individual countries' labor laws on their website "National Labor Law Profiles" found at www.ilo.org/Search5/search.do.

In the United States, airline labor relations fall under the Railway Labor Act (RLA). The RLA was enacted in 1926 to ensure railroads, then the economic lifeline of the country, would not be disrupted by strikes that could threaten commerce. Airline labor was added to the Act 10 years later. The RLA requires management and certified unions to collectively bargain in good faith. Strikes are permitted, but only after an extended process monitored by the National Mediation Board.

The European Court of Human Rights has held that while employees are free to form unions and collectively bargain, there is no requirement for an employer to enter into a collective bargaining agreement – called a voluntary system of collective bargaining. Trade unions that are not recognized by the employer can take steps, including industrial action like strikes, to gain recognition and persuade the employer to enter into a collective bargaining agreement. The right to strike is protected by the court but is not absolute and may be subject to regulation by various national laws. National laws, for example, may restrict strikes by workers in "essential services" (European Court of Human Rights [ECHR], 2022, p.44).

6.4.2 Types of Unions

Generally, labor unions are started through employees' internal communication and organization. Often a mutual need (like better working conditions) can be a driving force for workers' organization within a work group. Once deciding to form a trade union, workers must decide on the type of union to have. Several types of labor unions are typically found at airlines, including:

- *Craft unions* are made up of workers with the same craft. Example: the Association of Flight Attendants. In this example, flight attendants from any airline might join the association.
- *Industrial unions* are made up of workers in the same industry. Example: the Transportation Workers of America. In this example, there may be different crafts in the union – flight attendants, dispatchers, ramp workers, bus drivers, etc.
- *Enterprise unions* are made up of workers at the same airline. Example: Southwest Airlines Pilots Association.
- *Federations* are groups of unions. Example: the Coalition of Airline Pilots Associations (CAPA) which includes five associations: the Allied Pilots Association (American Airlines), the Independent Pilots Association (UPS), and three Teamsters Unions representing Atlas Air, Southern Air, Horizon Air, ABX Air, Cape Air, Omni Air, Silver Airways, and Republic Airways (CAPA, n.d.).

6.5 Union Formation

Each country has labor laws regulating the formation of unions. Unions have a constitution or bylaws in accordance with their national origin that outlines many of the functions of the union. The constitution will include membership restrictions, processes for the election of officers, duties of officers, duties of the board of directors, and compensation of officers, to mention a few.

6.5.1 Election and Certification

Laws and regulations concerning voting in and certifying that union vary by country. Generally, a simple majority vote is required; however, in some countries, like the US, specific voting procedures apply. (In the US, a majority of the entire membership is needed to form a union.) Certification and registration with government agencies are often also necessary. Generally, unions will be certified if their constitutions, including election processes, conform to national laws. Election and government certification, however, do not always require the company to recognize the union.

6.5.2 Election of Officers and Representatives

Union constitutions or bylaws specify the type and number of union officers and the officer election process. To support the officers, unions usually have a dozen or so committees staffed by members to help manage issues. The union constitution outlines the committees that typically include:

- Negotiating committee to work with the company on CBA updates and amendments.
- Safety committee to oversee safety issues reported by members.
- Rules committee (or contract administration committee) to oversee questions concerning the current CBA.
- Retirement committee to oversee retirement savings plans.
- Benefits committee to oversee medical and other benefit plans.
- Government affairs committee to oversee governmental actions that might impact union members.
- Communications committee to manage the union's website, and/or publish union news.
- Hotel committee to oversee (or help manage) the company's overnight accommodations for crewmembers.
- Military liaison committee to make sure special needs of employees who are National Guard or military reservists are represented.
- Crisis response team that can travel on short notice to an accident or incident to help members.

Committee members are generally volunteers; however, in some cases, members are paid by the union if they need to be pulled from work for union duties, or if they incur expenses related to the conduct of their duties on the committee.

6.5.3 Initial and Recurrent Collective Bargaining Agreement Negotiation

An airline union negotiates with the company on behalf of its members for terms of employment – a process called *collective bargaining*. The bargaining usually ends with an employment contract termed a *collective bargaining agreement* (CBA) or a *collective labor agreement* (CLA).

The American Federation of Labor and Congress of Industrial Organizations (AFL–CIO) defines collective bargaining as:

> Collective bargaining is the process in which working people, through their unions, negotiate contracts with their employers to determine their terms of employment, including pay, benefits, hours, leave, job health and safety policies, ways to balance work and family, and more. (2022)

The need to open negotiations for a collective bargaining agreement can occur for a couple of reasons. First, if a union is newly recognized by the airline, an initial CBA must be negotiated. Second, bargaining is required when an existing CBA has expired or is eligible for renewal. (Note: in the United States, CBA agreements at airlines are under the jurisdiction of the Railway Labor Act and never expire. Instead of expiring, they "become amendable" at a certain date, after which they are generally renegotiated. In the meantime, the status quo continues under the existing CBA.)

Negotiating a CBA, either for the first time or as an update, is a long and arduous process costing both the union and the airline time and money. Talks can go on for years with little happening. For example, Southwest Airlines flight attendants' contract became amendable in November 2018, and almost four years later in August 2022, SWA and the Transportation Workers of America had still not come to an agreement (Thomaselli, 2022).

6.5.4 CBA Elements

Collective bargaining agreements are typically for a term of three-to-five years and conform to labor laws pertinent to the country of representation. In the United States, the Railway Labor Act defines a specific collective bargaining process. The resulting CBA is detailed and complex, often several hundred pages long.

A CBA typically sets the terms and conditions of employment, such as:

- Company recognition, rights, and responsibilities.
- Pay and benefits (including retirement benefits).
- Employee work scheduling and work rules.
- Disciplinary procedures and appeal.
- Grievance and dispute resolution.

In the absence of a union and collective bargaining agreement, the employer unilaterally establishes work rules. For airlines, these work rules are often similar to union-negotiated CBAs at competing carriers. Keeping their work rules comparable to unionized competitors demonstrates to employees that unionization is not needed to obtain comparable terms of employment. This process has worked successfully for Delta Air Lines flight attendants for many years.

6.5.4.1 Company Recognition, Rights, and Responsibilities

Collective bargaining agreements usually start with a section that lays out the purpose and scope of the agreement. That section delineates:

- Recognition – the company recognizes the union as the representative of the employees. Most unions assert that their members have exclusive rights to work at certain jobs.
- Scope – outlines the purpose of the agreement and work covered by the agreement. The scope clauses would typically restrict the ability of the airline to wet lease aircraft and crews from another carrier.
- Merger protection – the company states that it will do its best to protect the seniority of its employees in the event of a merger.
- Procedures for amendment – the CBA will prescribe a method to amend the Agreement. Usually, amendments can be made with the agreement of both parties. Amendments then become part of the CBA. In some cases, amendments are called "*side letters*."

6.5.4.2 Pay and Benefits

The CBA specifies in detail the method and rates of pay, usually based on seniority. For example, the CBA would address pay for training, travel, vacation, and other non-normal situations, such as moving expenses and jury duty.

Airlines usually provide retirement benefits based on salary and other variables. Typically, airlines opt for one of two different types of retirement plans – *defined benefit plans* which are much like pensions, and *defined contribution plans* which are savings plans, such as the US' 401K programs, where employee contributions are usually matched to a predetermined level with company contributions. Some airlines use one plan or the other, while others use a combination of both. Over the past few decades, many airlines have transitioned their original defined benefit plans and now offer mostly defined contributions plans.

6.5.4.3 Employee Work Scheduling and Work Rules

The CBA attempts to define work rules for every potential circumstance; however, it is nearly impossible to put on paper every situation that might occur in a complex airline operation. The previous chapter (Flight Schedule

Development and Control) outlines different crew scheduling methods an airline might use. National aviation regulations set some minimum flight crew scheduling limitations that must be observed in the CBA. The US Federal Aviation Administration regulations, like 14 CFR Part 117, cover maximum daily flight times and duty-time periods. The CBA will expand on government regulations and often include further restrictions. Scheduling is a major section of pilot and flight attendant CBAs.

6.5.4.4 Disciplinary Procedures and Appeal

The CBA addresses disciplinary procedures, including documentation and investigation. Disciplinary actions range from verbal counseling to formal reprimand, suspension, or termination. Disputes over discipline are resolved via the dispute resolution procedures, also contained in the CBA. Employees in the initial probationary period may receive less protection.

6.6 Contract Maintenance

Managers might expect that the adoption of a new or amended CBA would resolve most management-labor issues until negotiations resume for a new contract. Rather, disagreements over the interpretation and implementation of a CBA arise frequently. In healthy labor relations, management and union representatives meet regularly to review outstanding disputes, staffing, hiring plans, new equipment acquisition, and other issues of mutual interest. To resolve unanticipated circumstances, the parties will sometimes agree to minor changes or clarifications to the existing CBA. Amendments can cover just about any aspect of the CBA and remain in effect until the next CBA. Agreements are recorded in a letter of agreement, sometimes called a "side letter" to the CBA. When agreements are not reached, formal dispute resolution methods are included in the CBA.

6.6.1 Grievances

A grievance is a formal complaint of a CBA violation. Disagreements arise over pay, working conditions, and a host of other issues. The CBA prescribes a multistep process to follow when an employee feels harmed because the CBA was violated. The US Railway Labor Act outlines the steps for resolution that the CBA will mirror.

- The employee and appropriate manager(s) try to work out a solution satisfactory to both parties. This often solves the problem, but if that does not work there are further options.
- The employee files a "formal grievance" with the union. If the union sees no merit in the issue, it is dropped. If the union agrees with the employee, it takes the matter to the airline for discussion and resolution. The services of

a third-party mediator are sometimes used. If no satisfactory remedy results, another option must be pursued.

- The grievance can become a formal "minor" dispute heard by a System Board of Adjustment (SBOA) as described in the Minor Disputes section below.

Convening a SBOA is expensive for both the airline and the union in time and money, so the sides have an incentive to settle disagreements before getting to an SBOA. While both the company and union benefit from efficient grievance processing, the process can, unfortunately, also become a means of obtaining an adversarial advantage. This can work both ways – employees can file grievances for anything and everything to backlog the company; or the company can refuse to settle and send grievances to arbitration, a long and costly process for both sides (Kahn, 1969).

6.6.1.1 Minor Disputes

Minor disputes result from disagreements on the interpretation of the collective bargaining agreement (or contract) that cannot be resolved through the grievance process (United States Government Accountability Office [GAO], 2006). For example, a minor dispute might be related to an employee who thought they were wrongfully fired or disciplined (GAO, 2013). If the union and management do not resolve the issue through direct negotiation, the company and the union jointly create a temporary System Board of Adjustment (SBOA) consisting of equal airline and union representatives and sometimes an arbitrator provided by the NMB (GAO, 2013; Southwest Airlines Pilots Association v. Southwest Airlines, 2021). The decision of an arbitrator is final and binding. Unlike major disputes, minor disputes cannot trigger self-help actions such as strikes or lockouts.

6.6.1.2 Major Disputes

Major disputes are related to the creation of a collective bargaining agreement or subsequent changes to one (National Mediation Board [NMB], n.d.). The NMB handles these disputes through a systematic process that includes collective bargaining and mediation (if necessary). Although not an actual dispute, a union's negotiations for an initial CBA with an airline are considered by the RLA a major dispute simply to kick off the collective bargaining process. The mediator will continue negotiations until determining that the sides cannot reach an agreement. At this point, both sides may agree to binding arbitration. Before engaging in strike actions, euphemistically called *self-help*, a cooling-off period must pass. Under the RLA, the US president can stop a strike and send the parties back to negotiations.

A recent example involves the determination of a minor versus a major dispute. A 2021 lawsuit filed by the Southwest Airlines Pilots Association

(SWAPA) against Southwest Airlines contended that the company's mandate for COVID-19 vaccines violated the RLA regarding their right to collective bargaining – essentially, they said that the vaccine mandate constituted a major dispute because it changed their collective bargaining agreement (CBA) work rules. The company argued that the mandate was an interpretation of the current CBA and thus a minor dispute. A federal judge agreed with the company, saying the issue constituted a minor dispute in which the court had no jurisdiction (Josephs, 2012; Pawlyk, 2021; Southwest Airlines Pilots Association v. Southwest Airlines, 2021).

6.6.2 Self-Help: Strikes and Lockouts

Once freed to engage in self-help, union members may strike their airline employer by refusing to work. Airline strikes are common in Europe, less so in other developed economies, and relatively rare in Southeast Asia and China where airline unions are weaker and represent a smaller portion of employees (CAPA, 2019). The Chinese government actively represses all types of social unrest, including strikes.

In the US, the NMB works to facilitate negotiations and avoid airline strikes. In Europe, however, strikes by airline employees are permitted in most countries – essentially, airlines are treated no differently than other industries. The European Court of Human Rights states that countries' granting the right to strike "represents one of the most important of the means by which the State may secure a trade union's freedom to protect its members' interests" (ECHR, 2022, p. 44). In 2019, for example, there were seven airline strikes in Europe. British Airways, Alitalia, Iberia, Lufthansa, Norwegian, Ryanair, and Scandinavian all experienced strike disruptions ("What is the Cost," 2020). Strikes at European LCC Ryanair present an interesting case study and is explored at end of this chapter.

Just the rumor of a strike can cost airlines millions. In August 2019, a strike threatened by London Heathrow airport workers cost British Airways €33 million, even though the strike was called off before it started.

Although less common than strikes, airlines may also engage in self-help – called lockouts. While union employees have the right to strike, an employer has the right to lockout its employees preventing them from working. In 2011, Qantas, frustrated by "extreme demands" by three of its unions, locked out its licensed engineers, ramp staff and baggage handlers, and Australian and international pilots, shutting down all Qantas domestic and international flights worldwide. This action was legal under Australia's Fair Work Act. The lockout stemmed from the unions' announcement of strikes that were then canceled after Qantas had attempted to reschedule passengers. This bait-and-switch game cost Qantas $68 million. Based on the cost of the lockout to Australian commerce, the Australian government had to intervene, ending the lockout after two days, but forcing the unions into arbitration (Smith & Howard, 2012).

6.7 Challenges to Management

The management of airline labor can be very complex, especially for airlines with crew bases in several countries, each of which may have different labor laws. Unions are generally affiliated with countries, like the Hungarian Airline Pilots' Association, or the Danish Airline Pilots Association, so a transnational airline might have multiple unions in various countries (CAPA Centre for Aviation, 2019). Europe has a very large number of civil aviation unions, with at least one pilot union in each European country (CAPA Centre for Aviation, 2019). Airlines that have crew bases across Europe, like Ryanair, are faced with multiple collective bargaining negotiations.

6.7.1 Restrictions of Ability to Manage

Managing an airline requires long-term strategic planning and short-term mission execution. However, even with good union relations, unions often restrict both planning and execution. Union opposition can derail an airline's strategic plans, as the next two examples illustrate.

The first occurred during Frontier Airlines' 2008 bankruptcy when the airline was to be sold at an auction (Seetharaman, 2009). Southwest Airlines was by far the highest bidder ($170.0 million versus Republic Airways' bid of $108.8 million), but Southwest's bid was contingent upon the two pilot labor groups – the Southwest Airlines Pilots Association (SWAPA) and the Frontier Airlines Pilots Association (FAPA) – agreeing to a pilot seniority integration plan (Southwest Airlines, 2009a; Southwest Airlines, 2009b). After lengthy meetings between the two unions, no agreement was reached. As a result, Southwest's bid was not selected. The sale went to Republic in August 2009 (Southwest Airlines, 2009c).

The second, more recent example, occurred in 2021 when British Airways was planning to re-open short-haul flying at London Gatwick Airport after the pandemic with a newly established LCC subsidiary. Early in the pandemic, British Airways (BA) dropped most of its short-haul flights from London Gatwick Airport (Schlappig, 2021). Because Gatwick (LGW) is slot-controlled, BA would forfeit the unused slots. In an attempt to protect the slots at Gatwick, in 2021, BA announced its intention to start a Gatwick-based low-cost, short-haul airline to compete with Ryanair and Easyjet (Ziady, 2021). A viable LCC subsidiary must have a lower cost structure than BA, so BA needed wage concessions from labor represented by the British Airline Pilots Association (BALPA) and Unite, representing cabin staff, engineers, and ground staff. The BALPA pilots rejected BA's plan. Although possibly as a bargaining technique, in September 2021, BA scrapped its proposed LCC removing flights from the summer schedule (Schlappig, 2021). The next month, BALPA reversed course and voted to approve pay cuts for pilots working for the short-haul subsidiary, claiming support for more pilot jobs (Maszczynski, 2021). Then Unite, the union representing cabin staff, engineers, and ground staff also voted in favor (Drum, 2021). The LCC subsidiary began operations on March 29, 2022, flying to 35 destinations from Gatwick. The airline, called "BA Eurofly," is still awaiting an operating certificate, but in the meantime is operating under the British Airways certificate. After getting its certificate, BA Eurofly will still display its British Airways brand to customers (British Airways, 2022).

Execution of the daily flight schedule is also a very complex task with flights often impacted by weather conditions or unpredictable mechanical failures, both of which can severely disrupt operations and be a nightmare for managers at the operations center. An often-repeated quote attributed to Italian airpower theorist General Giulio Douhet is that "Flexibility is the key to airpower." This holds true for airline operations, where flexibility can often help the airline avoid irregular operations or return to normal operations after a disruption. However, there are many factors hindering operations flexibility, one of which is the collective bargaining agreement. Contract provisions are binding on both management and labor even if a seemingly small deviation might solve an operational problem.

6.7.2 The Increased Cost of Labor Relations

The increasing complexity of negotiating collective bargaining agreements (CBAs) requires teams of managers and lawyers at the airline and the union. Often the cost of collective bargaining negotiations can run into millions of dollars. When needed, outside attorneys' fees can be exorbitant, sometimes $1,000 per hour. To compound the problem, airlines are often working on several CBAs at the same time. As an example, on December 31 2021, Southwest Airlines was in CBA negotiations with eight of its labor organizations (Southwest Airlines, 2022a).

6.7.3 Universal Management Opposition to Unions

Labor unions raise airline costs. Wages and benefits are usually higher for unionized employees but their productivity is often lower. Airlines also have to pay for expensive CBA negotiations and other contract administration, like grievances, mediation, and arbitration. Unions often restrict management's ability to react to changing business or operational conditions creating distrust between management and labor. In the US, if a majority of members of a craft vote for a union, the RLA requires the new union to represent all members of the craft. As a result, there may be employees represented by the union who do not want that representation (or subsequent union dues).

Union-management relations are sometimes harmonious, but tensions usually rise during CBA negotiations, which, if unsuccessful, can result in a strike or other action. It's no wonder that management is generally opposed to union representation for their employees.

6.8 Managing Labor Costs

An airline must actively manage its costs to remain competitive and earn sustained profits. Especially during the regulated era, labor costs were viewed as uncontrollable. With the consolidation of US major carriers in the first decade of the century, airline management focused on profits before market share. Controlling labor costs became part of the focus on profits.

6.8.1 Wage Concessions

Airline employment contracts, whether with individual employees or CBAs, remain in force for long periods – five years is common. If an airline negotiates a contract in good financial times (the peak of the business cycle), the contract may be very favorable for employees. However, when the business cycle turns down, airlines may find that costly labor is a heavy burden. Unions have no obligation to negotiate for lower wages and less restrictive work rules during the term of the contract. Airlines can appeal for wage concessions that unions may negotiate if the airline's existence is threatened. Such concessionary agreements usually include a "snapback" date at which time full wages would be restored. For example, in 2021, Ryanair was able to negotiate a 20% pay reduction due to the pandemic, however, the deal included restoration of pay over a three-to-five year period (Ryanair, 2022).

In 1983, American Airlines CEO Robert Crandall attempted to reduce pilot wage costs in what became a classic, if not infamous, labor relations gambit. To finance a post-recession growth spurt, Crandall didn't ask for across-the-board wage cuts but instead proposed a two-tier pay scale (Lieber, 1997). One scale for existing employees (A-scale), and another for new hires (B-scale). The B-scale was in some cases, 50% of the A-scale. There was no plan to ever merge the two scales (Hopkins, 2000). The plan was approved by American's Allied Pilots Association (APA) whose current members, of course, had nothing to lose (they all stayed on A-scale) and a lot to gain as the promised company growth would improve their relative seniority and result in career advancements, including job position and aircraft upgrades. Eventually, Crandall envisioned, all employees would be under the new B-scale wage as the A-scalers retired. American grew rapidly as promised, but the plan backfired when the B-scale pilots (referred to as the "Killer Bs") gained a substantial share of union membership. Crandall's new pilots resented their second-rate status. To avoid labor unrest, American agreed to merge the B and A scales by 1987 (Kaps, 1997). The B-scale was finally phased out about 10 years later (Cimini, 1997).

6.8.2 Outsourcing

Another method used to reduce labor costs is to outsource jobs to a third party or a company subsidiary. Outsourcing can affect many labor groups, including pilots, flight attendants, mechanics, ground handlers, reservation agents, accountants, and others. Outsourcing of aircraft heavy maintenance to third parties, often in countries with much lower wages, has become widespread. A 2018 Transport Workers Union study reported that "approximately 24% of heavy aircraft maintenance for US carriers is performed offshore, up from about 7% in 2003" (Reed, 2018, p. 1).

In another example of contentious labor relations at Qantas during the COVID-19 pandemic, in 2021 the airline outsourced its ground handling and aircraft cleaning services at Australian airports resulting in the loss of 2,500 Qantas jobs. The Transport Workers Union (TWU) took Qantas to court claiming that Qantas used the pandemic as an excuse to make significant cuts to its unionized workforce. The TWU claimed Qantas outsourced the jobs to avoid collective bargaining, in violation of Australia's Fair Work Act (Dowling, 2022). The court ruled in 2021 and again on appeal in 2022 that the decision to outsource was illegal (Dowling, 2022). Although siding with the TWU, the court also ruled that the jobs would not be reinstated, and the company may be ordered to pay significant compensation to the fired employees (Khadem, 2021).

FSNC outsourcing of some flying to regional airlines is explained in Chapter 4. Between 2000 and 2009, mainline carriers grounded almost 800 narrow-body aircraft, taking with them over 14,000 pilot jobs. During the same period, regional airlines' capacity went up 133%, more than doubling their domestic market share (Swelbar, 2010; Regional Airline Association, 2010).

Mainline pilot collective bargaining agreements attempt to limit this activity by including *scope clauses*. A scope clause places limitations on the network carrier's ability to transfer its capacity to the regional affiliate. Scope clauses usually limit the size of the regional jet based on seats or gross weight – the smaller the limit, the better for the mainline employees. For years, scope clauses restricted network carriers from transferring flying hours to regional carriers with aircraft larger than 50 seats. This allowed major carriers to use small regional jets in some markets but limited the use of larger regional jets which could be used to replace mainline aircraft in many more markets. Under bankruptcy restructuring or other financial disasters, scope clause limitations are often relaxed (Compart, 2013).

6.8.3 Atypical Employment

A new employment approach, called *atypical employment*, is growing in Europe, especially among LCCs. Atypical employment refers to employees considered self-employed rather than direct-employed (Miley, 2019). However, this self-employment of airline employees seems to skirt the standard definitions of self-employment, hence the pejorative "*bogus self-employment.*" Carriers who use bogus self-employment (also called *false self-employment*), can reduce tax liabilities, or other employers' responsibilities to a large extent – enough to be considered an unfair competitive advantage over other carriers (European Cockpit Association [ECA], 2020; Miley, 2019; European Commission, 2019). Self-employed crewmembers are typically paid by the hour with no minimum guarantees, and possibly no sick time or vacation time. Unions argue that the airline profession does not lend itself to "actual" self-employment, and although cases have been put before the European Parliament, no action has been taken (ECA, 2020).

6.8.4 *Flag of Convenience*

As part of the European Union's single market, airlines can now be owned and controlled by citizens of any country but operate under a certificate issued in another country where they might find more-liberal employment standards. This concept, called a *flag of convenience,* can be financially beneficial. Flag of convenience is not a new idea. For a century, the maritime industry has found that registering a ship in a country (including land-locked countries) other than that of the owner can be beneficial regarding taxes, regulations, environmental controls, and yes, labor as well (International Air Transport Association [IATA], 2019; International Civil Aviation Organization [ICAO], 2013). Currently, off-shore civil aircraft registries are growing in Aruba, Bermuda, Ireland, Malta, Georgia, and Lithuania where owners enjoy lower direct and indirect taxes, lower insurance, and less bureaucracy. "There is growing evidence that airlines under liberalization are increasingly restructuring their operations to reflect classic maritime 'flags of convenience' scenarios" (ICAO, 2013, p.7).

In 2012, Norwegian Air Shuttle announced it was starting a subsidiary called Norwegian Air International (NAI), to fly routes from the EU to international destinations, including the United States (International Transport Workers' Federation [ITF], 2014). To avoid strict labor regulations, labor costs, and taxes in Norway, NAI registered their aircraft and applied for an operating certificate in Ireland, which was granted in 2014 (ITF, 2014; ALPA v. Chao, 2018). While waiting for their certificate from Ireland, in December 2013, NAI applied to the US Department of Transportation for a foreign air carrier permit that was granted in 2016 despite the objections of several labor unions, including the Air Line Pilots Association (ALPA), Allied Pilots Association (APA), and Southwest Airlines Pilots Association (SWAPA) (ALPA v. Chao, 2018). These labor organizations, along with others, contended that NAI used Ireland as a "flag of convenience" to evade the strict labor laws in Norway. This allowed them to hire lower-paid pilots and cabin crews based in Thailand that work under Singapore labor contracts, a scheme permissible under Irish but not Norwegian law (ALPA v. Chao, 2018; Alder, 2017). US airlines and labor unions challenged the DOT's decisions in court but lost. A few years later, in 2019, NAI ceased the Ireland operations saying the routes were not commercially viable (United States House of Representatives, Committee on Transportation and Infrastructure, n.d.).

Additionally, other administrative functions, including accounting and reservations call centers, can be moved to countries with typically lower labor costs. This step is not limited to legacy carriers – in 2007, low-cost carrier Spirit Airlines closed its domestic reservation centers and outsourced all reservations to the Philippines and the Dominican Republic. Similarly, Norwegian Air International operated its administrative office in Latvia (ITF, 2014).

6.8.5 New Entrants

Newer airlines typically have lower labor costs than long-established carriers. There are a couple of reasons for this cost savings. First, new entrant airlines will initially have no unions. Unions must be voted-in, however, organizing and establishing a union may take years, if it happens at all. Examples of this include JetBlue Airways and Ryanair. Both had no unions for many years following their founding. In 2014, JetBlue recognized its first union, and in 2017, Ryanair followed suit. Also, airline employees' pay scales generally increase with company seniority. At start-up airlines, almost all employees begin at entry-level wages, which increase as time passes. Clearly, at long-established airlines, high-seniority workers will be earning top-level wages. FSNCs often attempt to take advantage of the lower cost structure of new entrants by establishing new LCC subsidiaries.

6.8.6 Bankruptcy

In the United States, airlines seeking Chapter 11 bankruptcy reorganization may petition the court to reject existing collective bargaining agreements. The airline must prove to the bankruptcy court that the rejection of the CBA is necessary to allow the carrier to further its reorganization (Sanders, 2007). This happened in 2004 with US Airways and United Airlines, and in 2012 with American Airlines, but one of the more recent examples was at Aeromexico in early 2021. Aeromexico requested the termination of its collective bargaining agreements with pilots and flight attendants in its attempt to exit Chapter 11 bankruptcy (Garbuno, 2021). In Aeromexico's case, just the threat of the CBA termination led both labor groups to accept wage concessions.

6.9 Mergers and Seniority

The relatively small number of carriers in the airline business makes it an oligopoly – like telecommunications companies, automakers, gun manufacturers, and so forth. In oligopolistic industries, executives often view merging with a competitor as a superior strategy to organic growth. From 1930 through 2019, 55 US airlines merged (Airlines for America, 2020). Such horizontal mergers are usually subject to governmental approval to avoid anti-competitive industry concentration and harm to consumers. Merging companies present a huge management challenge as executives attempt to meld different cultures preserving the best of each. Merging unionized labor is often difficult and contentious.

The compensation and employment rights of union members are usually determined by seniority within their craft. An employee's seniority governs many aspects of their work, including daily schedule, vacation times, salary, and work location (domicile). When airlines merge, one of the major concerns is how the pre-merger seniority lists are combined into one post-merger list – a process called *seniority list integration* or SLI. In some cases, the inability to merge the employees' seniority could end the deal, and in other cases, it can create strife among the workers.

6.9.1 Labor Protective Provisions

In the US, airline seniority list integration after mergers was historically managed by the Civil Aeronautics Board (CAB) as part of their economic regulation of air carriers (Jerman & Joshi, 2012). In 1950, the CAB first imposed what they called *Labor Protective Provisions* (LPPs) on airline mergers to ensure that employees of both pre-merger airlines were treated fairly" (Bakos v. American Airlines, 2010, pp. 2–3). Later in 1972, during their oversight of the merger between Allegheny Airlines and Mohawk Airlines, the CAB formalized the LPPs, specifically Sections 3 and 13. Section 3 states that provisions shall be made for the integration of seniority lists fairly and equitably, while Section 13 states that "if the parties cannot agree on a fair and equitable manner, any party may submit the dispute for binding arbitration" (Bakos v. American Airlines, 2010, p 3; Nesgos, 1982).

After that formalization, the LPPs were usually referred to as simply "Allegheny-Mohawk." Although a formalized policy, "Allegheny-Mohawk" was not always applied in mergers, especially after the closure of the CAB. In 2007, Senator Bond and Senator Claire McCaskill succeeded in enacting the McCaskill-Bond Amendment which formalized "Allegheny-Mohawk" Section 3 and 13 into statutory law (Bakos v. American Airlines, 2010). It was codified in 49 USC § 43112.

Under McCaskill-Bond, if the same union represents the merging employees, the union's internal integration policy applies. Otherwise, the SLI must abide by the law and merge seniority lists fairly and equitably. If seniority list integration is not successful, binding arbitration will be proffered.

6.10 Summary

With labor costs often the highest expense category for airlines, airline managers and employees should understand the various issues unique to airline labor. A highly-skilled, highly-paid, highly-unionized, and highly-regulated workforce is essential to airline operations. Poor labor relations can degrade customer service and imperil flight operations. Managing a unionized workforce is challenging and expensive. Management's ability to react to market changes is frequently restricted by collective bargaining agreements. Although unionization is falling in the developed economies, large airlines will remain highly unionized. Consequently, managers should recognize the costs and benefits of unions.

6.10.1 Pros

- Unions help employees collectively bargain for contracts. Individual workers might have a hard time dealing with huge airlines whereas unions have an advantage in being able to leverage the entire workforce. Vacations, sick time, and health benefits are among the many issues negotiated for the group as part of collective bargaining.
- Unions advocate for members in the event of discipline or termination. Union lawyers know the business and can provide a wealth of information to individuals who find themselves in tough situations.
- Unions can help improve the safety of the carriers' operations by ensuring strict standards are maintained. Unions will usually have committees to oversee operational safety.
- Unions can improve the overall flying experience by lobbying for both passenger and employee improvements. ("The Pros, Cons, and Future of Unions in Aviation," n.d.)

6.10.2 Cons

- The initial election of a union is often contentious. Unions can foster employee distrust in management, and sometimes pit employees against each other.
- Negotiations for both new and renewal CBAs are lengthy and expensive, often taking years to accomplish.
- The CBA usually reduces employee productivity and increases wage and benefit expenses.
- Ongoing contract management is time-consuming and expensive, requiring in-house and often outside labor attorneys. Airlines will pay $500 per hour for mid-level outside attorneys. Senior and partners' fees are higher.
- The CBA can reduce the ability of both management and employees to react to a changing business or operational environment. Airlines and employees must abide by the CBA, even in extremely irregular operations. The COVID-19 pandemic is a good example. Airlines were often unable to change CBA conditions to fit the rapid onset of the pandemic lockdowns. Instead, most changes that affect employee groups must be negotiated through the involved collective bargaining process.
- Personnel management is much more difficult. CBAs restrict and complicate management-employee relations. Unions are required (at least in the US) to represent all members of the group of employees, even when individuals within the group might prefer not to. This often manifests itself in objections and frequently in grievances. The grievance process requires management time, often by a costly dedicated staff. For issues that can't be resolved, arbitration is even more time-consuming and expensive.
- Managers who do not come from the labor group – which is usually the case, pilots being the common exception – are often unaware of how

constrained their management discretion is and how costly a contract violation can be. ("The Pros, Cons, and Future of Unions in Aviation," n.d.)

6.11 Case Study: Ryanair

The Irish airline market in the 1970s and early 1980s was stagnant and fares were
high. Aer Lingus cut routes between Dublin and provincial British cities; other
travel options were time-consuming. It was a perfect time for a new-entrant
carrier. Beginning operations in 1985 as a private company headquartered in
Dublin, Ryanair started as a full-service carrier operating Ireland–UK routes.
Ryanair's entry helped stimulate the market, attracting many passengers who
had not previously flown. Over-expansion and an unmanageable fleet, however,
soon brought mounting losses for Ryanair.

In 1990–1991 Ryanair re-structured as Europe's first LCC, adopting a cost
leadership model more recently termed an ultra-low-cost carrier. This restructuring, patterned largely after Southwest Airlines in the US, included reducing
staff, converting to single-class seating, and a single-type fleet – the Boeing 737
(see Figure 6.6).

The European airline deregulation of the 1990s created a single European
aviation market (SEAM) and spurred Ryanair's 1997–1998 expansion across

Figure 6.6 Ryanair Boeing 737 in Dublin.
Source: Photo from Wikimedia Commons by PvOberstein (2022) CC0 1.0 (cropped).

the continent. In 2000, Ryanair launched its website, simplifying reservations, ticketing, and distribution. Ryanair's fleet reached 100 Boeing 737s in 2007. Ryanair flourished under the cost leadership model despite growing competition from other LCCs, like EasyJet. In 2022, Ryanair operated 500 aircraft, 90% of which are unencumbered with debt.

Ryanair bases crews in many countries across the continent – the latest count, over 80 different crew bases. This strategy not only saves money on crew hotels (crewmembers are back in their base each night) but also gives them a transnational status making labor negotiations tricky. Ryanair classified many workers as self-employed under the atypical employment model that reduced Ryanair's tax and social security costs.

Ryanair's labor force is very productive, with one of the highest passenger-to-staff ratios among European carriers. Employees received productivity-based pay incentives. For its first 31 years of operations, Ryanair had no labor unions. Ryanair CEO, Michael O'Leary, known for his openly hostile relationships with organized labor, once stated he would close the airline before negotiating with unions. Despite a few brushes with unionization, Ryanair remained union-free.

The company went public in 1997, which was also the year O'Leary's first clash with unionization began. Dublin-based baggage handlers, supported by the Services, Industrial, Professional, and Technical Union (SIPTU) went on strike, ultimately closing the Dublin airport. After several self-help actions by both sides and a government mediation attempt, the case ended up in the Supreme Court. The High Court's finding in favor of Ryanair, and the fact that Ryanair had conceded nothing to SIPTU, sent a message to other potential labor groups that strikes were not effective against Ryanair.

Undeterred, the Irish Airline Pilots Association (IALPA) and the British Airline Pilots Association (BALPA) both also lost confrontations with Ryanair. Between 2007 and 2017, there were 10 key labor court cases in countries where Ryanair crews were based. Ryanair won most cases. The few that were not (cases in France, Norway, and Denmark), led O'Leary to close crew bases in those countries rather than accept unions. Ryanair's transnational makeup was at the heart of its ability to ward-off unions. Aircrews based in many locations and European countries diluted the effects of a strike by employees in one country. Employees in other countries were assigned to cover for striking employees, weakening or even quashing the strike. This was the case in 2019 when Ryanair pilots and cabin crews from UK, Belgium, Netherlands, Italy, Spain, and Portugal went on strike but caused little impact on customer service.

However, a staffing crisis in 2017 ultimately led O'Leary to recognize and negotiate with unions. The European Union Flight and Duty Time Limitations (FTL) Regulation (83/2014) that became effective in 2016 placed restrictions on flight crew hours, including 900 hours of flight time in any calendar year. Contrary to other countries, Irish authorities initially considered this period from April 1 – March 31, which benefited Ryanair, who typically flew crews hard in the summer months. However, in 2017, the Irish authorities acknowledged the period should be from January 1 – December 31. This was bad news for Ryanair. They had already flown crews hard during the summer, expecting to get through Christmas before slowing operations in the early part of 2018 to comply with the restrictions. Crews would now reach the time limit in the heart of Christmas holiday travel. As flights were canceled by the thousands, Ryanair pilots recognized the balance of power had shifted. In December, pilots in Ireland, Portugal, Spain, Italy, and Germany backed industrial actions. This impending transnational strike forced Ryanair to recognize unions for the first time. Flight attendants soon followed, and by September 2018, pilots and flight attendants in more and more countries were following suit. Within a year, Ryanair has signed deals with unions in eight out of the 21 countries where its crews were based.

Ryanair was successful in maintaining separate unions in each country rather than transnational unions. Working with single-country unions preserves the possibility of covering a strike in one country with crews from another. Unions from several countries can still unite and strike together, but that requires much more coordination and teamwork.

The European Court of Justice also got involved recently, ruling that disputes between Ryanair and its crew can fall under the jurisdiction where the worker is based (not necessarily Ireland), a decision that transport unions considered a significant setback for Ryanair.

Concepts: Unionization, strikes, cost leadership model, and international labor laws.

Sources: De Spiegalaera (2020), Golden & Erne (2022), O'Sullivan & Gunnigle (2009), Tungate (2017), PvOberstein (2022), Barrett (2009), Ryanair (2022), European Cockpit Association. (2017).

Notes

1 Fuel is generally the second-highest cost, but occasionally, depending on the current price, fuel will overtake labor as the highest cost category. In Asia, however, labor costs are typically about one-fifth of total operating costs, so fuel is often the highest cost area (Prokopovič, 2018; Blyton et al, 2001).

2 Indirect jobs include jobs created through the purchase of goods and services from companies in the air transport supply chain; induced jobs are created through the direct employees spending wages; and aviation-enabled jobs are tourism jobs created by aviation (ATAG. 2020).

3 The passenger airline wages are the average of 23 airlines used by the US Department of Transportation, Bureau of Transportation Statistics (n.d.-a) for comparison. In March 2021 that list included American Airlines, Delta Air Lines, United Airlines, Alaska Airlines, Southwest Airlines, JetBlue Airlines, Spirit Airlines, Frontier Airlines, Allegiant Airlines, SkyWest Airlines, Envoy Air, Piedmont Airlines, Republic Airlines, Endeavor Air, PSA Airlines, Horizon Air, Mesa Airlines, Air Wisconsin Airlines, GoJet Airlines, Hawaiian Airlines, Sun Country Airlines, Silver Airlines and Eastern Airlines.

References

Air Line Pilots Association v. Chao. (2018). United States Court of Appeals for the District of Columbia Circuit, No. 17–1012. https://law.justia.com/cases/federal/appellate-courts/cadc/17-1012/17-1012-2018-05-11.html

Air Line Pilots Association. (2017). ALPS fact sheet: Pilot retirement age facts. www.alpa.org/-/media/ALPA/Files/pdfs/advocacy/alpa-pilot-supply-retirement-age.pdf?la=en

Air Line Pilots Association. (2022). More than enough pilots to meet US airline demand: Debunking the pilot shortage myth. www.alpa.org/advocacy/pilot-supply

Air Transport Action Group (ATAG). (2020). Aviation benefits beyond borders. https://aviationbenefits.org/media/167517/aw-oct-final-atag_abbb-2020-publication-digital.pdf

Airlines for America. (2020). US airlines mergers and acquisitions. www.airlines.org/dataset/u-s-airline-mergers-and-acquisitions/

Airlines for America. (2021a). A4A presentation: Industry review and outlook. www.airlines.org/dataset/a4a-presentation-industry-review-and-outlook/

Airlines for America. (2021b). Labor per FTE data.

Alaska Airlines moots California crew base closure. (2022). *Ch-Aviation.* www.ch-aviation.com/portal/news/117364-alaska-airlines-moots-california-crew-base-closure

Alder, E. (2017). Fixing open skies: Policy proposals for the European Union and the United States in light of the Norwegian Airlines International dispute. *Aviation Law and Policy, 16*(2), 341–368.

American Federation of Labor and Congress of Industrial Organizations (AFL-CIO). (2022). Collective bargaining. https://aflcio.org/what-unions-do/empower-workers/collective-bargaining

Arnold, K. (2022). Southwest hits worker milestone before all the other major airlines. *Dallas Morning News.* www.dallasnews.com/business/airlines/2022/07/14/southwest-hits-worker-milestone-before-all-the-other-major-airlines/

Bakos v. American Airlines. (2010). United States District Court for the Eastern District of Pennsylvania. No. 17–402. www.jamhoff.com/wp-content/uploads/sites/3081/2017/07/45-Opinion-Bakos.pdf

Barrett, S. (2009). Deregulation and the airline business in Europe: Selected readings. Taylor & Francis Group. https://ebookcentral.proquest.com/lib/erau/reader.action?docID=431807&ppg=7

Becker, H. & Cunningham, C. (2017). Pilot retirement accelerate beginning in 2021 & peak in 2025. *Cowen and Company Equity Research.* https://static01.nyt.com/files/2018/business/Pilot_Retirements_Accelerate_Beginning_In_2021_Peak_In_2025_Cowen_and_Company.pdf

Blyton, P., Martinez, L., McGurk, J., & Turnbull, P. (2001). Globalization and trade union strategy: industrial restructuring and human resource management in the international civil aviation industry. *International Journal of Human Resource Management, 12*(3), 445–463.

Brady v. Allied Pilots Association and Amer. Airlines. (2003). Civil Action No. 3:03-CV-0984-D. *CaseText.* https://casetext.com/case/brady-v-allied-pilots-association-and-amer-airlines

British Airways. (2022). British Airways returns to short-haul European flying from Gatwick Airport. https://mediacentre.britishairways.com/pressrelease/details/86/0/13813

CAPA Centre for Aviation. (2019, 4 Aug). European airline labour relations: Multiple unions are a challenge https://centreforaviation.com/analysis/airline-leader/europ ean-airline-labour-relations-multiple-unions-are-a-challenge-484048

Cimini, M. (1997, Fall). Profile of an airline emergency dispute. Bureau of Labor Statistics: *Compensation and Working Conditions*. www.bls.gov/opub/mlr/cwc/prof ile-of-an-airline-emergency-dispute.pdf

Coalition of Airline Pilots Association (CAPA). (n.d.). About CAPA. www.capapilots. org/about-capa/

Compart, A. (2013). Scope clauses may affect use of new RJs. https://aviationweek. com/air-transport/scope-clauses-may-affect-use-new-rjs

De Spiegalaera, S. (2020). Transnational union action at Ryanair. *Transfer, 26*(2), 229–233.

Delta Air Lines. (2022). Annual report 2021 Form 10K. https://d18rn0p25nwr6d.clo udfront.net/CIK-0000027904/0c2d6e8c-ef47-43ad-8a88-da602e4b426a.pdf

Dowling, H. (2022). Qantas loses bid to overturn illegal outsourcing ruling. *Australian Aviation*. https://australianaviation.com.au/2022/05/qantas-loses-bid-to-overturn-illegal-outsourcing-ruling/

Drum, B. (2021). British Airways reportedly reaches agreement with Unite for a new LCC at London Gatwick. *World Airline News*. https://worldairlinenews.com/2021/ 11/07/british-airways-reportedly-reaches-an-agreement-with-unite-for-a-new-lcc-at-london-gatwick/

Dubois, T. (2022, January 24 – February 6). Flying solo? *Aviation Week & Space Technology*. https://aviationweek.com/air-transport/safety-ops-regulation/easa-european-oems-explore-single-pilot-operations

Enquoted. (n.d.). Top 19 quotes by Herb Kelleher. www.enquoted.com/herb-kelleher-quotes.html

European Cockpit Association. (2017). Ireland & EU flight safety rules on pilot fatigue. www.eurocockpit.be/news/ireland-eu-flight-safety-rules-pilot-fatigue

European Cockpit Association. (2020). Position paper: (Bogus) self-employment in avi-ation – ACP, ECA & EurECCA common views. www.eurocockpit.be/positions-publications/bogus-self-employment-aviation-acp-eca-eurecca-common-views

European Commission. (2019). Study on employment and working conditions of aircrews in the EU internal aviation market. www.eurocockpit.be/sites/default/ files/2019-06/Study%20on%20employment%20and%20working%20conditi ons%20of%20aircrew%2C%20EU%20Commission%202019.pdf

European Court of Human Rights (ECHR). (2022). Guide on Article 11 of the European convention on human rights www.echr.coe.int/Documents/Guide_Art _11_ENG.pdf

Garbuno, D. (2021). Aeromexico seeks to end crew collective bargaining agreements. *Simple Flying*. https://simpleflying.com/aeromexico-crew-collective-bargaining-agr eements/

Garbuno, D. (2022). KLM handlers strike leads to chaos in Amsterdam. *Simple Flying*. https://simpleflying.com/klm-handlers-strike-amsterdam/

Gerhardt, M. (2022). Why is there a pilot shortage? Start with ageism. www.nbcnews. com/think/opinion/pilot-shortage-start-ageism-rcna34831

Golden, D., & Erne, R. (2022). Ryanair pilots: unlikely pioneers of transnational col-lective action. *European Journal of Industrial Relations 28*(4) 1–19. https://journals-sagepub-com.ezproxy.libproxy.db.erau.edu/doi/full/10.1177/09596801221094740

Goldstein, B. (2022). Alaska Airlines and United Airlines launch training academies. *Aviation Week & Space Technology*.

Hopkins, G.E. (2000). *Flying the line, volume II. The line pilot in crisis: ALPA battles airline deregulation and other forces.* Air Line Pilots Association, Int'l.

Icelandair. (2021). Icelandair annual and sustainability report 2021. https://assets.conte ntstack.io/v3/assets/blt7c94950eb7f3edcb/blt4c78d9eb445c5cb1/628398e3747a0 167e7640db7/IcelandairAnnualReport_2021.pdf

Icelandic Airline Pilots' Association. (2021a). Bluebird Nordic strike. www.fia.is/en/ fr%C3%A9ttir-fundager%C3%B0ir/fr%C3%A9ttir/bluebird-nordic-strike/

Icelandic Airline Pilots' Association. (2021b). Pilots lay-offs ruled illegal: FIA wins labour court case against Bluebird Nordic and SA. www.fia.is/en/fr%C3%A9ttir-fundager%C3%B0ir/fr%C3%A9ttir/pilots-lay-offs-ruled-illegal-f%C3%ADa-wins-labour-court-case-against-bluebird-nordic-and-sa/

International Air Transport Association. (2019). Doug Lavin: False flag operation. https://airlines.iata.org/blog/2019/10/doug-lavin-false-flag-operation

International Civil Aviation Organization. (2013). The need for a strategy to address the negative consequences of continued liberalization: Would maritime style "flags of convenience" contribute to sustainable aviation? Worldwide Air Transport Conference (ATCONF) (Working Paper). www.icao.int/Meetings/atconf6/ Documents/WorkingPapers/ATConf.6.WP.099.2.en.pdf

International Labour Organization (ILO). (2003). The International Labour Organization's fundamental conventions. www.ilo.org/wcmsp5/groups/public/---ed_norm/---declaration/documents/publication/wcms_095895.pdf

International Labour Organization (ILO). (n.d.). Statistics on union membership. https://ilostat.ilo.org/topics/union-membership/

International Transport Workers' Federation (ITF). (2014). Flags of convenience in civil aviation. www.itfglobal.org/sites/default/files/resources-files/flags_of_convenienc e_in_civil_aviation.pdf

Jerman, T. & Joshi, A. (2012). United States: Seniority Integration and the McCaskill-Bond Statute. *Mondaq: Jones Day.* www.mondaq.com/unitedstates/aviation/164186/ seniority-integration-and-the-mccaskill-bond-statute

Jorens, Y., Gillis, D., Valcke, L. & De Coninck, J. (2015). Atypical forms of employment in the aviation sector, European social dialogue, European Commission. www.europarl. europa.eu/meetdocs/2014_2019/documents/tran/dv/report_atypicalemployment inaviation_/Report_AtypicalEmploymentInAviation_en.pdf

Josephs, L. (2012, October 26). Federal judge rejects Southwest Airlines pilots' request to block vaccine mandate. *CNBC: Airlines.* www.cnbc.com/2021/10/ 26/federal-judge-rejects-southwest-airlines-pilots-request-to-block-vaccine-mandate.html

Kahn, M. (1969) *Airline grievance procedures: Some observations and questions. Journal of Air Law and Commerce 35*(3) 313–323. https://scholar.smu.edu/jalc/vol35/iss3/2

Kaps, R. (1997). *Air transport labor relations.* Southern Illinois University Press.

Khadem, N. (2021). Qantas workers who lost their jobs due to outsourcing will not be reinstated, Federal Court finds. *Australian Broadcasting Corporation News.* www. abc.net.au/news/2021-12-17/qantas-twu-ground-crew-outsourcing-legal-federal-court-decision/100709612

Labor, 29 C.F.R. §1202.12 (n.d.) Retrieved 25 October 2021 from www.ecfr.gov/curr ent/title-29/subtitle-B/chapter-X/part-1202

Leff, G. (2021). British Airways starting a new short haul airline at London Gatwick. *View From the Wing.* https://viewfromthewing.com/british-airways-starting-new-short-haul-airline-at-london-gatwick/

Lieber, R. (1997). Bob Crandall's boo-boos: The fiery American Airlines Chairman faces labor strife that could create long-lasting scars at his company. Here's how he went wrong. *Fortune Magazine.* https://money.cnn.com/magazines/fortune/fort une_archive/1997/04/28/225523/index.htm

Maszczynski, M. (2021). British Airways pilots vote in favour of massive pay cuts to save Gatwick low-cost subsidiary. *Paddle Your Own Kanoo.* www.paddleyourownka noo.com/2021/10/07/british-airways-pilots-vote-in-favour-of-massive-pay-cut-to-save-gatwick-low-cost-subsidiary/

Miley, I. (2019). Bogus self-employment could be costing the state €1 billion per year. *Raidió Teilifís Éireann* (RTE). www.rte.ie/news/2019/1024/1085474-bogus-self-employment/

National Mediation Board (NMB). (n.d.). National Mediation Board. www.govi nfo.gov/content/pkg/GOVMAN-1995-07-01/pdf/GOVMAN-1995-07-01-Pg635.pdf

Nesgos, P. D. (1982). Call for Labour Protective Provisions in Canadian Aviation. *Annals of Air and Space Law,* 7, 127–160.

O'Sullivan, M., & Gunnigle, P. (2009). "Bearing all the hallmarks of oppression": Union avoidance in Europe's largest low-cost airline. *Labor Studies Journal 34*(2) 252–270. https://journals-sagepub-com.ezproxy.libproxy.db.erau.edu/doi/abs/10.1177/0160449X08319661

Pawlyk, O. (2021, October 26). Federal judge tosses out Southwest Airlines pilots' petition against vaccine mandate. *Politico.* www.politico.com/news/2021/10/26/southw est-airlines-petition-vaccine-mandate-517253

Prokopovič, K. (2018). Airlines under pressure from higher wages and increased fuel costs in 2018. https://aviationvoice.com/airlines-under-pressure-from-higher-wages-and-increased-fuel-costs-in-2018-2-201802151046/v

PvOberstein. (2022). Ryanair EI-DPP. *Wikimedia Commons, the free media repository.* Retrieved 21:51, August 1, 2022 from https://commons.wikimedia.org/w/index. php?title=File:Ryanair_EI-DPP.jpg&oldid=648285367.

Reed, T. (2018, April 6). Amount of outsourced offshore airline maintenance work has risen, report says. *Forbes.* www.forbes.com/sites/tedreed/2018/04/06/amount-of-outsourced-offshore-airline-maintenance-work-has-risen-report-says/?sh=ca411 8d26e2b

Reed, T. (2020, March 26). How labor unions won historic pay protection for aviation workers. www.forbes.com/sites/tedreed/2020/03/26/airline-workers-will-be-paid-until-sept-30-as-labor-unions-secure-historic-protections/?sh=52570aa4b806

Regional Airline Association. (2010). Fact sheet. https://members.raa.org/resource/res mgr/Docs/RegionalFactSheetDec2010.pdf

Ryanair. (2022). Annual report 2021. https://investor.ryanair.com/wp-content/uplo ads/2021/08/FINAL_Ryanair-Holdings-plc-Annual-Report-FY21.pdf

Sanders, T.G. (2007). The runway to settlement: Rejection of collective bargaining agreements in airline bankruptcies. *Brooklyn Law Review,* 72(4), 1401–1443. https:// brooklynworks.brooklaw.edu/cgi/viewcontent.cgi?referer=&httpsredir=1&arti cle=1345&context=blr

Schlappig, B. (2021). British Airways ends Gatwick short haul flights. *One Mile at a Time.* https://onemileatatime.com/news/british-airways-ends-gatwick-flights/

Seetharaman, D. (2009). *Republic wins Frontier auction over Southwest.* Reuters. www.reut ers.com/article/us-frontier-southwest/republic-wins-frontier-auction-over-southw est-idUSTRE57D03Z20090814

Smith, G. & Howard, L. (2012). The Qantas dispute: Employer's lockout, ministerial intervention and Fair Work Australia's decision. www.claytonutz.com/knowledge/2012/may/the-qantas-dispute-employer-s-lockout-ministerial-intervention-and-fair-work-australia-s-decision

Southwest Airlines Pilots Association v. Southwest Airlines Co. (2021). In the United States District Court for the Northern District of Texas, Dallas Division. www.flyert alk.com/wp-content/uploads/2021/10/177114656396.pdf

Southwest Airlines. (2009a). *Southwest Airlines submits bid to acquire Frontier Airlines in bankruptcy court proceeding*. www.swamedia.com/releases/release-0edcad2f7d6cf69a6 7bd6d104b6b1297-Southwest-Airlines-Submits-Bid-to-Acquire-Frontier-Airlines-in-Bankruptcy-Court-Proceeding?query=frontier

Southwest Airlines. (2009b). *Southwest Airlines developing bid proposal in Frontier Airlines bankruptcy court proceeding*. www.swamedia.com/releases/release-b663ff123945cbf43 e12b10a4b6c49f4-Southwest-Airlines-Developing-Bid-Proposal-in-Frontier-Airli nes-Bankruptcy-Court-Proceeding?query=frontier

Southwest Airlines. (2009c). *Southwest Airlines' bid to acquire frontier not selected at auction*. www.swamedia.com/releases/release-673cad0773b0f3630fb864d64b6b130e-Southwest-Airlines-Bid-to-Acquire-Frontier-Not-Selected-at-Auction?query= frontier

Southwest Airlines. (2022a). Form 10-K 2021. US Securities and Exchange Commission.

Southwest Airlines. (2022b). Destination 225o. https://careers.southwestair.com/D225

Swelbar, w. S. (2010, March). Mainline pilot scope: Will regional carriers be permitted to fly 90+ seat aircraft. 35th Annual FAA Aviation Forecast Conference "The Future of Scope Clauses." 35th Annual FAA Aviation Forecast Conference. Lecture.

The pros, cons, and future of unions in aviation. (n.d.). *Political Science*. https://polit icalscience-articles.blogspot.com/2018/01/the-pros-cons-and-future-of-uni ons-in.html

Thomaselli, R. (2022). Southwest flight attendants ask for federal mediation. *Travel Pulse*. www.travelpulse.com/news/airlines/southwest-flight-attendants-ask-for-fede ral-mediation.html

Tungate, M. (2017). A brief history of Ryanair. *Management Today*. www.managementto day.co.uk/brief-history-ryanair/food-for-thought/article/1449458

United States Bureau of Economic Analysis. (2021). National data. Table 6.6D: Wages and salaries per full-time equivalent employee by industry. https://apps.bea.gov/iTa ble/iTable.cfm?reqid=19&step=2#reqid=19&step=2&isuri=1&1921=survey

United States Department of Labor, Bureau of Labor Statistics. (2022). News release – Union members 2021. www.bls.gov/news.release/pdf/union2.pdf

United States Department of Transportation, Bureau of Transportation Statistics. (n.d.-a). US passenger airline employment down in March 2021 compared to February. www.bts.gov/newsroom/us-passenger-airline-employment-down-march-2021-compared-february

United States Department of Transportation, Bureau of Transportation Statistics. (n.d.-b). U.S air carrier traffic statistics through June 2021: System passenger – available seat miles (January 2000 – June 2021). www.transtats.bts.gov/TRAFFIC/

United States Department of Transportation, Bureau of Transportation Statistics. (n.d.-c). Employees at US scheduled passenger airlines month of January, 1990–2021. www.bts.gov/newsroom/employees-us-scheduled-passenger-airlines-month-janu ary-1990-2021

United States Government Accountability Office. (2006). National Mediation Board: Compensating neutral arbitrators appointed to grievance adjustment boards under the Railway Labor Act. www.gao.gov/assets/b-305484.pdf

United States Government Accountability Office. (2013). National Mediation Board: Strengthening planning and controls could better facilitate rail and air labor relations. www.gao.gov/assets/gao-14-5.pdf

United States House of Representatives, Committee on Transportation and Infrastructure. (n.d.). House report 116–636: Additional views. www.govinfo.gov/content/pkg/CRPT-116hrpt636/html/CRPT-116hrpt636.htm

Ziady, H. (2021). British Airways blames pilots as it ditches plans for new low-cost airline. *CNN Business*. www.cnn.com/2021/09/23/business/british-airways-low-cost-carrier-gatwick/index.html

Review Questions

1. Explain the unique characteristics of airline labor.
2. Describe why labor is an important consideration to an airline manager?
3. How do wages in the airline industry compare to other industries?
4. What are ways managers can reduce labor costs?
5. Describe so-called atypical employment in aviation.
6. Why do European airlines often have several different pilots' unions?
7. Why is airline labor in the US regulated under the Railway Labor Act?
8. What are Labor Protective Provisions and what aspect of an airline merger are they designed to protect?

7 Economics and Finance

Upon completion of this chapter, you should be able to:

- Characterize the profit history of the global airline industry.
- Compare the profitability of the airline industry with other major industries.
- Define several important airline financial and operational metrics.
- Using the profit equation, explain how airline managers can increase profits.
- Compare and contrast the growth of ancillary revenues at full-service network carriers and low-cost carriers.
- Identify and define the largest airline cost categories.
- List the typical steps in restructuring legacy carriers.
- Compare the effects of the COVID-19 pandemic on the airline industry with other 21st century shocks.

7.1 Introduction

Airlines are notorious for poor profitability. This chapter begins with a review of historical world airline profits and then examines the main sources of airline revenue and the interaction of the variables that affect total revenue. After considering revenue generation, the discussion then turns to airline costs. The many cost factors associated with running an airline can overwhelm managers as they attempt to maintain a competitive cost structure. Understanding these cost factors and their relationships is especially important to legacy carriers as they face ever-increasing competition from established and new-entrant low-cost carriers. Finally, the chapter examines government actions to save airlines from failure during the COVID-19 crisis.

7.2 Profit History

Since the end of World War II, airline profits have been mostly anemic and increasingly volatile. Figure 7.1 graphs the history of world airline profits and net profit margin. Real profits are adjusted for inflation measured in 2021 US dollars. In the early post-war years, airlines were subject to economic

DOI: 10.4324/9781003290308-7

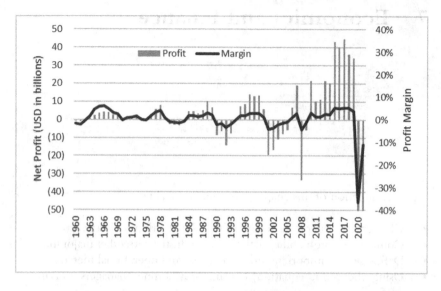

Figure 7.1 World Airline Real Net Profits and Margin 1960 through 2021.

Note: Real net profits are adjusted for inflation in 2021 dollars. The 2020 net loss of $148 billion is off the chart.

Data sources: A4A and St. Louis Federal Reserve Bank.

regulation and many were state-owned; the industry was modestly profitable in most years. However, as deregulation spread beyond the US, profits and losses grew larger and more cyclical. From the end of the Great Recession of 2008 until the beginning of the pandemic in 2020, the industry enjoyed its best stretch of profits leading some to claim that the years of steep losses were relegated to history. Nothing, however, was comparable to COVID-19 pandemic-induced losses in 2020 of nearly $140 billion, an amount that is off the chart in Figure 7.1.

Profits closely track the business cycle, the irregular fluctuations in economic activity over time. During periods of expansion, the economy grows in real terms (adjusted for inflation) with increases in jobs, industrial production, sales, and personal income. In recessions, the economy contracts, unemployment increases, production and sales fall. The effects of world recessions in the early 1990s, 2001, and 2008 are evident in large airline industry losses. In 2020, the world's airlines lost an unprecedented $139 billion due to COVID-19. Recessions cause businesses to hunker down reducing employee travel, thus airline revenue plummets as the highest price tickets go wanting. Leisure travel is less affected, but passengers look for bargains that airlines are forced to offer to fill otherwise empty seats. In response to declining demand, airlines reduce flights and cut costs wherever possible, but many costs are fixed in the short run and cannot be avoided. Aircraft ownership expenses, for example, remain. On its leased fleet, an airline must still make the same monthly payments even if aircraft utilization falls.

Airlines are also often victims of their own earlier success when profits were high. As economist Paul Clark and others have observed, some of the profit oscillation results from the time lag in adding capacity in response to high demand. In good economic times with rising corporate profits, business activity is vigorous with high-paying business travelers filling airline seats. Airlines, enjoying healthy profits, decide to increase capacity by ordering more aircraft that are delivered several years later. Unfortunately, the aircraft arrive just in time for the next economic downturn in the business cycle. Airlines then take delivery of aircraft they do not want or need; there are too many aircraft and too much capacity for the reduced demand. Large losses ensue (Clark, 2010).

7.2.1 Profit by World Region

Profitability varies by world region. In recent pre-pandemic years, the North American airline industry has been most profitable followed by substantially lower profits in Europe and Asia-Pacific. Airlines in the Middle East, Latin America, and Africa have struggled as Figure 7.2 illustrates. All airlines then suffered huge losses as countries restricted flights to combat COVID-19.

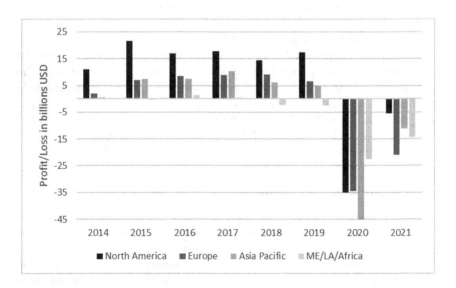

Figure 7.2 Profits by World Region.
Data source: Statista.

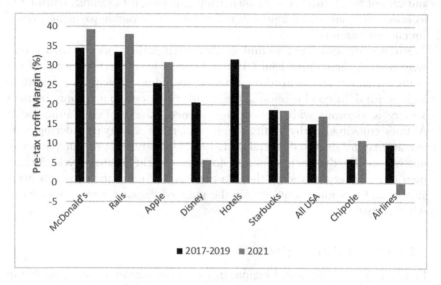

Figure 7.3 Profit Margin Comparison for US Airlines.
Data source: Airlines for America.

7.2.2 Net Profit Margin

By several measures, the airline industry has historically been a low-margin, low-profit business. Two common metrics for measuring profitability are *net profit margin* and *return on invested capital*.

Net profit margin is calculated by dividing either the before or after-tax net income (profit) by total revenue; it is a percentage of total revenue that remains after paying expenses. Net profit margin is useful for comparing the profitability of various industries and firms as shown in Figure 7.3. Before the pandemic, US airlines' before-tax profit margins were less than two-thirds of the average for all US industries. While US airlines lost billions during the pandemic, other industries remained profitable with some reaping higher profits.

7.2.3 Return on Invested Capital

Another common measure of profitability is the *Return On Invested Capital* (ROIC). ROIC compares profit to all invested capital – equity as well as debt-financed capital – providing a measure of management's efficiency at putting its capital to work in profitable investments. Although it seems easy to compute, airlines use several methods to determine ROIC. The obvious and most frequently used method is to divide net profit by total assets. For example, if an airline's net profit is $200 million and its total assets are $2.0 billion, the ROIC is 10%. The thorny part is that airlines do not always compute their profit in the same manner, for example, some use pre-tax figures and others use post-tax

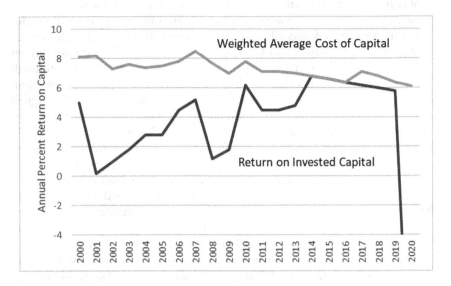

Figure 7.4 Return on Capital 2000 through 2020. Note: 2020 ROIC is off the chart at -18%.

Data source: IATA, 2021.

figures. Additionally, it is difficult to estimate the assets of an airline because of varied depreciation methods, varied proportions of leased equipment, and various government subsidies (Doganis, 2010). Like all financial measures, ROIC is not perfect but does provide another important measure of airline profitability.

Publically-owned firms whose stock trade in one of many stock markets must earn a competitive return for their shareholders or they will move their money to other firms thereby eliminating one source of funds for expansion. ROIC can be compared with the opportunity cost to investors; in other words, what return their money would earn if it were invested elsewhere at similar risk. This is opportunity cost is called the *Weighted Average Cost of Capital* (WACC). Figure 7.4 compares WACC with airline ROIC for the first two decades of this century. Except for a short period between 2014 and 2016 when lower fuel prices, controlled capacity growth, and high load factors led to record profits, airlines underperformed giving investors a good reason to withdraw their investments. Financial theory notwithstanding, airlines can usually sell stock to raise funds. Again, the effects of the pandemic are dramatic as airline losses drove ROIC highly negative.

7.3 Earning Profits

Despite historically low profits compared with other industries, most airlines survive – some since the birth of the industry – so it may be worth asking why profits are important. If airlines are viewed as a public service and state-owned or subsidized as was true for much of airline history, then profitability is not necessary for survival. Indeed, many state-owned airlines are not efficient or profitable, a topic addressed in Chapter 11. However, if airlines cannot rely on government support, then financing must be obtained in capital markets through a combination of stockholder-supplied funds and private loans. Investors shun firms that do not promise future profits, so airlines that are perennially plagued by poor profits are unable to obtain funding by selling new stock. A common perception of stock market investors is of a small, elite group of the very rich, but pension funds, insurance companies, and other financial institutions are the major sources of corporate equity financing. Airlines also borrow from commercial banks or by direct issuance of debt. Like investors, lenders will not make loans to firms with poor prospects for repayment that depend on future profits. Many airlines have survived surprisingly long periods without making profits but eventually find themselves unable to raise money to continue operations. Those airlines that do not successfully restructure go out of business.

An airline earns profits when total revenue exceeds total costs. Revenues for large airlines are many billions of dollars per year. American Airlines, the world's largest carrier, reported total revenue of $45.8 billion for 2019. However, billions in revenues are often accompanied by equally high costs resulting in scant profits or, as the previous section showed, large losses. The fundamental responsibility of airline executives, beginning from the chief executive officer down through middle managers, is earning profits for the airline's shareholders – so managers must ensure that revenues consistently exceed costs. This is a constant struggle in the competitive airline industry subject to many volatile outside forces.

7.3.1 Airline Finance Metrics

A few more measures of airline operations and finance are useful for appreciating the decisions that managers face in earning profits.

- The *available seat mile or kilometer* (ASM or ASK) is the basic unit of airline production, defined as one airline seat flown one mile (or kilometer) whether the seat is occupied or not. Just as an automobile company measures its production by the number of cars produced in a given period, an airline measures production by the number of seat miles produced in a period. For example, a 100-seat aircraft flown 1,000 miles contributes 100,000 ASMs to the airline's total production. With hundreds of aircraft flying 10 or more hours every day, a large airline's production in ASMs is a huge number ranging in billions per month. Using American Airlines again as an example, it reported producing an average of 23.75 billion ASMs per month in 2019. Two other metrics follow directly from the ASM, the revenue and cost per available seat mile.

- *Revenue per available seat mile or kilometer* (RASM or RASK) is the money an airline earns from a unit of production. RASM (pronounced ras-um) is computed by dividing total revenue by ASMs per period, either a month, quarter, or year. Because RASM measures the revenue for the airline's unit of product, it is also known as *unit revenue*. If a small airline has revenues of $100 million in a month in which it produced one billion ASMs, RASM is 10¢. RASM is often further refined by excluding revenues from cargo and other revenues above the base ticket price. The result is *passenger revenue per available seat mile* (PRASM). Care must be exercised with this metric because ancillary revenues are often excluded even though these are passenger-related and a substantial source of revenue, especially for LCCs. PRASM is measured in US cents or similar units in other currencies. PRASM for United Airlines, for example, was 13.9¢ in 2019 declining to 9.61¢ in 2020 amid the pandemic.

- *Cost per available seat mile or kilometer* (CASM or CASK) is similar to RASM, but measures the cost, rather than revenue, of producing an ASM. It is computed by dividing the airline's total cost per period by the ASMs. CASM (pronounced cas-um) is also known as *unit cost*. Because fuel prices are volatile, airlines often also measure CASM excluding fuel cost or CASM ex-fuel. Fuel prices are mostly beyond management's control, so CASM ex-fuel provides a better comparison over time of management's success in controlling costs. When RASM exceeds CASM, an airline is profitable.

- *Revenue passenger mile or kilometer* (RPM or RPK) is the basic measure of airline passenger traffic. It reflects how many of an airline's available seat miles were sold. For example, on a flight of 1,000 miles with 100 passengers aboard, the airline generated 100,000 RPMs (regardless of the aircraft seating capacity).

- *Load factor* (LF) is the portion of an airline's production that is sold. For any given flight, the load factor is the percentage of seats occupied or the number of passengers divided by the number of seats. For all flights, LF is computed by dividing RPMs by ASMs.

- *Yield* is the price a passenger pays per mile flown, so it's a direct measure of the airline's average price. To compute yield, divide total passenger revenue by RPMs. Typically measured in cents per mile, yield is useful in assessing changes in fares over time. Yield is easily confused with RASM. Yield does not incorporate or vary directly with load factor as RASM does, so an airline might have a relatively high yield but a low load factor. In this case, yield would be much higher than RASM.

Studying the evolution of a few of these metrics with the maturation of the airline industry provides an additional appreciation for how airlines can earn profits. Because of the ready availability, US airline data are used in the next sections; however, the basic trends hold for the world industry as well.

7.3.2 Yield History

Figure 7.5 is a graph of world real and nominal airline yield since 1960. Real yield is adjusted for inflation providing a valid comparison of prices over time. Except for occasional short periods, real yield declined over the period. In 2020, the real price of an average ticket fell by more than 60% from the 1960 price whereas the nominal price doubled. Over the long term, prices closely follow costs. During the period of economic regulation, the decline in costs was mostly the result of rapidly improving technology, especially more productive and efficient aircraft, but also from automation of other airline functions. Because regulated prices were set at cost plus a profit margin, declining costs led to lower prices. Prices are still set this way on some international routes subject to government control, but, with the spread of economic deregulation, competition has driven down fares. Aircraft technology has continued to improve, especially with the increased fuel efficiency of newer high-bypass jet engines. Real yield plunged during the 1960s with the introduction of jet aircraft. Nominal yield was stable during this period and has climbed modestly since. Automation of other areas of airline operations lowered costs.

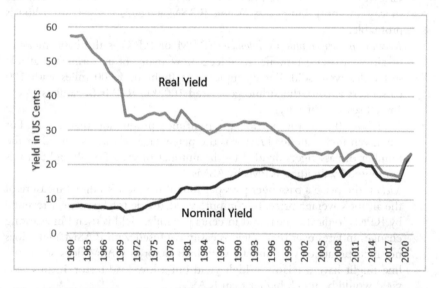

Figure 7.5 World Airlines Nominal and Real Yield.

Note: Because of data limitations, nominal yield is based on total operating revenue. Real yield is adjusted for inflation and measured in 2021 US cents. A graph of world airline yields is similar.

Data sources: A4A and Federal Reserve Bank.

7.3.3 Available Seat Miles

Airline managers must match capacity with constantly changing demand, so capacity has grown apace with demand over many decades. Figure 7.6 shows US airline capacity measured in available seat miles (ASM) since 1960. Demand for air travel falls during recessions (shown in gray) and capacity is correspondingly reduced. The effects of the recessions of 2001 and 2008 are especially prominent in the data. History, however, provides no comparison for the precipitous drop ASMs resulting from the COVID-19 shock of 2020. The US domestic market recovered more rapidly than in most world regions. The ASMs for 2021 indicate the restoration of flights in the early stage of the recovery.

ASMs and RPMs (or ASK and RPK) are easily confused, so a reminder may be helpful. ASM is the common measure of airline production. Each airline decides the number of ASMs to supply in its markets by changing the number of flights offered and the size of the aircraft assigned. RPM, in contrast, is a measure of airline demand that managers can influence, but not fully control. On a flight where every seat is filled by a paying passenger – 100% load factor – RPM equals ASM. Airlines attempt to match capacity with demand so ASMs are highly correlated with RPMs, so the graph in Figure 7.6 is very similar to Figure 2.1 in chapter 2. Figure 2.1, however, illustrates demand whereas Figure 7.6 is a graph of historical supply. The difference is important.

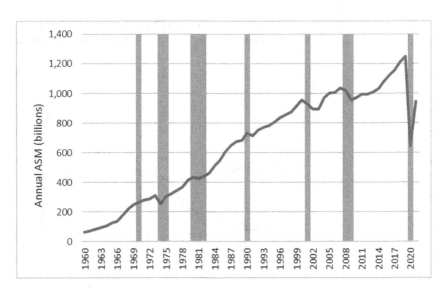

Figure 7.6 US Airline Capacity 1960 to 2021.

Note: Gray bars show economic recessions. A graph of world traffic is similar.

Data source: A4A.

7.3.4 Load Factor

The load factor (LF) is the ratio of seats sold to the seats available. On a given flight, it's just the percentage of seats occupied by paying passengers. To compute the load factor for the airline, divide RPM by ASM. Airlines have successfully increased load factors over the years filling more seats per flight even as capacity has grown. Higher load factors increase airline efficiency as fewer available seats go unfilled or spoil. This increased efficiency is the result of many factors, including intense competition, more efficient scheduling, the rise in non-refundable fares, and the increasing sophistication of revenue management systems.

Figure 7.7 shows the increase in world and US airline load factors over the past 60 years. Interestingly, US load factors lagged the world until reversing in the early 21st century. This reversal coincides with US major airline restructuring, mostly through bankruptcy, and the growth of LCCs.

High-demand flights will often be full, a frequent passenger complaint. However, some flights inevitably operate when demand is low and, hence, at lower load factors. The practical upper limit is about 85% which was reached before the pandemic. Most airlines have little opportunity to increase revenue with higher load factors.

Breakeven Load Factor (BLF) is the percentage of seats that must be filled for revenue to equal the total cost. BLF is determined using a simple equation: **BLF = CASM ÷ yield.**

If CASM is unchanged and yield goes down, the breakeven load factor goes up. Intuitively, if costs stay the same and fares go down, the airline needs more

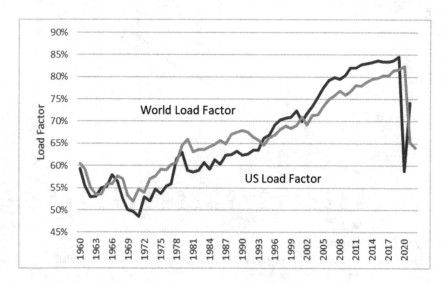

Figure 7.7 World and US Airline Load Factor from 1960 through 2021.
Data Source: Airlines for America.

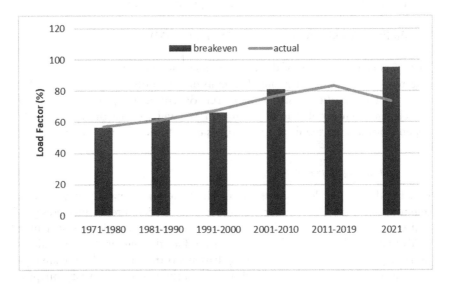

Figure 7.8 Breakeven Load Factors for US Airlines from 1971 to 2021.
Source: Airlines for America Industry Review, 2022.

passengers per flight to break even. Conversely, with a constant yield, if costs increase the breakeven load factor also increases. Figure 7.8 charts breakeven and actual load factors for each decade from 1971. Airlines are profitable when actual load factors exceed breakeven and vice versa.

During recessions, ticket prices fall as airlines struggle to keep seats filled, but the drop in yield raises the breakeven load factor. Concurrently, airlines reduce flights as demand falls, so ASM falls and CASM rises. The airlines are the victims of the "double-whammy" leading to a sharp increase in the breakeven load factor. The casual observer may fall victim to the fallacy that an airline must be profitable when flights are nearly always full. But, if ticket prices (yield) are low and costs are high, the airline will lose money even if every seat was full. During recessions, breakeven load factors usually exceed 100% as Figure 7.8 shows for 2021.

7.3.5 Profit Equation

With these metrics in hand, we can define a profit equation that will prove useful in understanding the challenge of earning consistent profits.

Profit = total revenue − total cost
Total revenue = yield x RPM; RPM = ASM x LF
Total Cost = ASM x CASM

Substituting these identities into the basic profit equation:

Profit = (Yield x ASM x LF) − (ASM x CASM)

Managers can adjust any of the four variables; however, a change in one variable will cause a commensurate change in one or more of the other variables. The effect on profit is usually uncertain. By considering how management can change each variable in this equation, we will see that the interaction or dependency among the factors sometimes thwarts the effort to increase profits.

7.3.6 Managing Revenue

Faced with losses or low profits, raising the ticket price may seem an obvious step to increase profit. However, raising the price, which directly increases yield, will reduce the load factor. The fundamental microeconomic law of demand holds that people will purchase less of a product the higher the price, all else remaining unchanged (ceteris paribus). Passengers traveling for leisure are especially price-sensitive. Many will choose an airline based primarily on price, so raising the price by as little as $10 will drive some potential passengers to a competitor. Then load factor will fall. Whether the price increase results in more or less revenue depends on the number of passengers who choose another airline as measured by the price elasticity of demand. In other words, raising the price may be self-defeating. The baffling complexity of airline pricing is the subject of Chapter 9.

Next, the manager might consider increasing the load factor and filling seats that would otherwise go empty. The problem here is the inverse of raising the price. The most direct means of filling more seats is to lower the price thereby attracting the most price-sensitive passengers. It may be possible to make discount fares available only to those passengers without lowering the fare of other passengers. However, if the price reduction results in a lower average fare, yield will fall. Total revenue may increase or decrease depending on how many additional passengers choose to fly at the lower price.

ASM is a variable in both the revenue and cost side of the equation. The airline, of course, can increase ASMs by adding flights or, if it has larger aircraft with more seating capacity available, assigning more of these aircraft to the flight schedule. Potential revenue increases by flying aircraft more hours per day. Unfortunately, as with the two previous examples, this tactic may or may not be successful. If aircraft operate more hours per day, then some flights will operate in periods of lower demand. Adding flights in the early morning hours or late in the evening or at night will substantially increase aircraft utilization and ASMs, but flights that depart during these hours are not popular with most travelers. To encourage passengers to choose a "red-eye" flight, prices must be very attractive. Yield falls and total revenue may or may not increase. Alternatively, an airline might consider reducing ASM to boost yield. However, if other airlines don't also reduce or, at least, maintain their capacity, the benefits for a single carrier are likely limited. The US airline profit streak before the pandemic was attributed to "capacity discipline," that is, reduced ASM growth by the largest airlines. Colluding to restrain capacity is illegal in most developed economies. In the fall of 2015, the US Department of Justice launched an investigation of the Big-4 airlines for illegally colluding to restrain capacity growth. Most analysts and industry observers felt the allegation was baseless (Karp and Walker, 2015). Indeed, the DOJ eventually dropped the inquiry.

Adjusting ASMs will also affect cost. One way to increase ASMs is to fly more hours per day. This spreads the fixed costs, such as monthly aircraft lease fees, across more ASMs. Unit cost (CASM) will fall but total costs will increase. The decrease in CASM will offset some of the lower yield required to fill more seats. Another way to increase ASMs is to put more seats in the aircraft, typically by decreasing the legroom in economy class. The trade-off here is between passenger comfort and price. Less comfort means the airline may have to lower prices. As noted in Chapter 4, LCCs employ high aircraft utilization and high-seat density to lower CASM.

An airline has many other tools to manage costs (and CASM). Before examining cost structure, we consider another facet of revenue generation.

7.4 Ancillary Revenue

Maximizing profits by manipulating the variables in the profit equation is a challenge that offers no single or easy solution. Especially since LCCs have become a major competitive force, airlines have turned to increase revenues beyond managing price (yield), load factor, and supply (ASM).

In addition to passenger ticket revenue, airlines also earn substantial revenue from ancillary sources. Airlines have long charged for beverages and meals in coach class. As with restaurants, sales of liquor are especially profitable. Low-cost carriers were the first to recognize the potential for ancillary revenue after Spirit Airlines began charging for checked baggage in 2007. LCCs aggressively tacked on fees for ever more services typically, including luggage (both checked and carry-on), priority boarding, assigned and preferred seating, and inflight entertainment. Airlines have also aggressively prompted sales of other travel-related services, including rental cars, hotels, and entertainment bundled with the purchase of the ticket.

Most airlines have adopted some form of *menu pricing* from which a passenger may choose to purchase services that were once complimentary. As a result, worldwide ancillary revenue increased rapidly from $2.45 billion in 2007 to nearly $110 billion in 2019 before falling precipitously. By 2020, ancillary revenues contributed a stunning portion of total revenues – more than 55% for Wizz Air and Spirit Airlines. Figure 7.9 shows ancillary revenue as a percentage of 2020 total revenues for a sample of airlines (Sorensen, 2021).

The world's largest carriers and a few others with large frequent flier programs, sell mileage points to third parties, mostly credit card companies, who then award the miles as customer incentives. The airline incurs a cost

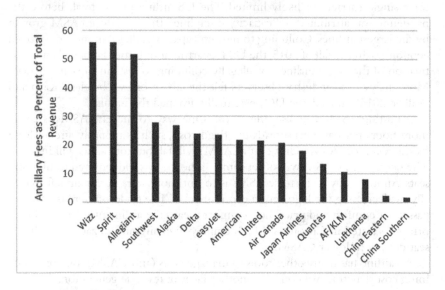

Figure 7.9 Ancillary Revenues Percentages for Sample Airlines.
Source: IdeaWorks.

when the miles are redeemed but the practice can contribute as much as 75% or more of the total ancillary revenue the carriers earn. Carriers that post a loss on passenger revenue alone can ameliorate that loss with ancillary revenue and improve their unit revenue (RASM). In recent years, many carriers have increased their reliance on ancillary revenue to improve their bottom lines, sometimes making the difference between an overall profit and loss for the carrier. In 2021, American Airlines' Advantage loyalty program contributed $2.167 billion to AA's total passenger revenue (about 7% of the total revenue), and another $2.166 billion in ancillary revenue (87% of the total ancillary revenue), overall, equating to 15% of their total operating revenue. Although this helped, their RASM was still slightly less than CASM and they posted an operating loss of about $1 billion (American Airlines, 2022).

Table 7.1 shows 2020 ancillary revenue data for the sample of airlines. FFM sales were not available for all airlines. Note that the Chinese airlines have taken little advantage of ancillary revenues.

In response to passenger objections to ancillary fees, government legislators and administrative agencies have periodically proposed regulations. In 2008, the European Commission enacted regulations requiring fare transparency (European Commission, 2013) pointing to Ryanair's opaque pricing. Not having entirely mended its ways, Ryanair was subsequently fined $521,000 by Italian regulators for failing to clearly and fully communicate "from the first contact with the consumer in such a way as to make the final price immediately clear" (Italian watchdog, 2013). Similar concerns resulted in an investigation by

Table 7.1 Ancillary Revenues for Sample Airlines

Airline	Ancillary Revenue (in mil)	Percent of total revenues	Average ancillary revenue per passenger	FFM Revenue (in mil)	Percent FFM revenue of total ancillary revenues
American	$3,782	21.8%	$39.81	$2,890	76
Delta	3,676	24	52.41	2,800	76
United	3,320	21.6	57.47	2,539	76
Southwest	2,552	27.9	46.63		
AF/KLM	1,262	10.7	43.70		
Spirit	1,009	55.8	54.72		
Alaska	961	26.9	53.59		
Air Canada	892	20.8	64.86	505	57
easyJet	881	23.5	13.31		
Japan Airlines	801	17.9	63.79		
Lufthansa	737	8.0	26.23		
Qantas	546	13.5	59.36		
Allegiant	513	51.8	59.49		
Wizz	465	55.9	45.69		
China Southern	224	1.7	2.31	168	75
China Eastern	189	2.3	2.53		

Note: FFM revenue is not available for several airlines.

Source: data for 2020 from IdeaWorks.

the US Senate Committee on Commerce, Science, and Transportation of US airline practices. The Committee made seven recommendations:

- Ancillary fees should be standardized and disclosed early in the booking process.
- Checked and carry-on baggage fees should have a clear connection to the costs incurred by the airline.
- Airlines should promptly refund fees for any checked bags delayed more than six hours on a domestic flight.
- Airline change fees should be limited to a reasonable amount tied to lead-time before departure and a maximum percentage of the original fare paid.
- Airlines should provide clear disclosures that "preferred seat" charges are optional.
- Airline and travel websites should have a clear and conspicuous link to the Department of Transportation's Aviation Consumer Protection website.
- The Department of Transportation should update its aviation consumer protection website to improve the consumer experience. ("The Unfriendly Skies," 2015)

Airlines for America president Nicholas Calio responded to the committee's report, saying that optional services are a positive development for both consumers and airlines have an incentive to transparently market their optional services (Calio, 2015).

More recently, US congressional Democrats introduced the Forbidding Airlines from Imposing Ridiculous Fees Act (FAIR) to limit fees placed on services such as checked bags, seat selection, and ticket changes. Although the FAIR Act didn't become law, proposals to regulate airline pricing arise periodically.

7.5 Cost Structure

Next, we look at airline costs in detail. Federal laws require that US airlines submit cost data to the Department of Transportation on a monthly, quarterly, and annual basis, so we have extensive data for US carriers. Airlines report costs by administrative area as shown in Figure 7.10 for the fourth quarter of 2021. *Labor* and *fuel* are the two largest expenses comprising 50% of all costs. *Rents and ownership* represent the capital cost of all of the airlines' physical assets of which aircraft are more than half. The fourth-largest cost category is titled *transport-related,* not a helpful label. The largest component of this category is the cost of regional airline flights operated on behalf of FSNCs. *Professional services* is a surprisingly large category that includes outside legal counsel, auditing, engineering, and consultant fees. All other expenses ranging from catering to advertising and insurance make up the remaining 15%.

Following the categorization in Figure 7.10, the next sections explore the tools available to manage labor, fuel, and ownership costs.

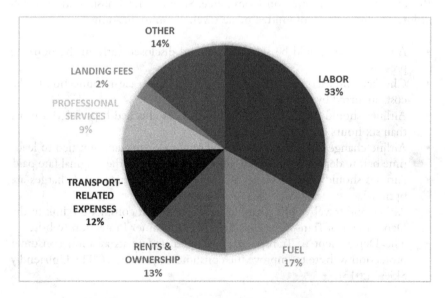

Figure 7.10 US Airline Cost Percentages for the 4th Quarter of 2021.
Data source: A4A Passenger Airline Cost Index, 4th quarter 2021.

7.5.1 *Labor*

Labor is the highest cost category although occasionally eclipsed by high and volatile fuel prices. A spike in oil prices in 2008 before the Great Recession drove fuel costs to nearly twice that of labor. Labor costs vary greatly by world region with western-European airlines generally facing the highest costs. Airlines in the developed economies are highly unionized and labor fiercely opposes attempts to increase productivity or reduce wages, so for much of airline history, management regarded labor costs as uncontrollable (see Chapter 6 for a discussion of airline labor relations). The recession of 2001 hit the US airlines harder than those elsewhere. Under dire circumstances, US legacy airlines restructuring altered the dynamics of labor/management relations. We begin by considering the actions of US airlines to reduce labor costs and improve productivity.

In any major recession, airlines reduce flights and furlough employees. Figure 7.11 shows a sharp drop in US airline employment beginning in the recession of 2001 and continuing to 2007. A slight recovery was then cut short by the Great Recession of 2008. Employment bottoms out in 2010, followed by robust hiring and employment growth until the pandemic of 2020 when employment again plummets. Available seat miles, however, do not follow the employment pattern but rather resume steady growth after both recessions, again until the pandemic of 2020 upends the entire industry. Airlines succeeded in producing more flights with fewer employees.

Producing more available seat miles with fewer employees increases employee productivity. In Figure 7.12, labor productivity is measured by the

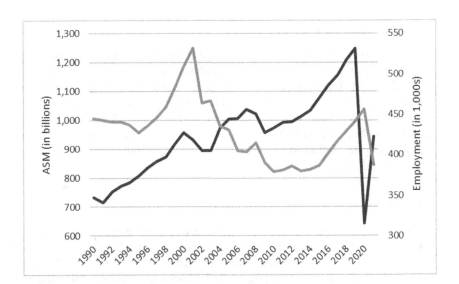

Figure 7.11 US Airline ASM versus Full-time Equivalent Employees.
Data sources: Airlines for America; Bureau of Transportation Statistics.

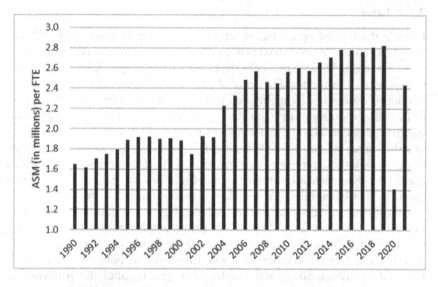

Figure 7.12 US Airline Labor Productivity Measured by ASM per Full-time Equivalent Employee.

Data source: Bureau of Transportation Statistics.

ratio of ASMs to employment. During the first two decades of the new millennium, airline labor productivity increased by 50%.

7.5.1.1 *Increasing Labor Productivity*

Airlines pursued many avenues to increase productivity. Automation, such as passenger check-in kiosks, replaced some employees. Other work was outsourced to third parties. Work rules were revised to increase the hours worked by limiting time off and vacations. The number of employees per task was reduced wherever possible. For example, to provide high-quality inflight service, flight attendant staffing at most FSNCs was higher than required by regulation. In the restructuring that followed the recession of 2001, staffing was reduced to the regulatory minimum.

7.5.2 **Fuel**

Global supply and demand determine the price of oil and, consequently, the base price of jet fuel. Local prices, however, vary considerably due to taxes and transportation costs. Following a decade of relative stability in the 1990s, fuel prices have been volatile since (Figure 7.13). Although difficult to predict, fuel prices tend to follow the world economy rising during economic expansion and falling in recessions. Note the precipitous drop during the Great Recession of 2008. As many economies quickly recovered from the pandemic, fuel prices

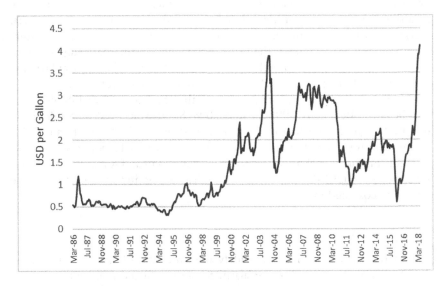

Figure 7.13 Historical Jet Fuel Prices.
Note: US Gulf Coast kerosene jet fuel spot price FOB in USD per gallon.
Data source: US Energy Information Administration.

leap to record highs. If efforts by central banks to reduce worldwide inflation in 2022 lead to a recession (as many economists predict as of this writing), fuel prices will drop again. Airlines can take some solace in this pattern because the lower prices often accompany a drop in demand for air travel. Nonetheless, airlines have limited ability to control this second-largest expense category.

7.5.2.1 Fuel Efficiency

As with labor, technology offers the best defense against rising fuel costs. Faced with rapidly escalating fuel prices in the first decade of the century, airlines ordered new aircraft with improved fuel efficiency in record numbers. When All Nippon Airways replaced older jets with the Boeing 787, it reported fuel savings of over 20% on long-range flights (ANA News, 2012). Jet engine manufacturers have steadily improved fuel efficiency. Fuel efficiency has improved by 70% over the past 40 years. The Airbus A320neo and Boeing 737 Max, introduced into service in 2016 and 2017, respectively, are 15% to 20% more fuel-efficient than the previous generation of the same types.

7.5.2.2 Fuel Hedging

The greater fuel efficiency of new aircraft lowers fuel costs. Fuel hedging, in contrast, is a financial tool for protecting an airline against fuel price volatility. Hedging locks in the price of the airline's future fuel purchases, essentially providing an insurance policy against fuel price increases. To hedge fuel prices, an airline purchases one or a combination of future contracts in the financial markets. Whatever investment instrument it uses, the airline incurs a fee (insurance premium, if you will). If the price of fuel goes up, the fuel hedge will save the airline money. However, if the price of fuel falls, the airline will still pay the higher hedged fuel price, as well as the cost of the instrument itself.

Southwest Airlines was for many years the most aggressive US airline in fuel hedging. Between 1999 and mid-2008, Southwest saved $3.5 billion on fuel by hedging. Then fuel prices plummeted in late 2008 (see Figure 7.13) costing Southwest a billion dollars on hedging contracts. Hedging is essentially a gamble on future fuel prices; sometimes airlines win and sometimes they lose. Lim and Hong (2014) found that hedging had no significant impact on airline operating costs and profits.

7.5.2.3 CASM Ex-fuel

CASM (or CASK) is the most widely used measure of airline costs. Rising CASM is sometimes a cause for alarm for airline financial analysts as it may indicate managers are unable to control costs. However, because fuel costs are mostly outside of management control, modified CASM metrics are commonly reported. The CASM excluding fuel cost is a better indicator to investors and other interested parties of how well the airline can manage costs that are within its control. Transport-related expenses can also be misleading for FSNCs, so these costs are often stripped from CASM data to provide valid comparisons.

Figure 7.14 compares the CASM less fuel and transport-related costs for US FSNC, hybrid, and LCCs from 2000 through to 2020. As we would expect from our study of these three business models in Chapter 4, FSNCs have the highest CASM and LCCs the lowest. The LCC cost dropped below the hybrid carriers as Spirit Airlines adopted its self-titled ultra-low-cost carrier model. As with most longitudinal financial data, the effects of the pandemic in 2020 are striking.

7.5.3 Ownership and Rental Expenses

Airlines require extensive capital goods – aircraft being the most obvious. Firms acquire capital goods with internally generated funds, by borrowing, or by selling equity. Capital structure is the combination of debt and equity used to finance operations and growth. Debt includes loans, some types of equipment leases, and other types of credit that must be repaid in the future with interest. Equity, obtained by selling corporate stock, whose value increases with the firm's potential future profits, does not obligate the firm to make payments to shareholders. Profitable firms often pay shareholder dividends.

A company's debt-to-equity (D/E) ratio is long-term debt divided by shareholder's equity. Airlines have continued to grow, but profits are cyclical and often meager. Without steady profits, the ability to finance growth through equity is limited, so how have airlines afforded to grow? The answer is often

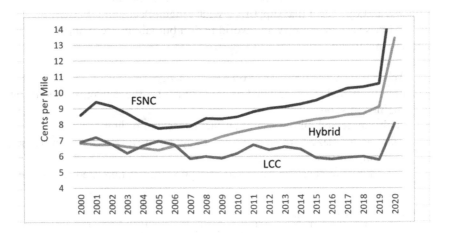

Figure 7.14 Cost per Available Seat Mile ex Fuel and Transport-Related.

Notes: FSNCs: American, Delta, and United; Hybrids: Southwest, JetBlue, Alaska, and Hawaiian; LCCs: Allegiant, Frontier, and Spirit.

Data source: MIT Airline Data Project.

ever-increasing debt. Adding debt increases the D/E ratio. A high debt-to-equity ratio, also known as high financial leverage, increases the volatility of earnings and is one of the reasons for the highly cyclical nature of airline profitability. Transportation companies typically have higher D/E ratios than firms in other industries and airlines have among the highest debt levels. Airlines can obtain loans and, thus, have higher D/E ratios than firms in other industries because aircraft can be mortgaged or used as collateral. Although it would be difficult for a creditor to repossess and resell a factory, highly mobile aircraft can be repossessed and sold or leased to new operators.

In periods of recession, firms with high debt levels may struggle to meet their repayment obligations. A firm that defaults on its debt payments can be forced into bankruptcy. As part of the bankruptcy proceedings, lenders usually agree to reduced loan repayments. The inability to make loan payments forced all major US airlines into bankruptcy at some time since deregulation. Because of the risk imposed by high D/E ratios, financial analysts look closely at an airline's ability to meet future payment obligations. Since emerging from bankruptcy, the US Big 3 airlines used their profits to pay down loans and reduce the D/E ratios until the pandemic struck.

7.5.4 Taxes

The airline industry is subject to heavy taxation adding to costs. Of course, some taxes directly support infrastructure and services essential for commercial aviation, but politicians often view the airline industry as a ready source of revenue to fund non-aviation projects. The International Air Transport Association argues that burdensome aviation taxes are counterproductive because they suppress air commerce that supports a wide range of economic activity (IATA, 2015). Taxes vary by country and world region but fall into several general categories, including:

- An air passenger duty leveled on each passenger.
- Environmental taxes intended to compensate for the damage caused by jet engine emissions.
- Fuel taxes levied on the sale of jet fuel per gallon or liter.
- Value-added taxes are applied in many countries to most purchases, including air travel.
- Special purpose taxes are imposed to support causes unrelated to aviation.

7.5.4.1 US Aviation Taxes

US taxes provide an example of the magnitude and bewildering array of airline taxation. Taxes have been increasing since the early 1970s and jumped dramatically in 2001 when the federal government assumed the responsibility for passenger screening. US airline ticket taxes fall under three main groups: Airport & Airway Trust Fund (AATF), Passenger Facility Charges (PFC), and Homeland Security Fees. Environmental Protection taxes, a fourth category, are minimal in the US but can be substantial elsewhere.

Airport & Airway Trust Fund was established in 1970 to fund the Federal Aviation Administration's (FAA) investments in airports and the airways system. The primary source is passenger taxes, but the administration also collects excise taxes from freight and fuel. The AATF provides nearly 95% of the FAA funding.

Passenger Facility Charges were added in the early 1990s. These taxes, currently capped at $4.50, are collected by airlines and paid to local airport operators to fund federally approved capital improvement projects at the airport ("Passenger Facility Charges," n.d.).

Before 2001, airlines directly funded airport security. The federal government assumed this responsibility after 9/11. Federal Security Fees fund the Transportation Security Administration's airport operations.

In many world regions, airports and air traffic control are privatized; airlines pay user fees rather than taxes for these services.

Table 7.2 illustrates the taxes on a hypothetical round-trip domestic flight on a US FSNC from Peoria, Illinois (PIA) to Raleigh/Durham, North Carolina (RDU) with a connection at Chicago O'Hare (ORD).

In this example, taxes add 28% to the base fare. The Department of Transportation prohibits airlines from advertising a flight's base fare only – advertised fares must include all taxes. However, when purchasing a ticket, the base fare and taxes are listed separately. In this example, connecting through ORD adds $18 to the total taxes, so an LCC with point-to-point service from PIA to RDU could have advertised an even lower fare.

Table 7.2 Taxes on a Hypothetical US Domestic Flight

Base Airline Fare	*$ 235.16*
Federal Ticket (Excise) Tax (7.5%)	17.64
Passenger Facility Charge (PIA)	4.50
Federal Security Surcharge (PIA-RDU)	5.60
Federal Flight Segment Tax (PIA-ORD)	4.50
Passenger Facility Charge (ORD)	4.50
Federal Flight Segment Tax (ORD-RDU)	4.50
Passenger Facility Charge (RDU)	4.50
Federal Security Surcharge (RDU-PIA)	5.60
Federal Flight Segment Tax (RDU-ORD)	4.50
Passenger Facility Charge (ORD)	4.50
Federal Flight Segment Tax (ORD-PIA)	4.50
Total Taxes	64.84
Total Ticket Price	$300.00
Taxes as a percent of the base fare	27.6%

Source: Airlines for America.

For an international flight, US Customs and Immigration collects additional taxes of more than $50. The foreign country also imposes similar taxes that vary greatly and are often much higher.

7.5.4.2 Cost of Carbon Emissions

The US does not currently impose a direct tax on jet engine carbon emissions, but the European Union does tax emissions and is pressuring other countries to follow suit. Since 2012, airline CO_2 emissions have been subject to the Emissions Trading System (EU ETS). All airlines operating in Europe, regardless of nationality, must report their emissions. Each airline receives a decreasing yearly allowance; if emissions exceed the allowance, the airline must purchase allowances to cover the excess through the ETS. In 2022, the EU adopted restrictions on airline emissions designed to help meet the goal of cutting net emissions by 55% by 2030 from 1990 levels. Among several steps, the policy would phase out the current free CO_2 allowances by 2026 resulting in a higher tax on emissions. The policy also requires airlines to begin blending sustainable aviation fuels with existing jet fuel beginning in 2025 reaching 63% blend by 2050. While these policies will evolve, airlines face substantially higher fuel costs as emission restrictions become more stringent in many world regions.

7.5.5 Fixed and Variable Costs

Depending on the type of management decision required, other methods for classifying costs can be useful. As the name suggests, *variable costs* change with the number of flight hours. Fuel cost, for example, increases when the airline flies more hours per day. Conversely, *fixed costs* do not vary with flight operations. Rents and ownership fees for aircraft and facilities are still due, whether aircraft fly a little or a lot. *Escapable cost* is a closely related concept that is particularly useful if the airline wishes to reduce costs, for example, during a recession. Faced with lower passenger demand, the airline will choose to reduce flight frequency and may drop some routes entirely. Fuel and some portion of flight-crew wages paid by the flight hour are immediately avoided; however, most other expenses must still be paid in the short run, perhaps for many months. Escapable costs increase with time. Employees can be furloughed relatively quickly but leased aircraft can be returned to the lessor only when the leases expire. Airport rental fees remain for the duration of the contract. If the airline cannot return to profitability, it may ultimately cease flying and liquidate all assets, a process of several years. At that point, all costs are escapable.

7.6 Economics of Scale, Scope, and Density

Three economic concepts provide a different perspective on airline economics: economies of scale, scope, and density.

7.6.1 Scale

Scale refers to the size of the business. In the airline industry, scale is measured by available seat miles and other similar metrics. Economies of scale exist when larger firms, those with more assets and greater production capacity, can produce at a lower unit cost than smaller rivals. If there are only a few small companies in an industry, this is strong evidence of economies of scale. Commercial aircraft manufacturers are one example. Only Airbus and Boeing produce large commercial aircraft, although China may enter the market soon. There were once several other manufacturers, but these either quit the business as did Lockheed Martin or were absorbed as was McDonald-Douglas by Boeing. In sharp contrast, the largest airlines usually have the highest unit costs. Most LCCs are smaller than FSNCs but enjoy lower CASM. This cost advantage has allowed LCCs to capture market share from competing FSNCs.

7.6.2 Density

Density refers to the demand per market or city-pair. Unlike economies of scale, airlines benefit from route density. On high-density routes, airlines can utilize larger aircraft with lower CASM. The hub-and-spoke system creates route density by combining passengers bound for many cities on a single flight, so network carriers benefit from density.

7.6.3 Scope

The third economy is that of scope. Economies of scope exist when the unit cost is reduced by producing several related products. This contrasts with economies of scale where unit cost is reduced by producing more of a single product. Airlines benefit from economies of scope in two ways. First, scope is the extent of geographical coverage and the number of destinations where each city-pair is a different product. Here, economies of scope are evident. Significant marketing advantages arise from offering many destinations, especially through a hub-and-spoke system. Passengers favor an airline with a large network, ideally allowing them to fly from any origin to any destination; hence, the attraction of the hub-and-spoke system. Airlines can increase their scope, especially internationally, by joining alliances with other airlines. Second, classes of service are also different products targeted to different passenger segments. FSNCs offer two or more classes of service on most flights. By combining demand from several passenger segments on a single flight, the cost of producing each product is reduced. Finally, another jointly produced product is belly cargo accommodated on the same flight potentially reducing the unit cost of cargo.

7.7 Legacy Carrier Restructuring

Since deregulation and liberalization of airline markets, legacy carriers have struggled to become more efficient and productive in the face of increased competition, often from new and rapidly growing LCCs. Although LCCs were not spared, the two severe recessions of the first decade of the 21st century were especially cruel to US legacy carriers.

The recession of 2001 hit North American carriers hard, forcing widespread restructuring culminating in mergers that created the Big-3 US FSNCs. American, Delta, and United plus Southwest controlled two-thirds of US domestic traffic in 2019 (MIT Airline data). US airlines had long focused on growth and capturing market share rather than profitability. With the increased concentration in the US airline industry, airline chief executives turned their attention to long-term profitability. In sharp contrast to the past, capacity growth was restrained to maintain higher prices. This *capacity discipline* led CEOs to predict that US airlines had broken the boom and bust cycle and would henceforth always be profitable. Doug Parker, then American Airlines' CEO, insisted in 2016 that the post-consolidation, post-bankruptcy US airline industry was structured for consistent profits (Karp, 2016). Although no one could have foreseen the pandemic, the industry optimism was premature; the historical cycle of profits and losses continues.

The 2001 recession was not as harmful to carriers outside North America, but the Great Recession of 2008 affected carriers worldwide. As in the US, growing competition from LCCs and the two recessions left three FSNC groups in Europe. Consolidation began with the merger of Air France and KLM in 2004. British Airways followed in 2011 in a merger with Iberia creating the International Airlines Group. Vueling and Aer Lingus have since been added to its stable. The Lufthansa Group is more complex and forever evolving. In addition to the flagship Lufthansa, it includes Swiss International Airlines, Austrian Airlines, Brussels Airlines, Eurowings, and other airline subsidiaries. In Asia, Japan Airlines successfully restructured through bankruptcy in 2009. It emerged as a more efficient airline having retired much of its older fleet and a smaller workforce. China's Big-3 is also the result of a series of consolidations. As of 2022, major mergers in Korea, Africa, and Latin America are possible.

Of course, restructuring is not new; every struggling airline attempts to revise its business model to stave off liquidation and return to profitability. The steps in airline restructuring are well known and similar to those undertaken by loss-making firms in other industries. Each airline-restructuring program is different but includes some or all of the following:

- Dropping unprofitable routes and simplifying the network.
- Reducing the number of fleet types and retiring older aircraft.
- Increasing utilization of the remaining aircraft sometimes by "depeaking" hubs.
- Outsourcing non-core tasks, particularly aircraft maintenance, and transferring thinner domestic routes to regional partners.
- Reducing staffing through employee buyouts, attrition, and involuntary furlough.
- Renegotiating or imposing concessionary work agreements with lower wage rates and improved productivity.
- Renegotiating aircraft lease agreements to obtain lower lease rates.
- Reducing debt through negotiation with lenders often facilitated by bankruptcy.
- Merger, acquisition, or partnership.

7.8 The COVID-19 Pandemic

The airline industry has survived repeated shocks in the 21st century, including terrorism, the SARS pandemic, volcanic eruptions, widely fluctuating oil prices, and two major recessions. Grave as these were, nothing compares with the industry ruination occasioned by the 2020 COVID-19 pandemic. Global flights fell by 50% in April and May. In several countries, 90% of flights were grounded for two or more months. Many figures in this chapter show the devastating results on profits (7.1 and 7.2), traffic (7.6, 7.7, 7.8, and 7.11), and cost (7.14). Recognizing that many airlines were in peril, governments provided unprecedented financial support ranging from just one airline of national importance to programs available across a nation's airline industry (Martin-Domingo and Martin, 2022). The magnitude of the disruptions required quick action, so it's not surprising that the type of government interventions varied widely. Adbate, Christidis, and Alloysius (2020) identify seven archetypes: (1) government-backed commercial loans and government guarantees; (2) recapitalization through state equity; (3) flight subsidies; (4) nationalization; (5) deferral and/or waiver of taxes and charges; (6) grants; and (7) private equity. Of these, government loans and guarantees were most widely employed.

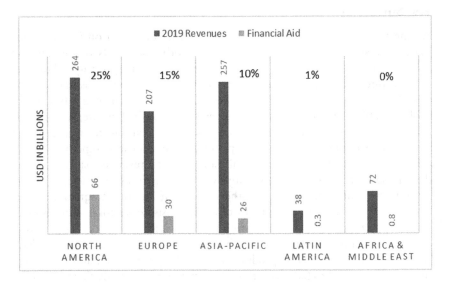

Figure 7.15 Government Airline Aid by World Region.
Data source: Pearce, 2020.

Figure 7.15 shows the amount of government support compared to 2019 airline revenues. Support was highest in North America at 25% of revenues totaling $66 billion dwindling to little or no support in Latin America, Africa, and the Middle East (Pearce, 2020). The aid usually came with restrictions. For example, France provided the highest percentage of support with about $8 billion going to Air France, and in return, the government increased its ownership stake from 14% to 30%. Air France was required to relinquish slots at Paris Orly Airport to competing airlines and banned from paying bonuses to executives, paying stock dividends, and share buybacks.

Despite the apprehension at the beginning of the pandemic, the number of large airline failures in 2020 and 2021 was not historically high. Skift (2021) lists nine, relatively small airlines that ceased operations in 2020 plus 10 others that filed for bankruptcy, numbers that do not differ markedly from previous years. Several of the significant bankruptcies are discussed in Chapter 4, including AirAsia Japan, LATAM, Aeromexico, Virgin Atlantic, and Norwegian. The US regional airlines Trans States and Compass and Singapore Airlines' subsidiaries Silk Air and NokScoot ceased operations as part of restructuring by their FSNC partners, actions that may have been taken even without the pandemic. Direct government aid to airlines and broader programs to support all industries were partly responsible for this better-than-expected outcome. Airline executives, however, quickly cut costs and downsized as the pandemic unfolded, timely steps that saved many airlines.

7.9 Summary

Airline profits as measured by net profit on sales or return on invested capital have historically lagged behind firms in other segments of the aviation industry and other unrelated industries. Profits, which generally follow the world economic cycles of boom and recession, have grown increasingly cyclical with large swings between profits and losses. The airline manager has several tools available to increase total revenue and thus profits but intense competition often makes higher prices counterproductive. Raising the average ticket price, for example, will lead some passengers to choose competing airlines offering lower fares. Load factor will fall and total revenue may rise or fall depending on the overall price sensitivity of passengers. The manager has more options for controlling costs, the other variable in the profit equation. Labor and fuel are the largest components of airline costs comprising about 50% of the total, so these receive great attention. Faced with two severe recessions in the new millennium, many airlines, beginning with US legacy carriers, have successfully increased labor productivity and lowered wages, although these improvements tend to fade with time. As they struggle with LCC competition, old-line, legacy carriers worldwide periodically restructure to survive. Fuel prices are determined in world markets and beyond the control of managers, but can be mitigated with new fuel-efficient aircraft and stabilized over the short term by fuel hedging. The COVID-19 pandemic forced some airlines to reduce schedules by 95% during spring 2020. Many governments, fearing the collapse of their national airlines, provided financial aid and support that cushioned the pandemic shock.

7.10 Case Study: The Turbulent History of Aer Lingus

The Irish national airline Aer Lingus was founded in 1936 to connect Ireland with major European cities. Like most other European flag carriers, Aer Lingus was either fully or partially government-owned and controlled for the next 70 years. Ireland's small population, predominately-agricultural economy, high marginal tax rates, and violence in Northern Ireland limited demand. The airline also suffered all the symptoms and characteristics of "Distressed State Airline Syndrome." These included strong unions, overstaffing, high fares, poor service quality, bureaucratic management, and constant government interference. Not surprisingly, the airline was rarely profitable.

With the support of the Irish government and limited competition allowed under international airline economic regulation, Aer Lingus expanded after World War II. By 1986, however, its high fares attracted public scrutiny. In response, Ireland became the first European country – 11 years before EU airline deregulation – to allow a new entrant carrier to compete with its national airline on its major route. Ryanair entered the Dublin–London route undercutting Aer Lingus' fare by more than half. With its low-cost carrier business model incorporating low-cost/low price, point-to-point route structure, no-frills product, high efficiency, enhanced productivity, and high asset utilization, Ryanair profited as Aer Lingus languished.

In the first of several severe economic shocks, the Gulf War of the early 1990s plunged Aer Lingus into a deep financial crisis. Losses mounted and debt ballooned. To staunch the losses, management sold or dissolved a host of non-core businesses and furloughed 1,500 employees. The Irish government injected 175 million Irish pounds while simultaneously discrediting the Aer Lingus board for their irresponsible management.

Less than a decade later, catastrophic events of the early 2000s included scares over foot-and-mouth disease in the United Kingdom and Ireland, the stock market crash in 2000, and the 2001 terrorist attacks. The heretofore-profitable North American operations plunged into disarray. Meanwhile, LCCs competed in over half of Aer Lingus' short-haul markets out of its Dublin hub airport. The airline was losing €2.5 million per day by November 2001; its financial situation was again rapidly deteriorating. Another restructuring was at hand.

Willie Walsh, later to head British Airways and the International Airlines Group, took the helm of Aer Lingus in October 2001. With prices driven down by LCC competition, it had to reduce costs. Walsh envisioned a "strategic fit" with Aer Lingus adopting as many low-cost features as possible while retaining some essential differentiators to compete more effectively with LCCs. Among other elements, his plan included fare reduction, higher labor productivity, a single fleet type, and moving to online distribution. Walsh cut employment by 2,000, earning the ire of the unions. This survival plan, which aimed to reduce costs by 30%, succeeded in transforming a loss of €139 million in 2001 to a profit of €69 million by 2003. Walsh also pushed for a public stock offering – an initial public offering (IPO) – with the proceeds going to finance a fleet expansion and a one-time contribution to the airline's pension fund. The government rejected the IPO proposal only to approve it one year later after Walsh resigned.

Meanwhile, Ryanair sought to rapidly expand by acquiring Aer Lingus. Public trading of the stock offered a means for the acquisition. Ryanair made three attempts to purchase Aer Lingus in 2006, 2008, and 2012, only to have each rebuffed by the airline's board of directors, the Irish government, and the European Commission, which ruled the merger would reduce competition.

Aer Lingus added new bases and routes in a period of intense growth only to be slammed once again, this time by the Great Recession that began in the US in 2008 and spread rapidly across the world. As in the US, permissive bank lending to the domestic property market had propelled the Irish economy. The bubble burst and the economy collapsed leading to a bailout from the International Monetary Fund, the European Central bank, and the European Commission. Aer Lingus suffered large net losses in 2008 and 2009. With its cash balance dwindling, Aer Lingus' chairman declared that it was facing the most difficult period in its history; the Irish flag carrier was again fighting for its life.

Another in the succession of CEOs, Chris Mueller, set about in mid-2009 to rescue the airline. He envisioned, not unlike Walsh, a new hybrid model combining the best elements of LCCs with the option to upgrade to features of a full-service airline. Cost savings included 863 voluntary job redundancies plus pay cuts and freezes. Management moved closer to the LCC product by unbundling fares and stressing ancillary revenues. For long-haul flights, however, upgraded business–class amenities, including lay-flat seats, were added to compete with American and Delta.

The biggest innovation was the development of a hub-and-spoke system connecting Europe and the US for which Dublin was geographically ideally suited. Laboring under the Irish government's direction, Aer Lingus for decades operated as a point-to-point airline to serve only Ireland. Aer Lingus was the Irish vehicle for immigration, tourism, and trade, but not to serve passengers traveling between Europe and the US as Finnair and TAP Air, similar midsized European airlines, were already doing. Following those examples, the new management built the Dublin hub to efficiently connect traffic between Europe and North America.

In 2015 after months of negotiations, the Irish government and Ryanair agreed to sell their remaining stock holdings to International Airlines Group (IAG). IAG's chairperson was then Willie Walsh who had joined Aer Lingu as a pilot cadet when he was 17. He added his first airline to the International Airline Group.

Turbulence returned as COVID-19 pandemic-related travel restrictions forced Aer Lingus to cut 95% of its flights by mid-2020. In a replay of earlier financial crises, another new CEO, Lynne Embleton, secured government-backed loans and closed its Shannon base. Embleton told employees in May 2021, that Aer Lingus had too many resources and needed to right-size and right-shape its structure. Shortly thereafter, Stobart Air, its regional airline partner, succumbed to the pandemic and ceased flying. By late 2021 the worst of the pandemic passed. Aer Lingus began adding back domestic and international flights. It selected a new regional carrier, Emerald Airlines, to support Aer Lingus hub-and-spoke network.

Sources: This case study draws from O'Connell and Connolly, 2017; other sources include O'Keeffe, 2004; "Aer Lingus," 2010.

Concepts: Early airline history (Chapter 1); air travel demand (Chapter 2); point-to-point and hub-and-spoke route structures (Chapter 3); full-service, hybrid, low-cost carriers, and regional airlines (Chapter 4); cost structure and control (Chapter 7), international airline regulation and state-owned airlines (Chapter 11).

References

American Airlines. (2022). Form 10-K. https://americanairlines.gcs-web.com/static-files/deb58cec-bf67-418f-a0f0-022fb75939fc

Aer Lingus – kissing goodbye to simplicity? (2010, March). *Aviation Strategy, 149*, pp. 5–9.

ANA News. (2012, November 12). *ANA Celebrates First Anniversary of Boeing 787 Services* [Press release]. www.ana.co.jp/eng/aboutana/press/2012/pdf/121101.pdf

Abate, M., Christidis, P., & Alloysius, J. P. (2020, October). Government support to airlines in the aftermath of the COVID-19 pandemic. *Journal of Air Transport Management*, 89. https://doi.org/10.1016/j.jairtraman.2020.101931

Calio, N. (2015). Letter to the Honorable Bill Nelson, August 12, 2015. http://airlines.org/wp-content/uploads/2015/08/8-12-15-Sen-Nelson-Report-Ltr.pdf

CAPA. (2021, December 16). *Airbus deal gives Qantas commonality and range advantages.* https://centreforaviation.com/analysis/reports/airbus-deal-gives-qantas-commonality-and-range-advantages-591244

Clark, P. (2010). *Stormy skies: Airlines in crisis.* Burlington, VT: Ashgate.

Doganis, R. (2010). *Flying off course: Airline economics and marketing* (4th ed.). New York: Routledge.

European Commission. (2013, March 13). Air passenger rights revision – frequently asked questions. http://europa.eu/rapid/press-release_MEMO-13-203_en.htm

Karp, A. (2016, July). Fundamentally changed. *Air Transport World*, 53(7), p. 51

Karp, A., & Walker, K. (2015). Claiming collusion. *Air Transport World*, 52(8), p.14.

Lim, S. H., & Hong, Y. (2014). Fuel hedging and airline operating costs. *Journal of Air Transport Management*, 36, 33–40.

Martín-Domingo, L., Martín, J. C. (2022). The effect of COVID-related EU state aid on the level playing field for airlines. *Sustainability*, 14(4). DOI:10.3390/su14042368

O'Connell, J. F., & Connolly, D. (2017). The strategic evolution of Aer Lingus from a full-service airline to a low-cost carrier and finally positioning itself into a value hybrid airline. *Tourism Economics : the Business and Finance of Tourism and Recreation*, 23(6), 1296–1320. https://doi.org/10.1177/1354816616683492

O'Keeffe, B. (2004, November 20). Aer Lingus restructuring to proceed as planned. *The Irish Times.* www.irishtimes.com/business/aer-lingus-restructuring-to-proceed-as-planned-1.1167289

Passenger facility charges. (n.d.). *Airlines for America.* www.airlines.org/Pages/Passenger-Facility-Charges.aspx

Pearce, B. (2020, May 26). COVID-19 government aid. *IATA.* www.iata.org/en/iata-repository/publications/economic-reports/government-aid-and-airlines-debt/[PPT]

Pension Benefit Guarantee Corporation. (2013). Retrieved from www.pbgc.gov/documents/apbletter/Decision--Delta-Pilots-Retirement-Plan-2013-27-09.pdf

Skift, T. L. (2021, September 4). The Airline Failures and Bankruptcies So Far in the Pandemic. *Skift.* https://skift.com/2021/09/04/the-airline-failures-and-bankruptcies-so-far-in-the-pandemic/

Sorensen, J. (2021). The 2021 CarTrawler yearbook of ancillary revenue. *IdeaWorksCompany.* https://ideaworkscompany.com/wp-content/uploads/2021/09/2021-Ancillary-Revenue-Yearbook.pdf

Swelbar, W. (2010). The future of scope clauses. *Massachusetts Institute of Technology International Center for Air Transportation.* www.faa.gov/news/conferences_events/aviation_forecast_2010/agenda/media/AF%20William%20Swelbar.pdf

Unfriendly skies: Consumer confusion over airline fees. *Committee on Commerce, Science, and Transportation, United States Senate.* (2015). https://fortunedotcom.files.wordpress.com/2015/08/8-6-15-final-airline-report.pdf

Review Questions

1. How do historical airline industry profits compare with other industries?
2. What financial metrics are useful in comparing the profitability of different firms?
3. Define ASM (or ASK), RASM (or RASK), CASM (or CASK), LF, and yield. How might a change in one affect another?
4. How have real and nominal airline prices changed over the past several decades?
5. What is ancillary revenue? What category of airlines realizes the highest percentage of ancillary revenue? Why?
6. What are the two largest airline cost categories?
7. What steps have airlines taken to control labor costs?
8. Can fuel hedging lower fuel costs? Why do some airlines hedge fuel prices?
9. What are the typical steps legacy airlines take in restructuring?
10. How does the EU Emissions Trading System affect airlines?
11. Do airlines benefit from economies of scale, scope, and density?
12. Compare the effects of the COVID-19 pandemic on the airline industry to earlier shocks.

8 Fleet Selection and Acquisition

Upon completion of this chapter, you should be able to:

- Discuss the operational and market factors an airline must consider in selecting aircraft to add to its fleet.
- Compare and contrast the aircraft requirements of a typical LCC and FSNC.
- Describe how aircraft acquisition and operating cost vary with aircraft size, age, and segment length.
- List and explain the various methods an airline may use to finance aircraft acquisition.
- Compare the advantages and disadvantages of aircraft ownership versus leasing.

8.1 Introduction

As an airline grows, its existing fleet ages and the route structure and markets evolve. As a result, the airline's fleet must adjust. In healthy economic environments, the fleet is expanded and upgraded; however, recessions may require downsizing. Commercial aircraft differ substantially in passenger and cargo capacity, range and payload capability, price, and operating costs. Airbus and Boeing dominate the production of larger commercial aircraft with seating capacities above 130 passengers. As of 2022, Embraer was the dominant regional jet manufacturer. In this chapter, we first explore how an airline selects the aircraft it operates to meet the demands of its route structure and passengers. Having selected aircraft that it will add to its fleet, the airline must determine how to finance the acquisition. Fortunately, there are several alternatives ranging from a purchase using operational cash flow to leasing. These options are explored in the second half of the chapter.

8.2 Aircraft Selection

An airline's fleet must have the performance capability to operate the airline's route structure and capacity to meet demand. The A380, the world's largest commercial passenger aircraft, was an engineering triumph for Airbus and

DOI: 10.4324/9781003290308-8

proved very popular with passengers. However, too many airlines concluded that with a capacity of more than 500 passengers, it was too big for the demand on all but a few international routes. Airbus decided in 2021 to end production for lack of sales. It delivered the last A380 to Dubai's Emirates Airlines in December 2021. Just over 250 were produced compared with the 1,000 or more once predicted. The next section considers aircraft performance and cost of acquisition and operations.

8.2.1 Range and Payload

As we have seen in Chapter 3, airline route structures vary in geographical scope. Some airlines serve a limited area such as the continental United States or European Union (plus the UK) whereas the route structure of full-service network carriers typically extends across much of the globe. Routes vary in length from less than a hundred miles and under an hour of flight time to more than 8,000 miles and 16 hours. Similarly, aircraft range capability varies greatly. For example, the Boeing 717, the last version of the venerable Douglas DC-9, is a short-haul mainline jet with an economic range of approximately 1,000 nautical miles. Despite the limited range, it is still in service with Delta Air Lines, Hawaiian Airlines, and QantasLink. In contrast, some models of the Airbus 330, 350, and 380 and Boeing's 787, 777, and 747 have a range capability of over 7,500 nautical miles (see Table 8.1).

An airline's fleet must match the requirements of its route structure. For most LCCs, a single aircraft type will suffice, but an FSNC will require several aircraft types from regional jets to wide-body, long-range aircraft. Conversely, the route structure may evolve with the improved performance of new aircraft. As Airbus 350 and Boeing 787 entered service, airlines expanded their route structure with non-stop international flights in thinner markets that would not previously have been profitable. The long-range model of the Airbus A321, the A321XLR, further extends the ability of airlines to serve even thinner medium-range markets.

Payload – the weight of passengers and cargo – and range vary directly with aircraft size. Larger aircraft can carry more passengers and cargo across greater distances but are more expensive to purchase and operate. There is, however, a trade-off between the weight of fuel carried, and thus range, and passenger and cargo weight. If an aircraft is fueled to capacity, the payload may be restricted by total takeoff weight limitations. Similarly, if an airline chooses high-density, single-class seating for its fleet, then the range may be restricted at high load factors because the fuel load is limited by weight restrictions.

Other factors also affect the payload range trade-off. Many airports are not capable of handling the largest commercial aircraft because of the limited terminal and taxi-way space, inadequate runway length, or weight-bearing capability. The Airbus 380, in particular, requires wider taxiways and tailored terminal facilities. Some airports such as New York LaGuardia and London City Airport are space and runway constrained. The high elevation of some cities such as

La Paz, Bolivia, and Bogotá, Columbia, impose performance restrictions that limit aircraft operating range and payload. Similarly, high ambient temperatures restrict allowable takeoff weight. Boeing designed the latest upgrades of the 777, the -8 and -9 models, to meet the high temperatures encountered in the Middle East without sacrificing range or payload. This capability garnered large orders announced at the 2013 Dubai airshow from Emirates, Qatar Airways, and Etihad Airways but certification delays have resulted in some cancelations. Finally, adverse en-route winds, often seasonal, will limit aircraft range.

8.2.2 Aircraft Operating Costs

Just as aircraft performance capability varies greatly by aircraft type, so does operating cost. Aircraft operations and maintenance are about half of the total airline costs, so understanding and managing these costs are critical to airline success. There are several ways to compute and categorize aircraft operating costs. The appropriate metrics depend on the information needed for specific decisions. One method is to divide costs into direct and fixed. Direct operating costs (DOC), also called variable costs, are incurred only when the aircraft is flown. Major direct costs are fuel, maintenance, and a portion of crew wages. Fixed costs such as depreciation or lease fees and insurance, on the other hand, are incurred each month regardless of how many hours the aircraft is flown. Suppose an airline is considering implementing a reduced mid-week schedule, then it would want to know what costs are avoided. For this decision, the division into direct and fixed costs is useful. For other purposes, however, costs are computed by block (or flight) hour or by cost per available seat mile (CASM).

8.2.2.1 Aircraft Size versus CASM

Block hour cost increases with aircraft size. Fuel consumption per flight hour increases with aircraft size and weight. For example, Boeing's smallest aircraft, the B-737–700 burns about 680 gallons per block hour whereas the B-747-400 consumes 3,300 gallons per block hour (Airline Monitor, 2015). Maintenance costs per block hour are similarly higher for the B-747.

While DOC per block hour is appropriate for some airline decisions, cost per available seat mile (CASM) is more important for many decisions as it represents the cost of providing air transportation for a single passenger (over one mile). Large aircraft cost more to operate per hour, but CASM decreases as aircraft size increases because maximum seating capacity increases faster than aircraft operating costs. Thus, the higher DOC of larger aircraft is spread over an even greater number of seats. An aircraft with 150 seat capacity has about a 17% lower CASM than one with a 100-seat capacity (Swan, 2002).

Airbus' very large aircraft, the A-380, has the lowest CASM because of its high seat capacity, long segment lengths, excellent fuel efficiency, and relatively low maintenance costs. It is, however, very expensive to operate per hour. In contrast, the regional jet has a lower cost per flight hour than larger mainline jets, but substantially higher CASM. Therefore, the yield for regional jet flights must be higher on average than for larger mainline jets. The CASM disadvantage for the smallest regional jets is leading to their rapid removal from service in favor of the new generation of larger regional jets.

Larger aircraft have higher seat capacity, but the number of seats installed on an aircraft is the choice of the airline. The number of cabins and seating density depends on the passenger segments targeted, markets served, and competition. This is a purchase option on new aircraft, but the configuration will likely be changed several times during the aircraft's service life. For example, when Delta established its low-cost Song subsidiary, it reconfigured the Boeing 757 aircraft to single-class, moderate seating density. Spirit Airlines and Ryanair configure their aircraft with high-density, single-class cabins whereas other airlines choose more legroom and two or more separate cabins. Using the Airbus A-320 in service with US airlines as an example, seating density varies from United Airlines' 142 seats to Spirit's 178. ASMs increase directly with the number of seats per aircraft, so CASM is lower for higher seating density. Spirit's CASM would be 20% lower than United's just due to seat density. The Airline Monitor (2015) computes United's A-320 operating CASM at 8.71¢ versus Spirit's 6.23¢. Spirit enjoys other cost savings as well, but seat density accounts for more than half of the difference.

8.2.2.2 Segment Length

Cost per available seat mile decreases with stage length (also called segment length) for several reasons. First, longer stage lengths result in proportionally more flight time spent at efficient cruise altitudes which increases cruise speed and fuel efficiency. Second, the time required for taxi and in-flight maneuvering for takeoff and landing is a smaller portion of the total block time on longer segments. A portion of total maintenance costs also varies with the number of takeoffs and landings, called aircraft cycles. Cycle and passenger handling costs are spread over more miles with longer segment lengths. Finally, the latest generation wide-body aircraft often operate the longest segments. The graph of CASM versus segment length in Figure 8.1 is computed by a formula often used in financial analysis to adjust CASM for stage length (cost per equivalent available seat-mile). The figure shows a representative airline with an average CASM of 11¢ and an average stage length of 1,000 miles. If the same carrier reduced its average stage length to 540 miles, they could expect their CASM to increase to 15¢; conversely, if the stage length were increased to 1,500 miles, the CASM would drop to 9¢. This adjustment is useful for comparing unit costs between airlines with different average stage lengths.

New generation aircraft, including the A350, A380, B777X, and B787 are capable of flights of more than 18 hours and 9,000 nautical miles. Singapore Airlines, for example, flies non-stop from New York to Singapore with the A350 covering 9,537 miles in 18 hours and 40 minutes. Ultra-long-haul flights, however, do not benefit from a reduction in CASM because the weight

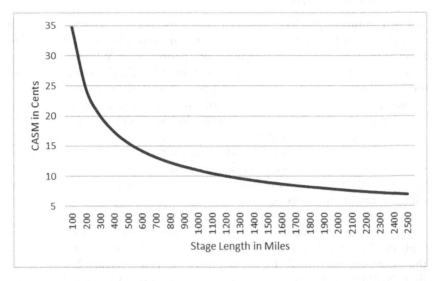

Figure 8.1 CASM Versus Segment Length.

Note: Curve generated from CASM = 11 x $\sqrt{\dfrac{1000}{STAGE\ LENGTH}}$ with 11¢ the assumed CASM at 1,000 miles.

Sources: Hamlin, n.d.; Stalnaker et al., 2018.

of the required fuel increases the total fuel burn. Consequently, the CASM is higher than a similar flight with an intermediate stop. These flights are only profitable on business routes that support high fares and potentially only when fuel prices are relatively low. For this reason, long-haul cargo flights such as from Asia to the US usually stop in Anchorage, Alaska, even if a non-stop flight is technically feasible.

8.2.2.3 Yield

The average price a passenger pays per mile on a flight is termed *yield*, a standard airline price metric. In competitive airline markets, the yield will closely follow CASM. For example, the base fare on Southwest Airlines for a mid-week flight in early 2020 from Dallas to Houston, Texas, a distance of 240 miles, was $154. The yield is 64¢ ($158/240). However, the fare for a flight on the same day from Dallas to Seattle, Washington was $198. At 1,668 miles or seven times the Dallas–Houston distance, the yield is 11.9¢. These two data points illustrate the fall of yield with segment length in the same way as CASM. There is limited competition in many short-haul markets – some are monopolies – which results in higher prices. If you are flying to or from a small city, expect to pay much higher fares.

8.2.2.4 New Versus Older Aircraft

Just as consumers purchase new and used cars, airlines can choose new or used aircraft for fleet expansion. Used aircraft are available for nearly all airline requirements in both seating capacity and performance. While most airlines have elected to upgrade their fleets with large orders for the latest-generation aircraft, there are exceptions. Delta Air Lines occasionally acquires used aircraft to supplement its new aircraft order book. In a notably large transaction, Delta purchased 88 used B-717s from Southwest Airlines. Southwest inherited these planes from its merger with Air Tran.

New aircraft have lower DOC than similar used aircraft. More fuel-efficient engines reduce fuel expenses and the longer design life of new components, improved reliability, and automated diagnostic tools lower maintenance costs. New aircraft components do not require overhaul or replacement for several years. State-of-the-art engines burn significantly less fuel than those of older technology. The latest Airbus A320neo and Boeing 737 MAX sport new engines with 15% better fuel economy than the models they replaced. Figure 8.2 shows the impressive gains in fuel efficiency over the past 60 years. Improvements from 2010 to 2019 are due to the introduction of the new fuel-efficient aircraft, including the Airbus A320neo and A350 and Boeing 737 MAX and 787 families. The only new aircraft on the horizon are the B777–8 and -9, so future improvements may be slow (Zheng & Rutherford, 2020).

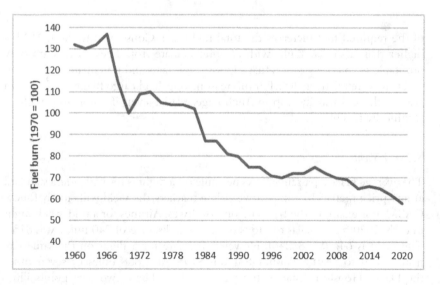

Figure 8.2 Aircraft Fuel Efficiency Improvement 1960 to 2019.

Note: Fuel burn index of new commercial aircraft jet aircraft from 1960 to 2019. Adapted from Zheng & Rutherford, 2020.

The lower acquisition price of used aircraft, however, may offset the lower DOC of new aircraft. Commercial aircraft purchase prices for both new and used aircraft vary with demand, with lower prices during economic recessions. Actual sale prices are proprietary but a 15-year-old model of today's most popular narrow-bodied aircraft can be bought for 40% of the price of a new aircraft (Airline Monitor, 2015). (This is the purchase price, not the list price. Like automobiles, the manufacturer's list price is usually negotiable, especially for good customers). Another benefit is that used aircraft are often available with little lead-time whereas airlines may have to wait years for the delivery of a new aircraft.

Older aircraft will incur higher maintenance expenses. Time-limited parts must be replaced or overhauled. Likewise, the airframe requires a periodic major overhaul. This extra maintenance limits availability for revenue service. Older aircraft also suffer a higher rate of mechanical failure that adversely impacts customer service. In turn, a higher level of spare parts is needed to maintain the required service levels. Allegiant's poor reliability and on-time performance rapidly improved as it replaced the original MD-80 fleet with Airbus A320 series aircraft even though these aircraft were used.

The US LCC Allegiant Air provides an illuminating example of trade-offs between new and used aircraft. Allegiant has preferred used aircraft for fleet renewal and growth. After an unsuccessful start, Allegiant emerged from bankruptcy in 2002 with a new management team. It began scheduled service the same year with well-used MD-80s that it acquired for as little as $4 million each. That fleet, which grew to more than 60 aircraft, was gradually replaced with used Airbus A320 series (A319 and A320) aircraft – the last MD-80 retired in 2018. By 2021, the Airbus fleet numbered 121, all but 13 of which were used, with more in the pipeline, including an agreement with Air Lease Corporation (see the section 'Leasing') to acquire 10 more A320 aircraft over two years. Allegiant Executive Vice President Gregory Anderson explained, "Since the onset of the pandemic, our fleet team has now signed up 24 A320 series aircraft – 21 of them since the beginning of this year (2021) – all at an average price discounted by 30% when compared to pre-pandemic levels" (quoted in Singh, 2021). Allegiant then surprised the industry in January 2022 with an order for 50 Boeing 737 Max aircraft and options for 50 more. Industry analyst Bret Snyder, aka the Cranky Flier (2022), observed that this violation of the single-type standard for LCCs isn't crazy. He noted that Boeing had recently lost two high-profile deals as Boeing narrow-body operators Qantas (see Case Study at the end of this chapter) and KLM defected to Airbus. So Boeing needed a win and Allegiant got a great deal. With an aggressive planned growth of 10 percent annually, Allegiant needed more aircraft than would likely be available in the used market. And with very high-seat density and a low purchase price, Allegiant projected the 737s will have a lower seat cost per departure (an uncommon metric) than its current A320s (Allegiant, 2022).

8.2.2.5 Commonality

Substantial cost savings result from operating a common aircraft type such as an all–Airbus A320 or Boeing 737 fleet as is typical for LCCs. The airline has to develop and maintain only a single training program, so crew-training costs are reduced. Similarly, mechanics must learn and qualify on a single aircraft. Both groups gain more experience and benefit from learning curve efficiencies. A smaller spare parts inventory is required with a single fleet type.

Airbus and Boeing both produce families for aircraft with high commonality. The Airbus A-320 series ranges from the A-318 typically configured for 120 seats to the A-321 with a maximum seating capacity of 220 seats. The competing Boeing 737 family offers a similar capacity range. Pilots are usually qualified to fly all models within the family. In contrast to the new families of aircraft, airlines operating used planes will face greater differences within a fleet. Most airlines could meet their operational needs with exclusively Airbus or Boeing fleets simplifying some aspects of their business even with several types in the fleet. On the other hand, bargaining power may be increased by having some aircraft from each manufacturer. This is one of several aspects of new aircraft acquisition illustrated in Case Study 8.4.

Most aircraft manufacturers offer a choice of engine manufacturers and engine models. Airlines generally prefer to minimize the number of engine types to obtain commonality cost advantages.

There are great advantages to commonality, but, as discussed earlier, FSNC route systems require more than one aircraft type to fit the market. Regional jets operated in partnership with regional airlines connect smaller cities with the network carrier's hub airport. Long-range, wide-body aircraft are best suited to many international routes. Operating a fleet with a mix of aircraft ages also offers some flexibility. Airlines can choose to operate older aircraft that are nearing the end of their economic lives when the economy is strong or these aircraft can be parked or scrapped at relatively low-cost during economic slowdowns.

8.2.2.6 Airbus and Boeing Product Line

Most airlines purchasing new aircraft will choose either Airbus or Boeing models or occasionally from both. Table 8.1 lists the aircraft in production and forthcoming models. For most capacity and performance requirements, Airbus and Boeing offer closely comparable models with a few gaps. With its purchase of the Bombardier C-series aircraft, rebranded as the A220, Airbus has an excellent smaller aircraft that Boeing does not match. Boeing tried to counter Airbus' advantage by purchasing Embraer's regional jet division, but the deal fell apart in 2020 with Embraer accusing Boeing of bad faith.

The narrow body market is the largest by the number of aircraft on order. The A320 and B737 series are both upgraded models of aircraft dating back more than four decades. Improved fuel efficiency is mostly the result of new

Table 8.1 Airbus and Boeing Product Lines

Airbus Model	Typical Seating	Range in NM	2021 List Price in $M	Boeing Model	Typical seating	Range in NM	2021 List Price in $M
Narrow Body Aircraft							
A220- 100	100–120	3,400	$81	B737 MAX 7	138–153	3,850	$99.70
A220-300	120–150	3,350	91.5	B737 MAX 8	162–178	3,550	121.6
A319neo	120–150	3,700	101.5	B737 MAX 9	178–193	3,550	128.9
A32Oneo	150–180	3,400	110.6	B737 MAX 10	188–204	3,300	134.9
A32lneo	180–220	4,000	129.5				
Wide Body Aircraft							
A33Oneo–800	220–260	8,150	259.9	B787-8	248	7,305	248.3
A33Oneo–900	260–300	7,200	296.4	B787-9	296	7,565	292.5
A350–900	300–350	8,100	317.4	B787-10	336	6,330	338.4
A350–1000	350–410	8,400	366.5	B777–200LR	317	8,555	346.9
				B777–300ER	396	7,370	375.5
Discontinued Models							
A380	575	8,000	445.6	B747-8	410	7,730	418.4
Forthcoming Models (planned to enter service in 2023)							
A321XLR	180–200	4,700	not available	B777-8	384	8,730	410.2
				B777-9	426	7,285	442.2

Sources: Airbus Commercial Aircraft, AXON Aviation Group, Boeing Commercial Airplanes, Simple Flying.

engines; Airbus' neo stands for *new engine option*. The two tragic B737 MAX crashes and subsequent grounding of the model for over a year afforded Airbus an advantage in this critical market segment.

Boeing's B757 has long been a favorite for thinner transatlantic and other mid-range international markets, but with production discontinued in 2003, the aircraft still in service are nearing the end of their useful lives. Boeing has considered a clean-sheet replacement, in industry argot, a *New Midsize Airplane*, but set aside the project in 2020 to correct widespread production problems. Airbus' forthcoming A321XLR (extra-long range) model is set to capture much of the segment.

At the other end of the spectrum, the Airbus A380 is the largest commercial aircraft. After first asserting that the market would not support a so-called very large aircraft, Boeing countered with an upgrade of its venerable 747. Unfortunately, for both, the market proved Boeing right – production of both aircraft ended by 2022.

8.3 Fleet Acquisition

Whether acquiring new or used aircraft, an airline must decide on how to pay for the acquisition. There are several alternatives, most of which are similar to the financing available for automobiles. A profitable incumbent airline could choose from all options whereas a new or financially struggling airline would have fewer choices. However, as shown in Table 8.1, and unlike automobiles, new aircraft prices are in millions of dollars. The A220-100, Airbus' smallest commercial aircraft, has a list price of $81 million. In contrast, Boeing's forthcoming B777-9 features a list price of $442.2 million. In 2021, United Airlines announced the biggest order in its history for 270 new aircraft – 70 A321neo and 200 B737 MAX planes. The order was valued at $35 billion at list prices, but airlines negotiate purchase packages that are well below list prices. Because United's order was made during the pandemic recession, CNN Business reported that United will pay less than half of the list prices (Isidore, 2021).

A large aircraft purchase is typically financed with a combination of methods that are discussed next.

8.3.1 Internal Financing

Despite a history of uncompetitive profit margins, airlines generate lots of cash that can be used to purchase aircraft. For example, in 2019, Delta Air Lines generated $8.4 billion in cash from operating activities (Delta, 2020). Over the longer term, cash from operations follows net profits. In the short run, however, non-cash expenses, primarily depreciation and goodwill, will cause cash flow and profit to differ. Economic recessions, for example, may reduce the market value of commercial aircraft. If an airline writes down the value of its fleet, this non-cash expense will lower profits but does not affect short-term cash flow. As an aside, cash flow statements are critical to evaluating company health. Profits can be misleading but cash flow often reveals the true health of operations.

Profitable airlines will usually choose to use internally generated cash for some portion of the fleet acquisition, but there are many competing uses for cash flow. From mid-2014 through to 2019, American Airlines spent $12.4 billion to buy back its own stock. Stock buybacks are one means of returning profits to shareholders, the owners of the corporation.

8.3.2 Equity Financing

Airlines may also choose to issue new shares of stock to finance expansion, including aircraft acquisition (equity financing). To sell new shares, a company must demonstrate a strong potential for earning future profits. Japan Airlines' stock offering in 2012 illustrates stock value is based on the potential for future earnings. Japan Airlines entered bankruptcy and had its stock delisted in 2010, but the new stock offering after exiting bankruptcy raised $8.5 billion, signifying a dramatic turn-around. Somewhat surprisingly, airlines were able to sell stock during the COVID-19 pandemic. American and United Airlines both sold new stock in 2020 to bolster cash reserves and ULCCs Frontier and Sun Country had successful initial public offerings (IPO) in 2021.

8.3.3 Debt Financing

Airlines will use debt and borrowing to finance all or a portion of a new aircraft purchase. Commercial bank loans are akin to a home or car loan but with the aircraft as collateral. Should the airline become unable to make loan payments, the bank can repossess and sell the aircraft. A lead bank will often sell a part of a large loan to other banks to diversify risk, a process known as syndicating a loan.

Larger airlines can bypass commercial banks by selling bonds directly to investors. These bonds, known as Enhanced Equipment Trust Certificates, offer deep capacity, attractive rates, long maturities, and a diversified investor base. Enhanced Equipment Trust Certificates generally include 20 or more aircraft with 20% financed by a lead investor who takes ownership with the remaining 80% financed in three tranches of progressively subordinated public debt.

Most commercial aircraft manufacturers are reluctant to finance aircraft but will often provide financing assistance to get a sale. Support varies from leasing to guaranteeing the aircraft's value at the end of a lease or loan. Airbus and Boeing have also taken trade-ins from an airline's fleet to obtain a new sale.

The US government supports the export of aircraft, mostly to boost employment in well-paying aircraft manufacturing jobs, by guarantying loans for foreign airline buyers. To facilitate the sale, the US Export-Import (EXIM) Bank will guarantee the loan to protect a commercial bank or other financing institution against nonpayment by an airline. In 2021 for example, Panama-based Copa Airlines applied to the Export-Import Bank of the United States to guarantee financing for some Boeing 737 MAX 9s on its order book (Bonnassies, 2021).

8.3.4 Leasing

Aircraft leasing, again similar to automobile leasing, is an alternative for an airline to finance the purchase of new aircraft. From less than 2% in 1980, aircraft leasing companies owned 46% of the worldwide commercial aircraft fleet in 2020 (Mazareanu, 2021). Agnew (n.d.) attributes this astonishing growth to airline deregulation that led many airlines to pursue market share over profits. Airlines struggled to finance rapid fleet growth with traditional financing creating an opportunity for specialized leasing companies. These companies are willing to lease aircraft to financially weak airlines because aircraft are excellent collateral that can be repossessed and resold or leased to another airline in the event of default on the lease payments. Or at least it seemed until Russia effectively commandeered commercial aircraft owned by western lessors during the Russian–Ukraine war. Although leasing companies are out billions of dollars, Russian airlines will not able to obtain aircraft from western leasing companies in the foreseeable future.

As with airlines, there has been a flurry of mergers and consolidations within the leasing business in recent years. In 2021, Dublin-based AerCap acquired GE Capital Aviation Services (GECAS) from General Electric creating the largest lessor. Although there are more than 150 leasing companies, the top seven lessors owned more than half of the leased fleet in 2018, a concentration that the AerCap acquisition will strengthen. By somewhat of a historical accident, leasing became concentrated in Dublin which still hosts four of the 10 largest lessors (see Table 8.2).

Although we tend to think of airlines as being Airbus' and Boeing's most important customers, the leasing companies usually place the largest early orders for new aircraft commanding large discounts. In addition to new aircraft, the companies will also have a large portfolio of used aircraft providing airlines with quick access to aircraft for fleet expansion.

Table 8.2 Commercial Aircraft Leasing Companies in 2020

Company	Aircraft	Country
GECAS	1,074	USA
AerCap	1,024	Ireland
Avolon Aerospace Leasing	575	Ireland
BBAM	516	USA
Nordic Aviation Capital	483	Ireland
SMBC Aviation Capital	470	Ireland
ICBC Leasing	454	China
BOC Aviation	398	Singapore
Corporation	396	USA
DAE Capital	364	Dubai

Note: After completing the purchase of GECAP in 2021, AerCap is by far the largest leasing company with over 2,000 aircraft.

Sources: Statista; corporate websites.

There are two basic types of leases – operating and capital – and several variations. These are the next topics.

8.3.4.1 Operating Lease

An operating lease is similar to most car leases. The title remains with the lessor and the aircraft is returned at the end of the lease, so the lessor bears the risk of changes in the aircraft's value. The lease period is relatively short and much less than the useful life of the aircraft varying from just a few months to several years – five years is typical. The leasing company might place an aircraft at several different airlines over the aircraft's economic life.

Operating leases are more expensive than aircraft ownership, but there are several benefits. An operating lease requires less upfront cash and little aircraft trading experience. Aircraft are available in a shorter time than with a purchase where lead times can be several years. The airline gains flexibility in managing its fleet as it may return the leased aircraft to the lessor at the end of the lease or it may negotiate a lease extension. The operating lease is often the only option available to small and new-entrant airlines.

The lessor bears the risk of aircraft resale value and premature economic obsolescence but does so with the promise of attractive returns on invested capital. The aircraft is depreciated rapidly yielding tax benefits to the leasing company and its investors. Lease payments are greater than the interest on borrowed funds. With an operating lease of five to seven years, the airline would pay the leasing company 9% to 12% of the aircraft's value annually. One percent of the original price per month is a common industry rule-of-thumb. In good times, at the end of the lease, the used aircraft's value is 85% of the original price, but there is a market risk. The price of older aircraft collapsed in the recession of 2001. If all goes well, a $100 million aircraft brings in $84 million over the seven-year lease and then is sold used for $85 million – a pretty good business proposition.

8.3.4.2 Capital Lease

A capital lease, on the other hand, is just an alternative form of long-term purchase financing. Financing is provided by the leasing company rather than from a commercial bank or by issuing debt. Through the capital lease, the tax benefits from aircraft depreciation can be transferred to individuals or companies with higher tax liabilities than the airline, thus making the lease attractive to both the airline and investors. Capital leases range from 15 to 25 years, with the title transferring to the lessee at the end of the lease.

8.3.4.3 Sale and Leaseback

Leasing companies purchase the bulk of their aircraft inventory directly from Airbus, Boeing, and other manufacturers, but obtain some aircraft through sale and leaseback agreements with airlines. Airlines frequently place orders for new

Table 8.3 Aircraft Financing Options

	Operating Lease	Capital Lease	Purchase
Term	Few months to 7 years	12–18 years	Aircraft life
Capital usage	Low	Medium	High
Repayment	Rental	Principal & interest	Balance
Asset exposure	None	Some/all	full
Fleet flexibility	High	Low	Low
Entry cost to latest technology	Low	Medium	High
Lead Time	Short	Short to long	Long
Balance sheet	Off	Off or on	On
Deposit	3 month's rent	10%	25% or more

aircraft directly from the manufacturer with the first delivery several years into the future. At the time of placing an order, the airline may not have arranged for financing or even determined if it wishes to purchase or lease the aircraft on order. With a sale and leaseback, the airline subsequently sells some or all of its new aircraft to a leasing company while agreeing to lease back the same aircraft from the new owners. The lessor pays the airline for the aircraft providing the airline with a cash infusion. Most leasebacks are on operating leases with some aircraft on capital leases. Table 8.3 summarizes the various aircraft financing options.

The last variation on financing is an agreement with the manufacturer or large third-party maintenance provider whereby the airline pays a set amount for every hour of operation, so-called *power-by-the-hour*. Power-by-the-hour agreements most often cover engine operation. The airline pays a negotiated fee and the counterparty provides all maintenance and overhaul. The airline can accurately forecast a major operating cost and avoid having to inventory engine parts and accessories. The maintenance provider benefits from economies of scale and assumes the risk of repair in exchange for a long-term guarantee of future work.

8.3.4.4 Wet Lease

If an airline needs the use of an aircraft for a short period, perhaps to meet seasonal demand or for some operational difficulty, a wet lease is an option. Another airline, often a specialist in the business, provides an aircraft, crew, maintenance, and insurance (aircraft, crew, maintenance, and insurance – ACMI). The lessee provides passenger and ground handling and, critically, fuel. A wet lease provides the airline with immediate capacity without long-term obligations. Wet leases are subject to government approval and often restricted by labor agreements. Amazon Air is an interesting example of an airline in name only; its fleet of 96 aircraft comprised of ATR-72, B-737, and B-767 all of which are operated by seven different airlines. Amazon does not have an Air Operators Certificate or any crew of its own (Bailey, 2022).

8.3.5 Financing Portfolio

Just as an airline may benefit from having a range of aircraft ages in its fleet, it can also benefit from financing its fleet with various financial vehicles. Operating leases are more expensive than capital leases, but the airline may choose to return an aircraft to the lessor at the end of the lease. This option provides flexibility to change the fleet composition or easily downsize in the face of a recession. The optimal mix of financing is a complex decision; however, a large airline might structure its financing so that it purchases 40% to 50% of the fleet with internal funds or directly issued debt, acquires 30% to 40% of the fleet using capital leases, and the remaining 20% to 25% on operating leases.

8.4 Summary

Airlines continually assess their aircraft fleet requirements in light of changing macroeconomic conditions and opportunities and threats within the industry. Aircraft reaching technical or economic obsolescence must be replaced and additional aircraft may be required for an expanding or revised route structure. Commercial aircraft vary greatly in capacity and performance so the airline planner must carefully consider the fit between aircraft capabilities and the airline's requirements. Most LCCs find that a single aircraft type, usually either the Airbus 320 or Boeing 737 families, meets their operational needs. Operating a single fleet type yields substantial cost savings. The typical FSNC with routes extending across the globe, on the other hand, requires several types ranging from regional aircraft to long-range wide-body aircraft. Though most airlines opt for new aircraft for fleet replacement and expansion, used planes that meet the operational needs are usually available at prices far below those of new aircraft. However, the operating costs of used aircraft are higher and maintenance reliability is often lower than for new aircraft. The trade-off between lower capital cost and higher operating cost compared to new aircraft is complex involving business risk. Risk avoided by purchasing new aircraft may be the determining factor in fleet selection. Several methods are available for an airline to finance the purchase of aircraft. Some airlines enjoy substantial cash flow from operations and can use this cash to purchase all or a portion of a fleet acquisition. Alternatively, the airline can borrow funds either from a commercial bank or, for large airlines, by issuing debt directly in the capital markets. Nearly half of all newly acquired aircraft are leased from one or more of many aircraft-leasing firms. Leases can be relatively short-term, usually five to seven years, but can extend over much of the useful life of the aircraft. With an operating lease, the aircraft is owned by and returned to the leasing company at the end of the lease. An aircraft financed with a capital lease, on the other hand, remains with the airline at the end of the lease. A large airline, whether LCC,

FSNC, or hybrid, gains flexibility with a portfolio of new and older aircraft funded by several financing methods.

8.5 Case Study: Qantas Orders New Narrow-bodies from Airbus

Financially struggling from the effects of the COVID-19 pandemic, many airlines deferred aircraft on order for later delivery, but Qantas saw an opportunity to get a good deal. It sealed a large order with Airbus in late 2021 with firm orders for 40 A321XLRs and A220s and purchase rights for an additional 94 aircraft. Qantas, which has 75 Boeing 737s and 20 B717s in its fleet, described the deal as renewing the domestic fleet with the A220s replacing the 20-year-old 717s and with the A321s replacing the slightly younger B737s. Deliveries begin in 2024 extending over more than 10 years. A few months later, it added 12 Airbus A350-1000s with deliveries starting in 2025. The A350's long range will allow Qantas to begin flying the Kangaroo route from Sydney to London non-stop.

The A220s have 20% lower fuel burn and greater range than the 717s they replace. The A220's range will allow Qantas to serve secondary markets such as Brisbane-Broome that are beyond the reach of the 717. Similarly, the XLR's range extends beyond the domestic and short-haul international routes served by the 737s opening new route opportunities. The higher fuel efficiency will also allow Qantas to improve its relatively poor emissions profile.

Qantas requested proposals from Airbus, Boeing, and Embraer. Qantas CFO Vanessa Hudson said that all the models considered met Qantas' technical requirements and that the selection was "an incredibly competitive process," and was "neck-and-neck almost right to the end" (Quoted in CAPA, 2021). Jetstar, a Qantas wholly owned LCC, has more than 100 Airbus narrowbodies on order. Combining the new order with the Jetstar orders allows Qantas to choose between the entire A320neo and A220 families.

With the replacement of the Boeing 717 and 737 aircraft, Airbus will dominate the Qantas Group fleet leaving Boeing with only the 787–8 and–9 models reversing Boeing's dominance from the earlier era. CEO James Joyce does not believe that Airbus' dominance will reduce the airline's future bargaining power arguing the losing manufacturers will continue to compete intensely for future orders.

Concepts: new versus old aircraft; fuel efficiency; range capability; commonality; carbon emissions; financing methods; operational cost

Sources: "Airbus Deal," 2021; Qantas, 2022.

References

Agnew, R. F. (n.d.) The birth and growth of the aircraft leasing business. *World Leasing Yearbook.*

Airbus deal gives Qantas commonality and range advantages. (2021, December 16). *CAPA Centre for Aviation.*

Airline Monitor. (2015). *Traffic, fleet & financial data for leading 80 airlines of the world.*

Allegiant. (2022, January 6). *Form 8-K, Fleet update.* https://sec.report/Document/000 1362468-22-000002/a1622fleetdeckvfinal.htm?utm_source=newsletter&utm_med ium=email&utm_campaign=cranky_flier_allegiant_s_turn_toward_boeing_isn_t_ as_dramatic_as_you_might_think&utm_term=2022-01-10

Baily, J. (2022, February 25). Amazon Air: A risk to FedEx and UPS? *Simple Flying.* https://simpleflying.com/amazon-air-a-risk-to-fedex-and-ups/

Bonnassies, O. (2021, September 27). Copa seeks more Ex-Im financing. *Airfinance Journal* (Online).

Cranky Flier. (2021, January 10). *Allegiant's turn toward Boeing isn't as dramatic as you may think.* https://mail.yahoo.com/d/folders/1/messages/AHsi920jMgfOYdwhFgjK AHKyGTQ

Delta. (2020, January 14). *Delta Air Lines announces December quarter and full year 2019 profit.* https://ir.delta.com/news/news-details/2020/Delta-Air-Lines-Announces-December-Quarter-and-Full-Year-2019-Profit/default.aspx

Hamlin, G. W. (n.d.) Adjusting operating cost for segment length. *SpeedNews.* https://speednews.com/article/7233

Isidore, C. (2021, June 29). United Airline orders 270 jets, its biggest aircraft purchase ever. *CNN Business.* www.cnn.com/2021/06/29/business/united-record-jet-order/index.html

Mazareanu, E. (2021, April 21). *Share of leased aircraft in the aviation industry worldwide 1970–2020.* Statista.

Singh, J. (2021, July 29). Allegiant secures 10 more Airbus A320s as it sheds pandemic impact. *Simple Flying.* https://simpleflying.com/allegiant-10-a320s-air-lease/

Stalnaker, T., Usman, K., Taylor, A., & Alport, G. (2018). Airline economic analysis. *Oliver Wyman.* www.oliverwyman.com/content/dam/oliver-wyman/v2/publications/2018/January/Airline_Economic_Analysis_AEA_2017-18_web_FF.pdf

Swan, W. (2002). Airline route developments: a review of history. *Journal of Air Transport Management,* 8(5), 349–353.

Zheng, X. S. & Rutherford, D. (2020, September). Fuel burn of new commercial jet aircraft: 1960 to 2019. *International Council on Clean Transportation.* https://theicct.org/sites/default/files/publications/Aircraft-fuel-burn-trends-sept2020.pdf

Review Questions

1. What operational factors must an airline consider in selecting new aircraft?
2. How do aircraft direct operating costs vary with aircraft size, segment length, and age?
3. Why does an FSNC typically have several aircraft types in its fleet?
4. How does an LCC benefit from operating a single-fleet type?
5. Why has Delta Air Lines added a mix of new and used aircraft to its fleet?
6. Why is airfare per mile flown much less for long international flights than on a regional airline flight from a small city to a nearby FSNC hub?
7. List each Airbus aircraft type and the competing Boeing type.
8. What financing methods are available for an airline to acquire aircraft for its fleet?

9. What is the difference between borrowing funds from a bank for fleet expansion and leasing aircraft?
10. One aircraft type in an airline's fleet is grounded. How might it maintain its flight schedule for a short time?
11. How can an airline benefit from financing its fleet with a variety of financing methods?

9 Pricing and Revenue Management

Upon completion of this module, you should be able to:

- Describe the types of airline pricing and the reasons an airline might use each type.
- Identify the several product characteristics needed for effective revenue management.
- Explain the objectives of revenue management.
- Discuss why airlines sell more seats than they have on their aircraft.
- Identify the various types of revenue management an airline might use.
- Recognize the value of revenue management to an airline's bottom line.

9.1 Introduction

Unlike products and services in many other industries, the price of an airline ticket can vary greatly for any given flight. Two passengers sitting next to each other in economy class may discover they paid greatly varying fares even though both receive the same service. The price on different routes of about the same distance may also vary greatly. As we know from Chapter 2, demand varies by time of day, day of the week, season, and the business cycle – and so do fares. Of course, ticket prices differ by class of service whether first, business, premium economy, or economy, but prices also vary by many other factors such as the time of purchase and payment, and whether the booking was made online or through a travel agency. Restrictions will apply to many fares, with discounted tickets usually, including a minimum length of stay at the destination, a requirement to stay over a Saturday night, or other restrictions (Hanlon, 2007; Doganis, 2019).

This chapter explores this puzzling, seemingly random, and senseless variation in airline pricing. We will find that airline pricing is based more on the demand for travel than the cost to the airline and is designed to extract the highest price each passenger is willing and able to pay to maximize the revenue per flight. Airlines control pricing with sophisticated software applications developed after US airline deregulation and known now as *revenue management systems*.

DOI: 10.4324/9781003290308-9

9.2 Regulated Prices

Before the spread of airline liberalization across the world, ticket prices were usually set by government agencies – either singly for domestic operations, or jointly on international routes, based on cost-plus-profit. A mileage-based formula promulgated by the International Air Transport Association, the global airline industry trade association, was widely used to set international fares and remains in use in some markets today. Domestic markets were often dominated by a single state-owned airline with government-set fares based on both economic and political considerations, often with little profit margin. In the US, which has never had state-owned airlines, the Civil Aeronautics Board (CAB) set ticket prices for scheduled air carriers based on the length of the flight and the airline's cost to operate the flight. Except for some discounted fares (e.g., youth fares, family fares, etc.), passengers paid the same fare on all airlines flying the same route with different fares allowed only for the class of service, such as first-class and coach (Belobaba, 2009; Vinod, 2021).

The CAB's fares should have allowed airlines to make a profit; however, in the early to mid-1970s airlines were losing money. During the same period, charter airlines, largely left alone by the CAB, were making profits by flying to many vacation and leisure destinations like Las Vegas and Orlando. Charter operators benefited not only by flying less expensive older aircraft (aircraft sold off cheap as scheduled airlines took delivery of newer models) but also by the "unscheduled" nature of their operation which allowed them to fly an aircraft only after it was fully booked. As a result, they enjoyed very high load factors and could charge lower fares suited to leisure passengers. The CAB could have added operating restrictions to the charters to help the scheduled airlines fill their planes, but they chose not to. As a result, the scheduled airlines lost some of the leisure traveler market to charters (Petzinger, 1995).

To enhance the ability of scheduled airlines to compete with charter carriers, in February 1977, the CAB approved Texas International Airlines' so-called "Peanut Fares" designed to enhance load factors on flights in selected markets (Vinod, 2021; Curry, 1977; Bailey et al., 1991). However, American Airlines had a better idea – although they were occasionally flying charters in addition to their scheduled service, American wanted to put the two together and offer charter fares for the unsold seats on their scheduled flights. The CAB approved their request in March 1977 and American began the new "Supersaver" fares between New York, Los Angeles, and San Francisco. These fares included discounts as much as 45% below the standard CAB-set coach fares (Wensveen, 2011; Petzinger, 1995; Bailey et al., 1991). The number of discounted seats was limited, and sales were restricted to advance purchase round-trips with seven to 45 days at the destination (Bailey et al., 1991). The new fares proved hugely popular, stopping the market share loss to charters while filling seats that would otherwise have been empty ("History of AMR," n.d.; Petzinger, 1995). The Supersaver fare process was the beginning of a system originally called *yield management* now known by the more descriptive name of *revenue management*.

Today's complex and sophisticated revenue management systems employed by most airlines utilize high-tech computers and software to maximize revenue and profitability (Petzinger, 1995).

9.3 The Objective and Rationale for Revenue Management

The purpose of revenue management is fairly simple – maximize the revenue earned for each flight. Although a little counterintuitive, maximizing revenue is not the same as maximizing the load factor or the average yield – sometimes the process of maximizing revenue will lead to some empty seats and some below-average fares (Doganis, 2019). Revenue management uses *demand-related pricing* to exploit the passengers' willingness to pay. Purchase restrictions, called *fences*, attempt to prevent passengers willing to pay higher prices from purchasing discounted seats.

The sales of most goods and services are not well suited to revenue management – retailers stock products on shelves with prices defined by their cost – called *cost-based pricing*. As inventory sells, they just restock. All buyers pay the same price, regardless of how much they may want or need the product, or how much they are willing or able to pay. For example, two shoppers go to a department store looking for a sweater, and they both like one priced at $50. The first shopper bought the sweater for $50 and walks out thinking she got a steal – in fact, she liked the sweater so much she would have paid $100 if she had needed to. The second shopper, who also liked the sweater, had a limited budget and couldn't afford the $50 price tag, so she left with nothing. If the sweater cost the department store $25, the store made only $25 on the sale of just one sweater.

Using price and inventory management, a seller sets multiple prices for the same product or service and allocates a specific amount of inventory for each price. The retailer in the example above could have made more revenue by pricing the sweater at $100, then offering deep discounts to $35 with restrictions on the sale, like maybe a special coupon or early-bird sale. The budget-conscience buyer would go out of their way to abide by the special restrictions required of the lower price, while the other buyer would simply pay the $100. In that situation, the retailer might have sold two sweaters: one for $100 and one for $35, for total revenue of $135 and a profit of $85. In the airline business, the use of even the most basic revenue management can typically see revenue gains of 2%–5% over airlines that do not use revenue management (Belobaba & Wilson, 1997).

9.4 Revenue Management Product Characteristics

Revenue management was first developed in the airline industry, but it has since been applied in many other industries, including railroads, shipping, bus lines, rental cars, hotel rooms, and others. The products sold in each of these industries share the following product characteristics needed to employ revenue management (Hellerman, 2006; Cramer & Thames, 2021).

- *Perishable inventory.* The airline product is not storable and inventory cannot be stockpiled for sale at a future date. An empty airline seat spoils soon as the door is closed; the potential revenue is lost. Hotel rooms are similar.
- *Fixed short-term capacity.* The number of airline seats for a flight is fixed once the schedule is in place, which may be as long as six to 12 months before the day of the flight. In the short run, an airline cannot easily vary the number of seats to meet variances in demand. Only in the long-term can significant capacity be added or reduced to match demand.
- *High fixed costs.* The cost to fly a commercial aircraft from one city to another is high. Once the airline schedules a flight, it is committed to that fixed cost of operation. Adding payload (passengers and cargo) to the aircraft increases the weight resulting in only a small additional cost. Consider the cost of driving your car from Texas to California. Once you decide to make the drive, adding another person or suitcase to the car adds a minor additional cost, but little in comparison to the total driving cost.
- *Very low marginal costs.* The cost of carrying one additional passenger (marginal cost) is very low relative to fixed cost. The marginal cost varies with the length of the flight, but adding one additional passenger to a 500-mile flight probably costs the airline less than $10 (the cost of a couple of gallons of fuel, a soda, and a bag of peanuts). The combination of high fixed costs and low marginal costs results in a situation where it is better to sell the product for a substantial discount rather than let it spoil (Weatherford & Bodily, 1992; Belobaba, 2009).
- *Uncertain demand.* Airlines can forecast the demand for each flight with surprising accuracy but some error is inevitable. With certain demand, a company can adjust capacity and price to maximize revenue. With uncertain demand or fluctuating demand (called stochastic demand), periods of excess or insufficient capacity are unavoidable.
- *Segmentable market.* The airline's seat inventory on a flight can be segmented with different sales restrictions applied to each segment. This segmentation is obvious with first-class versus coach, but "fences" can be used to establish several price categories in a single cabin class. Additionally, purchasing behaviors can indicate passengers' willingness and ability to pay for air travel.
- *Advance sales/bookings.* Tracking advance ticket sales allows the airlines to fine-tune seat allocation and overbooking rates as the day of departure approaches. Advance booking requirements segment passengers by willingness to pay.
- *Historic sales data.* Airline's computer reservation systems contain historical data needed to forecast demand.

Revenue management depends on the ability to forecast demand for each flight. Airline computer reservation systems developed in the latter part of the 20th century (and covered in detail in the next chapter), not only provided airlines with a faster and more efficient reservation process but also captured demand information. Revenue analysts mine historical sales to forecast future demand and thereby increase profits.

Revenue management allows the airline to exploit passengers' willingness to pay. "In its 1987 annual report, American Airlines broadly described the function of yield management as 'selling the right seats to the right customers at the right prices'" (Smith, et al., 1992. P. 8).

9.5 Revenue Management Components

The characteristics of the airline product allow the use of several revenue management techniques, including the following:

* *Overbooking.* The practice of accepting more reservations than the aircraft has seats.
* *Seat allocation.* Offering the same seat for several different prices by controlling the seat inventory and purchase restrictions.
* *Fare nesting.* Ensuring that seat allocation restrictions do not deny the sale of a higher-priced fare as long as a lower-priced seat remains available.
* *Network inventory allocation.* Managing seat inventory not just for a single flight, but for all flights connecting the passenger's origination to their destination.

Each characteristic is discussed in the following sections.

9.6 Overbooking

Overbooking has gotten a bad reputation over the years. We have all heard horror stories about people getting kicked off aircraft – possibly the most infamous was United Express passenger Dr. David Dao, who had to be removed from a flight in 2017. His forcibly being dragged, seemingly unconscious, from the plane, made headlines across the US (Zizka et al., 2020). Although Dao's situation was not really an overbooking issue,[1] it does exemplify the fact that denying people the transportation for which they have paid is problematic at best. Involuntarily denying passengers boarding, however, does not happen often. According to the Department of Transportation, in 2021, among the largest US carriers, only about 11,000 out of over 632 million passengers were involuntarily denied boarding, or "bumped" off their flights (US Department of Transportation, Bureau of Transportation Statistics [BTS], 2022). With about 7.3 million departures in the US in 2021 (BTS, n.d.), only one person out of every 660 flights is involuntarily denied boarding. In this section, you'll learn why airlines overbook, and why it's good for most passengers.

Overbooking started in the late 1960s during CAB economic airline regulation (Hellerman, 2006) and was the first type of airline revenue management. Most airlines had liberal no-show and last-minute cancelation policies enabling passengers to book duplicate reservations, then receive full refunds for the unused reservations or a last-minute cancelation. Having missed a flight, a passenger would be accommodated on the next available flight or receive a full refund at no additional charge. (In comparison, have you ever missed a concert or football game and tried to get a refund on your ticket? Good luck!) The downside of these liberal policies, however, is that they invite more no-shows or late cancelations – sometimes so late that they cannot be re-booked before the flight's departure.

In the 1980s and 1990s, no-show rates averaged 10%–15% with peaks as high as 20% (Toh & Raven, 2003; Belobaba, 2009). Recent US no-show rates have declined to 5% (Zizka et al., 2020). Not surprisingly, leisure customers, who normally book their reservations far in advance, are less apt to cancel or no-show because of penalties they might incur canceling hotel, cruise, or other vacation plans. Business travelers, on the other hand, usually have higher no-show and cancelation rates (Walczak et al., 2012).

Always struggling with low profits, airlines cannot afford to lose 5%–10% of their potential revenue because of no-shows. To offset no-shows or late cancelations, airlines *overbook* by selling more reservations on a flight than they have seats. If, for example, historical data shows a specific flight will have a 3% no-show rate, the airline might overbook by 2% to fill the seats left unoccupied by the no-shows. Without overbooking, there would be empty seats on a fully booked flight (called spoil). The empty seats are lost potential revenue that can never be recovered.

During the booking period, the lowest-priced seats sell first, typically to leisure passengers who plan well in advance of travel. The late-booking passengers, generally business people who travel on short notice, pay the highest fares. Without overbooking, some late booking, high-paying business travelers will be denied a reservation and instead fly on a competing airline. The industry term for a passenger denied a reservation is called *spill*. This is a double whammy for the airline – they turned down their highest-paying passengers because the aircraft was booked full, but then the aircraft departs with empty seats due to no-shows. As a result, airlines judiciously overbook reservations based on historical no-show rates. They are betting that the potential lost revenue is greater than the potential denied boarding cost. The overbooking rate, which varies by flight, is computed using complex mathematical algorithms that weigh the costs incurred due to overbooking against the opportunity cost of empty seats (Belobaba, 2009).

If the overbooking algorithm works perfectly, the overbooked passengers will exactly offset the no-shows. That's a win-win-win situation. The late booking, high-paying passengers, who would have been turned down if it were not for overbooking, get seats on the aircraft (wins for both the passenger and the airline). The passenger who missed a flight due to traffic on the freeway benefits from the airline's liberal refund or rebooking policy (Vinod, 2021).

Perfect overbooking, however, is rare. Some overbooked flights will depart with empty seats, termed *spoil*. Spoiled or empty seats are an opportunity cost for the airline (Smith et al, 1992). Overbooking also risks *overselling* – a situation

One or more of your flights today may be overbooked.

Are you interested in maybe taking a later flight in exchange for a $200 travel certificate?

Figure 9.1 Check-In Kiosk Screenshot.
Source: Photo by B. Billig.

when more passengers than forecasted show and the airline doesn't have seats for all of them. The cost of an oversale comes in two ways.

First, if a flight looks to be oversold, the airline is required to solicit volunteers to give up their seats for compensation and take a later flight ("FAA Oversale Policy," 2013; "European Commission (EC) Regulation 261/2004," 2004). Incentives for volunteers might include ticket reimbursement for that leg, vouchers for future travel, hotel or restaurant vouchers/discounts, or a combination of these (Belobaba, 2009). Figure 9.1 is an example of an offer from United Airlines. This screen appears on the check-in kiosk if there is a chance of an oversell on your flight. The goal is to get enough volunteers to avoid involuntarily bumping passengers off the flight. In 2021, 183,000 passengers volunteered to give up their seats (BTS, 2022).

The compensation paid to volunteers reduces the potential revenue from overbooking, so overbooking algorithms account for that possibility. Oversale costs increase with the number of volunteers needed. Here's an example: passengers flying to Venice for a cruise might not volunteer to give up their seats even for a huge voucher. Fans going to a sporting event or concert are also unlikely to volunteer. Leisure travelers with flexible schedules are the best candidates for oversale volunteers. About 90% of the time, oversell situations are resolved by passengers voluntarily giving up their seats (Belobaba, 2009).

A more detrimental cost of overselling arises from passengers who are involuntarily denied boarding or "bumped" from a flight. If the agent cannot find enough volunteers to give up their seats, some passengers must be involuntarily denied boarding. Although the passengers bumped depends on the airline's policies, the US Department of Transportation's guidelines state that a person who has boarded the aircraft cannot be removed unless there is a "safety, security, or health risk, or due to behavior considered obscene, disruptive, or otherwise unlawful" (US Department of Transportation [DOT], n.d., p. 1). The last passengers arriving at the gate will usually be the ones involuntarily bumped.[2] An involuntarily bumped passenger in the US has legal rights under 14 CFR. § 250 (Oversales, 1982) to compensation in cash (not vouchers). Similar rights protect European passengers through European Commission (EC) Regulation 261/2004 (2004), with financial compensation varying by the distance of the flight. The EU regulation also requires compensation for certain late flights. In both jurisdictions, bumped passengers have the option of cash or voucher compensation. Regulators can fine airlines for non-compliance. In 2019, the US fined Spirit Airlines $350,000 for categorizing passengers who were involuntarily denied boarding as "volunteers" (Department of Transportation, 2020). To avoid being bumped, follow the airline's check-in procedures and get to the airport early (Zizka et al., 2020).

Oversales can also be costly to an airline's reputation (Lindenmeir & Tscheulin, 2008). Word-of-mouth reputation suffers when bumped passengers tell their friends they will never fly on the airline again. Additionally, the Department of Transportation tracks and reports numbers of passengers denied boarding. These and other statistics are the basis for reports such as the annual Airline Quality Rating (Bowen & Headley, 2015). Passengers have won lawsuits resulting in reimbursements for non-refundable prepaid vacation expenses lost because of involuntarily denied boarding (Maull, 2005).

To illustrate the overbooking process, refer to Figure 9.2. Note how potential revenue climbs with the number of seats booked. If the reservations stopped at capacity and all passengers showed up, revenue is R. But with refunds to no-shows and late cancelations, the revenue drops to R_1.

In Figure 9.3, bookings continue past the aircraft's capacity. If more passengers show up than the capacity of the aircraft, the airline incurs a cost beginning with the first passenger above capacity. This escalating cost is shown by the short curve at the bottom of the graph. The overbooking cost curve is not linear; little

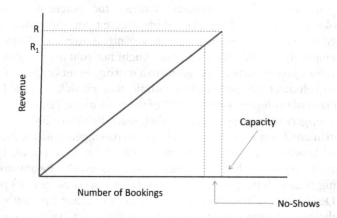

Figure 9.2 Revenue with No Overbooking.

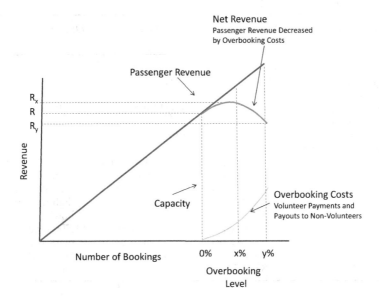

Figure 9.3 Revenue with Overbooking.
Note: These curves are exaggerated to illustrate the concept.

incentive may be sufficient for the first few volunteers, but further incentives will increase to attract more volunteers. The total potential revenue line is reduced by this cost. If the flight is overbooked to y%, the cost of oversales outweighs the potential revenue so that actual revenue drops to R_y. The highest point Rx is maximum potential revenue and the optimal overbooking level at x% (Belobaba, 2009).

The value of overbooking is demonstrated in Table 9.1 for three notional flights. In the first column, the flight was insufficiently overbooked. Although the historic no-show rate was 3%, the actual no-show rate for that flight was 9%. The result is seven empty seats that would have been filled with higher overbooking. The airline incurs an opportunity cost of $3,367. Flights in the other two columns have lower no-show rates than projected. The airline pays oversell costs in vouchers for future travel.[3] In the latter two oversold flights, note that the oversale costs were less than the opportunity cost of the under-sold flight.

9.7 Seat Allocation and Pricing

This section first explains the microeconomic theory behind airline revenue management. Demand curve #1 shown in Figure 9.4 (left panel) indicates customers' willingness and ability to buy a product or service at different prices. Each consumer values the product or service differently. Not surprisingly, the

Table 9.1 Overbooking Examples

Flight Origination – Destination	SAT-BWI	SFO-LAX	DEN-PDX
Aircraft Capacity	143	137	175
Average Historic No-Show Rate	3%	10%	7%
Overbooking rate (authorized bookings)	2% (146)	8% (148)	5% (184)
Total seats booked	146	148	182
Passengers that showed	136	142	181
Result	7 empty seats Opportunity cost of spoiled product	5 denied boarding 5 volunteers moved to other flights	6 denied boarding 4 volunteers moved to other flights, 2 sent to hotels
Cost	7 empty seats worth $481 each. Total lost $3,367	$1,200 in flight vouchers to volunteers	$1,500 total for flight vouchers, hotels, and meals

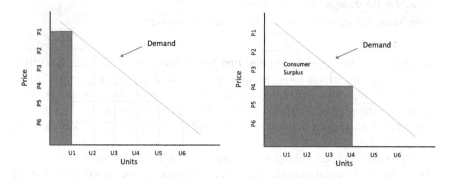

Figure 9.4 Demand Curves #1 and #2.
Note: Left panel: one unit sold at price P1. Right panel: four units sold at price P4.

lower the price, the more of the product or service consumers are willing to buy. In Figure 9.4 (left panel), at price P1, only one unit would be sold; in other words, only one consumer was willing and able to buy the product at that price. At price P4, however (right panel), four units would be sold, presumably to four different consumers.

In competitive markets such as the US and Europe, most products sell for a single, cost-based price. This is because prices are easily obtained and compared, several convenient sellers are available to the consumer, and there is generally no lack of supply. If one seller raises the price above the others, consumers move to the lower-priced sellers. Over time, consumer choice tends to even prices.

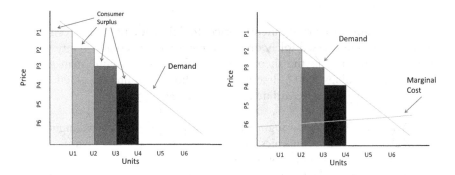

Figure 9.5 Demand Curves #3 and #4.

Note: Left panel: one unit is sold at each price P1, P2, P3, and P4. The total revenue is indicated by the total of the four shaded areas. Note the smaller area of consumer surplus.

With just one price, total revenue is simply the price times the quantity sold as indicated by the shaded areas in demand curves #1 and #2 (Figure 9.4).

In Figure 9.4 (right panel), the price is reduced from P1 to P4. Buyers who were unwilling to pay P1 now purchase a total of U4 units of the product. Even though several buyers were willing to pay more, all pay P4. In a sense, some got a good deal which is measured by *consumer surplus*, the area above the P4 and below the demand curve.

Total revenue can be significantly increased if each consumer can be charged the maximum price he or she is willing and able to pay. In Figure 9.5 (left panel), each consumer pays a different price; one unit is sold at price P1, the second unit at P2, and so on. The total of the shaded areas representing total revenue is much larger than the area when the four units are sold at a single price (Figure 9.4, right panel). The airline revenue manager's challenge is to devise a system that will require, or at least encourage, each passenger to pay the maximum price he or she is willing and able to pay.

One of the important airline product characteristics is low marginal costs. Marginal cost is the additional cost incurred by selling an additional unit; for airlines, adding one more passenger to an aircraft. In Figure 9.5 (right panel), the seller might be willing to sell the fifth unit at a price P5; but is not willing to sell the sixth unit at price P6 because the price is lower than the marginal cost of producing this last unit. The seller would lose money selling at price point P6. To illustrate this point and provide an example of low airline marginal costs, in 2010, JetBlue offered a promotion selling seats to several US destinations for only $10. The fare applied only to seats unsold just before departure that would have presumably spoiled if not for the promotion ("JetBlue Offers," 2010).

9.7.1 Price Discrimination

Selling the same product at different prices to different customers is called *price discrimination* and is based solely on a customer's willingness and ability to pay.[4] Price discrimination is the theoretical basis for airline revenue management. Dunleavy & Phillips (2009) describe six conditions required to practice price discrimination and, hence, airline revenue management:

- A market that can be segmented by varying degrees of willingness and ability to pay (in economics, different price elasticity of demand).
- Independent demand between segments.
- Barriers, also called fences, must prevent customers from buying down (buying a discounted seat when they are willing to pay for a more expensive seat).
- Arbitrage, the practice of purchasing products or tickets for resale to others, can be prevented.
- Customers' access or knowledge of other fare levels is limited.
- Marginal demand exceeds marginal supply.

9.7.2 Market Segmentation

Airlines must segment passengers to price discriminate. They do this by attaching various restrictions called *segmentation devices* to an aircraft's seat inventory. In Figure 9.5, the revenue manager wants to segment the seats into four booking classes sold at four different prices. The highest price ticket, P1, will probably have no sales restrictions, while the P2, P3, and P4 will have increasing levels of restriction. If the segmentation devices work as designed, the four buyers would each pay the highest price they are willing to pay and the resulting revenue is the shaded area. If the segmentation devices do not work and there was no control of the sales in the booking classes, all four buyers would each pay the lowest price offered, P4, and the resultant revenue would be shown in Figure 9.4 (right panel).

The challenge for airline revenue managers is determining which customers are willing to pay the higher prices and somehow devising appropriate restrictions so that these passengers cannot pay a lower price. Historically, people traveling for business are willing and able to pay more than those traveling for leisure. Most business travelers do not pay for the ticket out of their own pocket and the cost of the travel to their employer is probably a small fraction of the ensuing business deal. Business travelers usually have time-restricted schedules, and often make travel arraignments on short notice to meet their needs. They value fully refundable tickets. Further, business passengers want to be home for the weekend. For these reasons, business travelers are less price-sensitive and more time-sensitive (Donovan, 2005).

Leisure travelers, on the other hand, exhibit elastic demand – they are very sensitive to price changes. Leisure travelers pay for their tickets and often travel with family. Their travel costs are possibly the biggest expense of their trip. Leisure passengers will often adjust their travel plans for a cheaper airfare.

9.7.2.1 Segmentation Devices

Segmentation devices are designed for a specific market. For example, movie theaters sometimes use a customer's age as a segmentation device to price discriminate. They might sell tickets for a movie for three different prices – student, adult, and senior – although the products they are selling are the same, a seat in the theater (Boyd, 2007).

Airline segmentation devices are designed to separate business and leisure passengers with the overall goal of "fencing the population into different segments based on their willingness to pay" (Boyd, 2007, p. 13). With perfect segmentation devices, customers who are willing and able to purchase a certain fare are dissuaded from purchasing a lower fare by restrictions (Botimer & Belobaba, 1999). Look at the list of typical segmentation devices below and consider how they might differentiate between business and leisure passengers.

- *Advanced booking.* Flight must be booked in advance, typically 7, 14, and 21 days out from the day of departure.
- *Advanced purchase.* Pay for a ticket a specified number of days in advance of the flight or a minimum number of days after making the reservation.
- *Minimum stay.* The time between outbound and return flights must be at least a certain number of days (usually seven).
- *Round trip.* The ticket must be a round trip without intermediate stops.
- *Saturday night stay-over.* There must be a Saturday night between outbound and return flights.
- *Refund penalty or no refund.* If the booking is canceled, the passenger forfeits all or part of the ticket price.
- *Rebooking fee or no rebooking privilege.* Fees apply to any change of itinerary.
- *Limited or no stopover privileges.* No extra time may be spent at connecting or stopover points.
- *Limited or reduced service.* Reduced in-flight meal and beverage service.
- *Limited time of day.* Price not available at popular times.
- *No interline privilege.* A ticket is not accepted by another airline.
- *Tickets are not transferable.* This prevents an entrepreneur from buying discounted tickets and reselling the tickets later at a profit but for less than the airline's business fares.

The cheapest fares for each cabin will have the most restrictions while the most expensive ticket is unrestricted. The highest-price ticket is fully refundable with no advance purchase or stay-over requirement. The cheapest ticket, on the other hand, is not refundable, requires an advance booking and payment, requires a Saturday overnight, must be booked on the airline's website, and is limited to certain times of the day and days of the week.

Each fare level is given a single letter designation code (also known as reservation booking designator, RBD) (Vinod, 2021). The International Air Transport Association (IATA) defines standardized fare codes, but airlines stray from those guidelines and come up with their own lettering schemes. For example, generally, P, F, and A are first-class fares, and J, C, D, I, and Z are business-class fares. A Y-class ticket is the full-fare coach ticket and a Q-class ticket is the most discounted coach ticket. Many other classes of economy/ coach class fares will each have a letter designator (Airline Tariff Publishing Company, n.d.). Most airlines have 10 to 12 booking classes, but some go much higher – in 2017, British Airways offered 40 booking classes (Doganis, 2019). Airlines sometimes name their most popular booking classes. Qantas has three business classes: "Business Flex" (fare codes C and J), "Business Saver" (fare code D), and "Business Sale" (fare code I) (Qantas, 2020).

9.7.3 Estimating Demand

If the airline knew exactly how many late-booking business passengers wished to travel on each flight, revenue management would be relatively easy. Seats would be held in inventory for those business travelers who will book and purchase tickets a few days before the flight at high prices. The remainder of the seats is offered for earlier sale at discount prices with one or more segmentation restrictions attached. The segmentation restrictions would generally prevent the business traveler from taking advantage of the discounted tickets (prevent price diversion). Although not every passenger pays the maximum price he or she is willing to pay, this pricing method does significantly increase revenue over a single price for all seats –sometimes as much as 9% or more (Boyd, 2007).

For the largest airlines, even a small increase in revenue per flight is substantial in aggregate. For example, in 2019 (pre-pandemic), Southwest Airlines averaged over 3,700 flights a day throughout the continental United States and the Caribbean. Up-selling just one passenger per flight from a discounted seat to a full-fare seat (perhaps just $100) would increase the company's annual revenue by over $136 million (Southwest Airlines, 2022).

Revenue management, of course, is not that simple because the number of late-booking business passengers and early-booking, price-insensitive passengers cannot be known with certainty; the number can only be estimated. From earlier chapters, we know air travel demand varies with time of day, day of the week, season, and other seemingly random reasons that cannot be fully identified in advance. Because of this fluctuating demand, the optimal number of seats held in inventory for late sale can only be estimated.

A good start for estimating the demand for any fare class on a given flight is historical reservation data. Figure 9.6 shows an example of ticket purchases over an extended period for a particular flight with a B (economy) $600 fare. The lowest number of seats sold at the B fare is 30; this happened just once. Similarly, 50 seats were sold only one time. The average number of seats sold was 40.

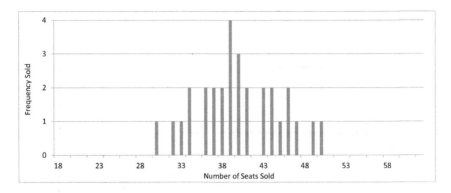

Figure 9.6 Historical Reservation Data.
Source: From a specific flight for the $600 B fare.

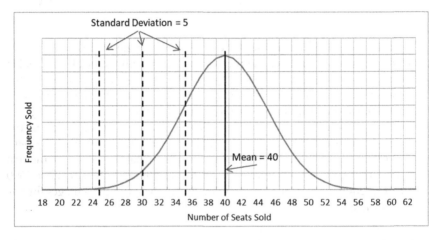

Figure 9.7 Probability Density Function for $600 B Fare.

Analysis of historical booking data shows that a normal probability density function (bell curve) accurately models the demand for a given flight. The mean and standard deviation, which define the curve, are computed from the raw data (Belobaba, 1989). In Figure 9.6, the mean is 40 and the standard deviation is 5.0. The normal distribution function can be generated from these two statistics (Figure 9.7). Notice how the plot resembles the raw data in Figure 9.6. For those who have not studied statistics (or who have forgotten everything you once learned), not to worry, just follow the general logic.

The probability that a given number of seats will be sold at the B Fare can be computed from the probability density function. The probabilities are more easily seen in a revised curve called the survivor function. The survivor function

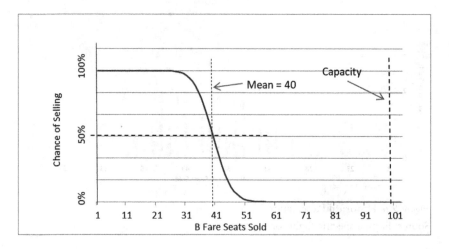

Figure 9.8 Survivor Function for $600 B Fare.

is the complement of the area under the bell curve. Figure 9.8 shows the survivor function for the B fare. The mean, or the average number of seats sold at the B fare, is 40. What's the probability of selling the 40th seat? Well, half the time, fewer seats are sold, and half the time more are sold, so the answer is 50% – as Figure 9.8 shows. It follows that the chance of selling the 41st seat is just a little lower, the 42nd seat a little lower still. On the other side of the mean, the chance of selling the 39th seat is just slightly higher than 50% and the 38th seat a little higher still. Taken to the extremes, the probability of selling the first couple seats at $600 is 100% while the chance of filling up the aircraft at $600 is zero. Revenue management software calculates the exact percentages for any point on the curve.

9.7.4 Expected Marginal Seat Revenue

The *expected marginal seat revenue* (EMSR) is the single most important concept in airline revenue management. EMSR is the estimated (or expected) revenue from the next (marginal) seat sold at a given fare – in other words, it's the fare times the probability of selling the seat. Once the analyst knows the probability of selling the next seat (from the survivor function in Figure 9.8), he or she can multiply that by the fare to determine how much the airline can expect to make selling that seat. Going back to our earlier example, at the B fare, the 40th seat has a 50% probability of selling. At the $600 B fare, the airline will sell the 40th seat on half the flights; on average the airline makes $300 selling the 40th seat ($600 x .50 = $300). The plot of the EMSR is the same as the survivor function, except the vertical axis now is labeled in dollars (Figure 9.9).

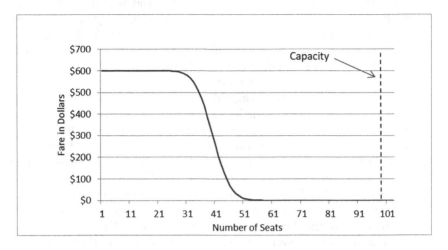

Figure 9.9 Expected Marginal Seat Revenue Plot for the $600 B Fare.

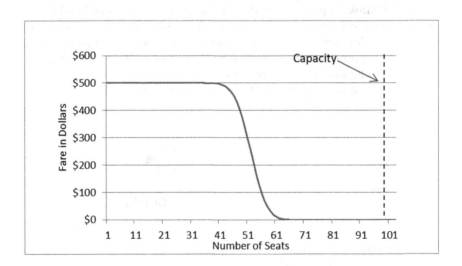

Figure 9.10 Expected Marginal Seat Revenue Plot for the $500 M Fare.

The EMSR curve can be calculated for every available fare on a given flight (assuming sufficient historical data). Now let's try another fare – an M fare at $500. The sales of $500 seats would produce a similar-looking EMSR plot, with different values, maybe something like Figure 9.10. Notice it has a shape similar to the $600 EMSR but starts at $500. It also has a slightly higher mean at about 52, which makes sense—chances are that more people will want to purchase the $500 seats than the $600 seats. Also note from the plot that even at this lower fare, the 100-seat aircraft will never sell out.

Now let's add in two more fare levels, our lowest and most restrictive, which we'll call a Q fare for $300, and an unrestricted fare called a Y fare at $800 (the highest). The two EMSR plots are shown in Figures 9.11 and 9.12. As the plots show, the Q fare is very popular, and if it is offered without fences, it would occasionally sell out the aircraft to capacity. On the other hand, the Y fare is not very popular, rarely selling the 30th seat.

By analyzing these EMSR plots, airline revenue managers can determine "how many seats *not to sell* in the lowest fare classes and to retain for *possible* sale in higher fare classes closer to the departure day" (Belobaba, 1989, p. 186). They do this by superimposing the plots on top of each other. Figure 9.13 shows the four EMSR plots from Figures 9.9 through 9.12 together. From the graph, notice that the airline can expect to sell a few Y fare ($800) seats, but the EMSR of the 21st seat (Point 1) sold at $800 is only $600. In other words, the average revenue from selling the 21st seat is the same whether it's sold at the Y fare or the B fare. If the airline tried to sell a 22nd seat at $800, its EMSR would be only about $530. However, the EMSR of the $600 tickets for the 22nd seat is still $600. As the plot shows, between Point 1 and Point 2, the EMSR of the $600 seats is the highest. At the 35th seat sold (Point 2), the $600 EMSR drops to about $500. Seats sold between Point 2 and Point 3 (the 51st seat) get the highest revenue from the $500 M fare tickets. Beyond Point 3, the plot shows the most discounted $300 seats are all that are selling. This booking limit process is known as "Littlewood's Rule," after Kenneth Littlewood who first described the process in 1972 while working for British Overseas Airways Company (BOAC) (Littlewood, 2005; Phillips, 2005).

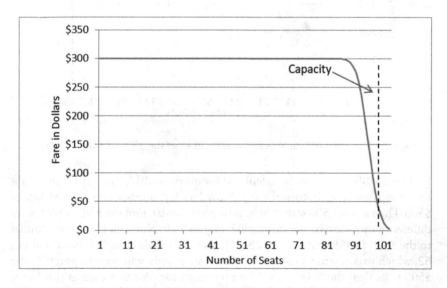

Figure 9.11 Expected Marginal Seat Revenue Plot for the $300 Q Fare.

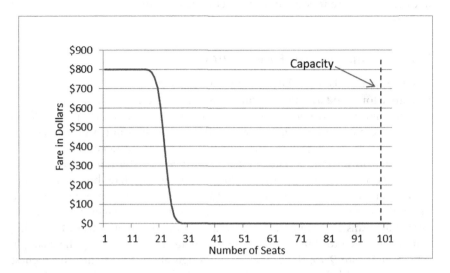

Figure 9.12 Expected Marginal Seat Revenue Plot for the $800 Y Fare.

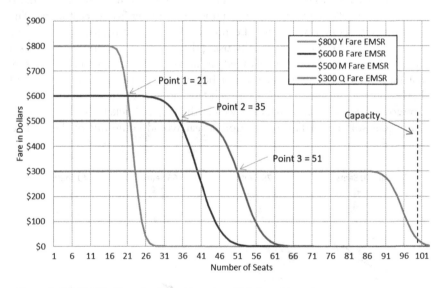

Figure 9.13 EMSR Plots for Four Fares in a 100-Seat Aircraft.

Based on this example, the airline would save 21 seats for $800 buyers, 14 seats (seats 22 through 35) for $600 buyers, 16 seats (seats 36 through 51) for the $500 buyers, and allow the 49 remaining seats (seats 52 through 100) to be sold at $300. The total expected revenue using this method can be calculated as the area under each respective EMSR curve between the respective limits (a calculus exercise if you are so inclined). This is similar to the area calculated in Figure 9.5.

9.8 Fare Buckets and Fare Nesting

From these calculations, the revenue management software allocates the seat inventory for each fare category on future flights. If forecast demand is perfect, the seat allocation would not change throughout the booking period. With imperfect demand forecasting, however, the airline does not want to get into a situation where a higher fare has sold out, but discount tickets are still available. Can you imagine telling a Y-fare customer that she can't book a reservation because you have already sold 21 seats at that fare, but you are still selling seats at the discounted B, M, or Q fares? If the Y-fare passenger wants a ticket, the revenue manager wants to ensure that the passenger can get a seat, even if it means using one of those allotted to the discounted fares. To do this, seats are assigned to "buckets" for reservation agents to sell. Using the example shown in Figure 9.13, the airline would assign all 100 seats into the highest Y-fare bucket – it is willing to sell as many Y-fare tickets as there are buyers. The Y fare is available until the aircraft is sold out. Similarly, the B fare bucket contains only 79 seats because the airline is willing to sell all of the seats at the B fare *except* the 21 they are saving for the late-buying Y-fare customers. The M-fare bucket contains 65 seats (the aircraft capacity minus those saved for the B-fare and Y-fare customers). Finally, using the same process, the Q-fare bucket has only 49 seats. This process is called *fare nesting* and is designed to make sure that higher-priced seats never sell out before lower-priced seats. Figure 9.14 shows how the buckets are filled for each fare category. As tickets are sold, a seat is removed from the sales bucket as well as the higher buckets. In this example, if an agent sold an M-fare reservation, a seat would come out of the M-fare,

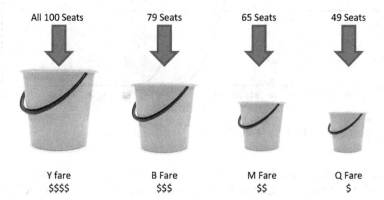

Figure 9.14 Fare Buckets.

B-fare, and Y-fare buckets. It follows that selling a Q-fare seat would reduce each of the buckets by one, and a Y fare would decrease only the Y bucket. The bucket concept might seem like an additional unnecessary step in the process, but it comes in handy as we progress to the next step where seats are allotted across a network rather than a single fight.

Fare nesting is often evident on airline reservation websites. Table 9.2 is an example website offering four different economy fare categories, decreasing in cost from left to right. Notice that in the example, the cheaper seats *always* sell out first. Because the fare buckets are nested, a higher-priced bucket will never sell out if there are still seats in a lower bucket. Sometimes you might notice an anomaly in fare nesting, but this is probably due to the physical limitations of the aircraft. For example, a flight might sell out the first-class seats (P, F, and A codes), but still have economy seats available because there are physically a limited number of seats installed in the first-class cabin.

9.8.1 Updating Expected Bookings

Revenue management as described to this point allocates specific numbers of seats to each fare bucket based on historical information, current trends, and other circumstances that influence the demand of various passenger segments. No forecasting model will be 100% accurate – many issues impact the forecast, both positively and negatively. For example, if the economy dips into recession, business demand will fall as will business ticket sales. Holidays, promotions, or other special events, on the other hand, can increase bookings. As bookings increase over time, the originally estimated demand can be refined so, that as the day of flight approaches, the demand comes more accurate.

The upward-sloping expected booking curve for a fare category shown in Figure 9.15 is generated from the sales history. It shows how a market segment is expected to book over time, starting the day the flight is opened for sale and ending on the day of departure.

Actual bookings are continuously compared with the expected booking curve allowing demand estimates to be refined and seat allocations adjusted. A brisk sales pace exceeding the expected booking curve for a certain fare may indicate demand is underestimated for that fare level. In that case, the number of seats allocated to lower fare categories would be reduced and those seats reallocated to the higher fare. Returning to the example, there are 21 seats in the Y-fare category and 14 seats in the B fare category. On a given day, maybe 10 days before departure, you expect to have sold 15 Y fares, 10 M fares, 8 M fares, and 45 Q fare seats. If the Y fare has sold 19 seats (five higher than projected for that day), the manager would want to reduce the lower fare seat allotments, anticipating increased Y-fare sales. On the other hand, if a higher fare category isn't selling as fast as it should, seats might be reallocated to lower-fare categories – in some cases re-opening previously closed discount fares.

Table 9.2 Typical Airline Online Reservation Web Page

Departure Time	Arrival Time	Flt #		Travel Time	Flex-Economy	Premium Economy	Classic Economy	Discount Economy
5:00 am	7:05 am	5757	Nonstop	2 +05	$280	$250	$220	Not Available
6:35 am	8:35 am	2387	Nonstop	2 +00	Not Available	Not Available	Not Available	Not Available
8:15 am	11:15 am	23	1 Stop	3 +00	$380	Not Available	Not Available	Not Available
10:30 am	12:35 pm	543	Nonstop	2 +05	$430	$400	$350	Not Available
12:45 pm	2:45 pm	980	Nonstop	2 +00	$380	$350	$320	Not Available
3:50 pm	7:00 pm	784	1 Stop	3 +10	$440	$410	$350	$250
4:55 pm	6:55 pm	555	Nonstop	2 +00	$500	$470	Not Available	Not Available
6:30 pm	8:35 pm	2990	Nonstop	2 +05	$400	Not Available	Not Available	Not Available
8:00 pm	11:30 pm	3125	1 Shop	3 +30	$300	$270	Not Available	Not Available

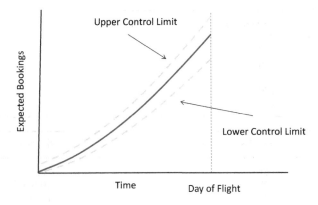

Figure 9.15 Expected Booking Curve.

Seat allocation among the fare categories is adjusted frequently, more often as the day of the flight nears. Thus, a discount fare that had previously sold out its allocation, may be reopened later if bookings in higher fares are slow. If you have a reservation, it pays to check frequently for a lower fare. The airline, however, wants to avoid price diversion and may or may not refund the difference if a lower fare becomes available. Policies differ among airlines and are subject to change.

9.8.2 Selling-Up

Selling-up occurs when a customer who meets all of the restrictions for a discount fare instead purchases a higher fare. In a way, it's the opposite of price diversion. Selling-up can occur because the lower fare category is sold out, or the customer is enticed into a higher fare by marketing or advertising. As an example of selling-up, if the seats are still available, Southwest Airlines offers a walk-up upgrade to their highest class (Business Select) as customers are waiting in the gate area. The target customers have already met the restrictions and purchased the lower-category fare, but might be inclined to upgrade at the last minute as an impulse.

9.8.3 Revenue Enhancement with Revenue Management

To show how revenue management can increase an airline's bottom line, Table 9.3 provides examples using the four fare classes from the examples above (with no oversales).

9.9 Network Allocation

Our discussion of revenue management thus far has been limited to maximizing revenue for single-leg flight segments. We now move to another refinement in

Table 9.3 Revenue Management Example for a 100-Seat Aircraft

Class	Fare	Unconstrained Sales[1]	Unconstrained Revenue
Y	$800	0	$0
B	600	0	0
M	500	0	0
Q	300	100	30,000
Total		100	$30,000

Class	Fare	Initial Seat Allocation[2]	Sales	Revenue
Y	$800	15	8	$ 6,400
B	600	20	18	10,800
M	500	25	25	12,500
Q	300	40	40	12,000
Total		100	91	$41,700

Class	Fare	Updated Seat Allocation[3]	Sales	Revenue
Y	$800	10	8	$ 6,400
B	600	20	18	10,800
M	500	30	30	15,000
Q	300	40	40	12,000
Total		100	96	$44,200

Notes:
1. These sales indicate no fences (segmentation devices, or sales restrictions) and everyone would simply buy the cheapest ticket they could (price diversion). In this case, everyone buys the Q fare and the aircraft goes out full, but only earns $30,000 in revenue, which might be below the airline's cost to fly the aircraft. Many of those $300-seat buyers would probably have been willing and able to purchase higher-priced seats but instead, they buy the discounted seats – just because they can.
2. Based on an EMSR analysis, the airline sets seat allocations (buckets) and derives segmentation devices (like early purchase) to prevent buying down. However, actual sales were less than forecast. As it turns out, 9 seats spoiled as the Y- and B-fares did not sell as predicted, and no adjustments were made. Even with 9 spoiled seats, the total revenue was significantly higher than the unrestricted sales.
3. The updated column shows a situation where a revenue manager noticed that the Y fare for this flight was not selling as planned and they reallocated 5 of the Y-fare allocations to the M-fare (more seats in the M-fare bucket). As a result, more M-fares were able to book seats. The spoil dropped to only 4 seats and the revenue went up $2,500.

revenue management, that of *network revenue management* (also called O&D *revenue management*).

The passengers' goal is to travel from their origin to their destination (O&D). The passenger would prefer a non-stop flight; however, most itineraries will require a connection at the airline's hub. In hub-and-spoke systems, demand is not based on the individual flight legs, but rather on the O&D market demand. Passengers purchasing O&D travel on a network purchase tickets for a single

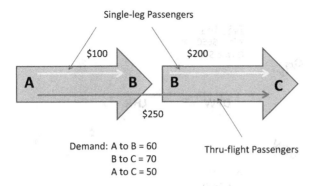

Figure 9.16 Network Allocation Example.

class – in other words, tickets are not issued for travel in a Y fare for one leg and a Q fare for the other (although a passenger could book separate flights using different fare classes) (Walczak et al., 2012).

With high demand, airlines have to be sure they allocate seats optimally to maximize revenue. In network travel, this is not an easy process as complications arise even in a simple single-fare level linear system such as the one depicted in Figure 9.16. In this example, the airline operates a single flight that originates at point A, flies to point B, and then continues to point C. Some passengers boarding at A are traveling just from A→B, but others are traveling from A→C with a short layover at B. These through-flight passengers get a slight break on the price compared to the two single legs combined. Additionally, more passengers will board at B who are traveling only from B→C. The challenge here is to allocate seats to maximize the revenue for the two legs. If the revenue manager allows tickets to sell without restrictions, there is a chance A→C tickets ($250) will take up seats that could be sold to individual leg buyers earning higher revenue. With accurate demand forecasting, the best combination can be determined, although as you may appreciate, even the solution to this very simple problem is not obvious (try it for yourself).

The maximum revenue is obtained by first meeting the demand for the two single (higher paying) legs. Sixty seats can be filled for both single legs (there will be some spill in the second leg but, as you will see, spill is better than spoil). The remaining 40 seats are sold to the A→C passengers. If the airline allocates 60 seats to A→B passengers (60 x $100 = $6,000), 60 seats from B→C (60 x $200 = 12,000) and 40 through-flight seats from A→C (40 x $250 = 10,000). The total revenue is $28,000 and the aircraft operates full on both legs.

Alternatively, if the airline accepted all the B→C demand (70), they could accept only 30 A→C through-flight passengers for that leg. That would result in the A→B having only 90 filled seats (60 A→B + 30 A→C) and 10 empty seats. This would net only $27,500 in revenue. Five hundred dollars might not sound like much, but a large network system that includes thousands of

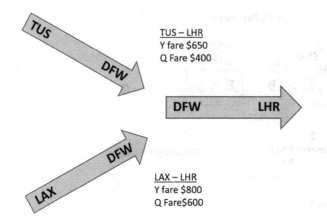

TUS – LHR
Y fare $650
Q Fare $400

LAX – LHR
Y fare $800
Q Fare$600

Figure 9.17 Origination/Destination Revenue Management.

city-pair combinations and several possible daily connecting itineraries for each can add up fast. That $500 can easily snowball into $150 million a year in additional revenue enhancement.

Figure 9.17 extends the problem to the hub-and-spoke network which is vastly more complex. Passengers connecting in DFW will be arriving from other spokes (Tucson [TUS] and Los Angeles [LAX] in this example), each with various fares. For large networks, the variability for fares could become very large – in the hundreds (Smith et al., 1992). One way to solve this problem is by using the fare buckets described earlier. To sell a connecting flight from any spoke to London (LHR) at a given fare level, each leg would need a seat in the applicable fare bucket available. In this example, a Q-fare passenger wanting the connecting flight from Tucson to London through Dallas (TUS→DFW→LHR) would need the Q-fare level available in both legs' Q-fare buckets. However, with not all Q fares the same, it becomes obvious that some Q fares would be better than others in maximizing revenue.

A technique to solve network revenue management problems is known as *virtual nesting*. Virtual nesting, pioneered by American Airlines in 1983, creates virtual buckets that correspond to O&D revenue values instead of single-leg revenue values. These virtual revenue buckets include combinations of different fare classes with similar values. In Figure 9.17 again, one bucket might include TUS→LHR Y fares as well as LAX→LHR Q fares, while another bucket might contain only LAX→LHR Y fares (Smith et al., 1992). While this might seem rather simple for the two-spoke network shown in the figure, in reality, there might be hundreds of fare combinations to sort into virtual buckets. The reservation process with virtual buckets works like the single-leg bucket process discussed earlier in this chapter – lower-value reservations are limited in number, while the highest revenue reservations are limited only by the capacity of the aircraft.

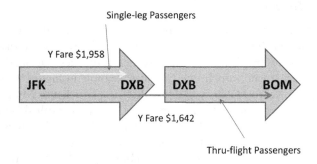

Figure 9.18 Demand-Driven Fares.
Source: From Emirates (n.d.).

The last example (Figure 9.18) emphasizes a fact that many students find puzzling: although airfares follow costs, demand is the driving factor. In this example, Emirates Airlines serves New York (JFK) to its Dubai (DXB) hub with non-stop service and connecting service from JFK to Mumbai (BOM) (with a stop in DXB). The fare for the non-stop flight from JFK to DXB is $1,958 whereas the connecting flight from JFK to BOM (through DXB) is only $1,642 (Emirates, n.d.).

Origin and destination markets vary greatly by the number of passengers wishing to travel, the purpose of travel (broadly business or leisure), and the competition. Emirates' cost to fly from JFK to BOM is considerably higher than from JFK to DXB, yet the fare is less. Emirates faces lower demand in New York to Mumbai market, so it must charge a lower fare to compete with other carriers. Were Emirates to charge a higher fare based on its cost of providing service, many passengers would choose competing airlines. Its total revenue would then be less than with the lower $1,641 fare. Of course, Emirates offers many fares in each market with various restrictions attached. The $1,958 and $1,641 fares are just one example.

A similar odd pricing scheme happened for a while in the San Antonio-Dallas-Washington DC market (SAT→DFW→DCA). American Airlines competed with Southwest in this O&D market (Southwest provides nonstop service from SAT to Baltimore/Washington). As a result, American listed an inexpensive connecting[5] fare (competitive with Southwest's) from SAT→DCA with a stop and aircraft change in DFW. Passengers originating in DFW for the non-stop leg to DCA paid a much higher fare – so much higher in fact, that some DFW→DCA travelers "commuted" to San Antonio to begin their trip there and pay the significantly lower fare. They would book an additional flight from DFW→SAT, get off the plane, and re-check in for the SAT→DCA flight. They would then re-board the same aircraft they were just on and fly back to DFW for the connection to DCA. Only in the airline business could this craziness happen!

9.9.1 Hidden City Ticketing

As illustrated in a couple of the examples above, there are occasional odd fares. Travelers have discovered that they can circumvent the airlines' revenue management in some instances by planning a trip past their actual destination, then bailing out at the intermediate stop. In the Emirates example in Figure 9.18, a passenger wanting to travel to Dubai, would buy the cheaper ticket to Mumbai but then disembark when the aircraft landed in Dubai. This ploy is called *hidden city* or *point beyond ticketing*. While not illegal, this purchasing trick is discouraged by airlines and generally forbidden in their contracts of carriage. For example, American Airlines' Conditions of Carriage prohibits "[p]urchasing a ticket without intending to fly all flights to gain lower fares (hidden city ticketing)" (American Airlines [AA], 2022, p.3). Penalties for violation could include cancelation of unused parts of the ticket, refusal to check bags, no refunds, or even added charges (AA, 2022). American and United have threatened to delete frequent flyer accounts for passengers who do not pay differences in fares (Puckett, 2019).

Nonetheless, the practice of hidden-city ticketing is still prevalent among fliers. An online website called "Skiplagged" finds hidden cities and boasts exposing "inefficiencies in airline pricing, such as hidden-city, to find you deals you can't get anywhere else" (Skiplagged, n.d., p. 1). Although United and Lufthansa sued Skiplagged, the lawsuits were thrown out (Puckett, 2019). Undeterred, airlines vow to use corporate security to find abusers and enforce the contract of carriage penalties. Passengers caught attempting hidden-city ticketing could have other problems. Aircraft diverts, reroutes, last-minute checked baggage, customs, and return flights could all be problematic for hidden-city travelers.

9.10 Revenue Management in Air Freight

Cargo carriers can be categorized into three groups. First the integrated carriers FedEx and UPS that provide door-to-door service with fleets of aircraft, sorting hubs and ground delivery vehicles (hence the term "integrated"). Second, the all-cargo carriers operating under contracts with various shippers and cargo forwarders provide airport-to-airport cargo service. And finally, the combination carriers, passenger airlines that also carry cargo in the aircraft baggage compartments (see Chapter 4).

About half of all airfreight is carried on scheduled passenger flights. In contrast to the integrators and all cargo carriers, these airlines have little market power to control pricing. Belly capacity is a "joint product" with passenger service. There is no accepted or theoretically correct method to determine the portion of total flight cost attributable to freight. If airline managers view belly capacity as a by-product of the passenger flight, then any cargo revenue exceeding the direct handling cost of the freight is considered a contribution to profit. This view leads to low freight pricing. Other airlines may account for the cost of belly capacity differently, allocating more of the total cost of the flight to freight. With this view, low freight rates which just cover handling costs result in a loss.

There are several key differences between passenger revenue management and cargo revenue management. First, freight varies greatly in volume, weight, and number of containers. With passenger carriers, each passenger has a definitive volume (defined by the seat pitch), and a definitive average weight (defined by the carrier's approved weight and balance program). However, cargo can come in many size-weight configurations requiring cargo revenue managers to manage all three variables. Second, much of the aircraft's cargo space is generally sold under long-term contracts, also called allotments, with the remaining capacity open on a general free-sale basis. Third, the major shippers who purchase capacity allotments are often unsure of the exact weight and/or size of their shipment until just before the cargo is loaded. As a result, the carrier has a difficult time planning space/weight combination requirements. Sometimes the negotiated capacity is exceeded and the carrier has the difficult task of determining which cargo is removed (Amaruchkul, Cooper, & Gupta, 2007; Hellerman, 2006). Fourth, itinerary control is not as important as passenger revenue management. Cargo shippers don't care about the routing of their shipment as long as it gets to the right destination on time (Kasilingam, 1996).

While the goal of passenger revenue management is to maximize the number of high-paying business passengers versus discount-paying leisure passengers, the goal of freight revenue management is to maximize higher-paying contracts versus lesser-paying free-space shipping.

9.11 The Future of Revenue Management

In this chapter, we discussed the development of passenger and cargo revenue management systems in the latter decades of the 20th century. In the 21st century, ancillary revenue has now become a major part of total airline revenue, especially for LCCs. Ancillary revenue (covered in detail in Chapter 7) is obtained from selling passengers services and amenities not included in the base ticket price such as fees for baggage, extra legroom, meal service, retail sales, boarding priority, and others. The percentage of revenue varies by carrier, but the clear winners in the ancillary revenue business are Spirit Airlines in the United States and Wizz Air in Europe. In 2020, the two carriers lead the industry with each collecting nearly 56% of revenue from ancillary charges (IdeaWorks, 2021). Airlines will benefit from enhanced ancillary revenue management systems.

As with passenger tickets, revenue management for ancillary sales is heavily dependent on the availability of sales data. However, legacy airline distribution systems do not have the capability to provide the data needed for improved revenue management. The idea is to correlate the passenger's fare class (or even the passenger's name) with ancillary purchases. It's thought that the fare class' (or passengers') history of ancillary purchases could predict their willingness to purchase ancillaries in the future and allow "smart pricing." In doing so, some think airline revenues could be raised by as much as 1% (Canaday, 2015). Better reservation systems like the International Air Transport Association's New Distribution Capability (NDC), covered in the next chapter, will provide some of the needed data.

As an example of revenue management's use in ancillary charges, Spirit Airlines carry-on bag fees vary by date of travel, route, and the time of purchase (Spirit Airline, n.d.). If you add the carry-on bag fee when you first book the flight, the fee might be $42, but if you wait until adding it when you check in, the fee goes up to $48–$53. If you show up at the gate with and have not paid for the carry-on bag, it'll cost you $71 (Spirit Airlines, n.d.).

With advances in data processing, revenue management as we know it today could change significantly. Where we now see a dozen or more fare categories on a 250-seat aircraft, future data processing may set a different price for each seat.

9.12 Summary

Unlike the pricing of many other products and services, airline ticket prices are not directly based on the cost of providing travel. Rather, prices are derived from demand and based on an estimate of the passenger's ability and willingness to pay. Appreciating this difference is important to understanding revenue management.

Airline revenue management attempts to maximize revenue through a combination of several practices:

- *Overbooking.* Overbooking offsets no-shows. Using this revenue management technique, the airline sells more seats than are available on the aircraft. The overbooking rate is based on the predicted number of no-shows or late cancelations.
- *Seat allocation.* Seat allocation is an attempt to get the passenger to pay the most they are willing to pay for their seat. Airlines using this revenue management technique will assign segmentation devices, like an advance purchase requirement or stay-over requirement, to place restrictions on the sales of limited discounted seats. In addition, limited seats are nested to ensure higher-priced seats do not sell out with lesser-priced seats still available. Seat allocations are continually reviewed and adjusted as the time of departure nears.
- *Network allocation.* Network allocation expands the single-leg seat allocation idea to include the whole origination-destination market where passengers are flying multiple legs to get to their destination within a large network system.

Is revenue management fair? Price discrimination and overbooking may run afoul of many peoples' sense of fairness; however, it does have two positive aspects. First, it assures business travelers the opportunity to reserve seats very near the departure date, and second, it allows some travelers discount fares that would not be available under other pricing systems.

9.13 Case Study: Southwest Airlines Dynamic Pricing and Seat Allocation

Southwest Airlines, often a maverick in the airline industry, not surprisingly uses a rather unique pricing method that includes a combination of dynamic pricing and seat allocation. What looks to be a fairly simple pricing scheme is more complex as you look deeper. There are four branded fare families as shown in the table. The highest price and least restrictive is Business Select (BUS), and the most restrictive and lowest price is Wanna Get Away (WGA). There are two additional fare families in-between: Anytime (ANY) and Wanna Get Away Plus (WGA+). The four families are partitioned by benefits as shown in Table 9.4.

The "base fare" is the WGA fare. The other three fare categories with incremental increases above the WGA fare. The WGA+ is $30 higher than WGA, the ANY fare is $50 above the WGA+, and the BUS is $40 above the ANY fare. Sales of the most-discounted fare, WGA, are limited in numbers based on historic booking trends of the higher categories. The WGA fare is generally the first to sell out. The same for the next two higher categories, WGA+ and ANY that are similar to the M and B fares (respectively) in Figure 9.13. The twist comes with the BUS category. Following the fare nesting and fare bucket process, the highest fare category sells out only when the aircraft is full. However, Southwest limits the sale of the BUS category. The reason is the definition of BUS itself. The company states BUS provides a "better boarding position" so as to be "one of the first customers to board the plane" (Southwest, n.d.). Once

Table 9.4 Southwest Airlines Branded Families

	Business Select™ (BUS)	*Anytime* (ANY)	*Wanna Get Away Plus*™ (WGA +)	*Wanna Get Away*™ (WGA)
Rewards Points	12 x fare	10 x fare	8 x fare	6 x fare
Two Free Bags	✓	✓	✓	✓
No Change Fees	✓	✓	✓	✓
Cancelation Credit	✓	✓	✓	✓
Same-day Changes	✓	✓	✓	
Same-Day Standby	✓	✓	✓	
Refundable	✓	✓		
Priority Boarding	✓			
Priority Security Lanes	✓			
Premium Drink	✓			

Notice how, moving from right to left, as the fares increase from WGA to BUS, more benefits are added. From Southwest Airlines (n.d.-a).

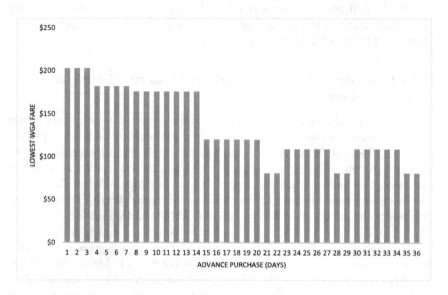

Figure 9.19 WGA Base Fares Plotted in Relation to Advance Purchase.
Note: Fares from Southwest.com (Southwest Airlines, n.d.-b) from SAT to HOU.

reservations hit the limit (probably about 15), the category closes. If they sold an unlimited number of BUS fares, they would reduce the incentive to buy the fare in the first place. For this reason, the Southwest website might show the BUS fare sold out before lower-priced categories.

The company's manipulation of the WGA base fares presents an example of dynamic pricing – in other words, pricing based on seats remaining in inventory, advance purchase, or future demand (Wittman & Belobaba, 2018). In this example, two things happen with the base fare regarding the amount of advance purchase – Figure 9.19 provides illustrations of both. First, a shorter advance purchase results in a higher base fare. The graph indicates five distinct price increases over the period displayed based simply on advance purchase. Second, the chart also shows deep discounts for two days of the week if reserving a seat more at least three weeks in advance – in this case, Tuesdays and Wednesdays, two days that often see less demand from business travelers. These deep discounts are added to attract leisure passengers to fill empty seats left by reduced business demand.

Considering the total of six different fare levels shown in Figure 9.19 for the base WGA fare, and the four fare levels that are derived from each base fare (WGA, WGA+, ANY, and BUS), a total of 24 different fares can be sold for a flight from San Antonio to Houston Hobby ranging from $81 to $323.

Concepts: Dynamic pricing, fare buckets, and seat allocation.

Major sources: (Southwest, n.d.-a, n.d.-b, n.d.-c) and (Wittman & Belobaba, 2018).

Notes

1 United says the flight was full, but not overbooked. Just before departure, United made the decision to add four crew members who needed to be repositioned for a flight the next day. That necessitated the removal of four passengers who had already boarded (Mutzabaugh, 2017).
2 Be sure to check the airline's contract of carriage for specifics concerning denied boarding policies.
3 On the average, vouchers for future travel do not cost the airline the full face-value of the voucher because not all vouchers are redeemed.
4 Seats sold with varying purchase and/or travel restrictions are not identical products. Most economists argue, however, that setting restrictions on seat sales does not rise to the level of creating a separate product and, therefore, is still price discrimination.
5 In industry terms, "non-stop" means just that, "direct" means a stop with no aircraft change, and "connecting" means a stop with an aircraft change.

References

Airline Tariff Publishing Company (ATPCO). (n.d.). www.atpco.net/sites/default/files/2017-10/RBD-Answer-Table_factsheet.pdf

Airlines for America. (2022, April 10). Data and statistics. www.airlines.org/dataset/a4a-presentation-industry-review-and-outlook/

Amaruchkul, K, Cooper, W., & Gupta, D. (2007, November). Single-leg air-cargo management. *Transportation Science* 41(4), 457-469.

American Airlines. (2022). Conditions of carriage. www.aa.com/i18n/customer-service/support/conditions-of-carriage.jsp?anchorEvent=false&from=footer?

American Airlines. (n.d.). Hidden city sample letter. www.aa.com/i18n/agency/Booking_Ticketing/Ticketing/hidden_city_ltr.jsp&locale=de_DE

Bailey, E., Graham, D., & Kaplan, D. (1991). Deregulating the airlines. The MIT Press.

Bailey, J. (2007). Bumped fliers and no plan B. *New York Times.* www.nytimes.com/2007/05/30/business/30bump.html?_r=0]&adxnnl=1&pagewanted=all&adxnnlx=1382065847-0odybQGfDbNbKl4U/VDBcg

Belobaba, P. P. (1989). Application of a probabilistic decision model to airline seat control. *Operations Research, 37*(2), 182-197.

Belobaba, P. P. (2009). Fundamentals of pricing and revenue management. In P. Belobaba, A. Odoni & C. Barnhart (Eds.), *The global airline industry* (pp. 73–111). Chichester, UK: Wiley.

Belobaba, P. P., & Wilson, J. L. (1997). Impacts of yield management in competitive airline markets. *Journal of Air Transport Management, 3*(1), 3–9.

Botimer T. C. & Belobaba, P. P. (1999). Airline pricing and fare product differentiation: A new theoretical framework. *The Journal of the Operational Research Society, 50*(11), 1085–1097.

Bowen, B. D., & Headley, D. E. (2015). Airline quality rating 2014. http://commons.erau.edu/cgi/viewcontent.cgi?article=1025&context=aqrr

Boyd, A. (2007). *The future of pricing: How airline ticket pricing has inspired a revolution.* New York: Palgrave Macmillian.

Canaday, H. (2015, February 2-15). Priceless optimization. Airlines are becoming increasingly sophisticated in maximizing yields. *Aviation Week & Space Technology,* pp. 40–41.

Cramer, C. & Thams, A. (2021). Airline revenue management: Current practices and future directions. Springer Gabler.

Curran, A. (2021). Airlines enjoy an ancillary revenue bonanza in 2020. *Simple Flying*. https://simpleflying.com/airlines-ancillary-revenue-2020/

Curry, B. (1977). Texas airline says nuts to D.C.-Houston fares. *Washington Post*. www.washingtonpost.com/archive/business/1977/12/07/texas-airline-says-nuts-to-dc-houston-fares/04a3faa8-0a8a-4aea-b619-b46c46dc1206/

Department of Transportation. (2020). Spirit Airlines, Inc. violations of 49 USC. §§ 41708 and 41712 and 14 CFR Part 250. www.transportation.gov/individuals/aviation-consumer-protection/spirit-airlines-order-2020-6-5

Doganis, R. (2019). Flying off course: Airline economics and marketing (5th ed.). Taylor and Francis Group.

Donovan, A. W. (2005). Yield management in the airline industry. *The Journal of Aviation / Aerospace Education & Research*, 14(3), 11–19.

Dunleavy, H. & Phillips, G. (2009). The future of airline revenue management. *Journal of Revenue and Pricing Management*, 8(4), 388–395.

Emirates. (n.d.) Make a reservation. http://fly4.emirates.com/CAB/IBE/SelectPrice.aspx

European Commission Regulation 261/2004. (2004). *Official Journal of the European Union*. http://eur-lex.europa.eu/resource.html?uri=cellar:439cd3a7-fd3c-4da7-8bf4-b0f60600c1d6.0004.02/DOC_1&format=PDF

FAA Oversale Policy, 14 C.F.R. § 250 (2013).

Hellerman, R. (2006). *Capacity options for revenue management: Theory and applications in the air cargo industry*. Berlin: Springer-Verlag.

History of AMR Corporation and American Airlines. (n.d.). www.aa.com/i18n/amrcorp/corporateInformation/facts/history.jsp

IdeaWorks. (2021). 2021 CarTrawler yearbook of ancillary revenue – Report. https://ideaworkscompany.com/wp-content/uploads/2021/09/2021-Ancillary-Revenue-Yearbook.pdf

International Air Transport Association. (n.d.). Resolution 728: Code designators for passenger ticket and baggage check. https://kupdf.net/download/rbdiatares728_5a12d1d6e2b6f52f1233f653_pdf

Jansen, B. (2015, May 4). Judge throws out United Airlines case against Skiplagged. USA Today. www.usatoday.com/story/todayinthesky/2015/05/04/united-skiplagged-lawsuit-federal/26864961/

JetBlue offers $10 seats to 10 spots. (2010). Travel Deals on NBCNews.com. www.nbcnews.com/id/35762207/#.UmbciVOgSVo

Kasilingam, R. G. (1996). Air cargo revenue management: Characteristics and complexities. *European Journal of Operational Research*, 96, 36–44.

Lawrence, R.D., Hong, S. J. & Cherrier, J. (2003). Passenger-based predictive modeling of airline no-show rates. The ninth ACM SKGKDD International Conference on Knowledge Discovery and Data Mining, August 24–27, 2003. http://citeseerx.ist.psu.edu/viewdoc/download?doi=10.1.1.15.2513&rep=rep1&type=pdf

Lindenmeir, J., & Tscheulin, D. K. (2008). The effects of inventory control and denied boarding on customer satisfaction: The case of capacity-based airline revenue management. *Tourism Management*, 28, 32–43.

Littlewood, K. (2005). Forecasting and control of passenger bookings. *Journal of Revenue and Pricing Management*, 4(2), 111–123.

Martin, H. (2010, April 28). Southwest fined for bumping. *Los Angeles Times*. http://articles.latimes.com/2010/apr/28/business/la-fi-southwest-fine-20100428

Maull, S. (2005). Bumped passenger wins suit. *Houston Chronicle*, 17 Nov, 2005.

Mutzabaugh, B. (2017, April 11). United "clarifies" that Flight 3411 was not oversold. *USA Today*. www.usatoday.com/story/travel/flights/todayinthesky/2017/04/11/united-clarifies-flight-3411-not-oversold/100331782/

Oversales, 14 CFR § 250 (1982). www.ecfr.gov/current/title-14/chapter-II/subchapter-A/part-250

Petzinger, T. (1995). *Hard landing: The epic contest for power and profits that plunged the airlines into chaos*. New York: Times Books.

Phillips, R. L. (2005). *Pricing and revenue management*. Stanford, CA: Stanford University Press.

Puckett, J. (2019). United Airlines is cracking down on the hidden-city ticket trick. *Conde Nast Traveler*. www.cntraveler.com/story/united-airlines-is-cracking-down-on-this-trick-for-cheap-flights

Qantas Airways. (2020). Qantas class types. www.qantas.com/us/en/frequent-flyer/qantas-class-types.html

Skiplagged. (n.d.). What is skiplagged? https://skiplagged.com/

Smith, B. C., Leimkuhler, J. F. & Darrow, R. M. (1992). Yield management at American Airlines. *Interfaces, 22*(1), 8–31.

Southwest Airlines (n.d.-c). Fare information. www.southwest.com/html/air/fare-information.html

Southwest Airlines Co. (2022). 2021 annual report. https://otp.tools.investis.com/clients/us/southwest/SEC/sec-show.aspx?FilingId=15534609&Cik=0000092380&Type=PDF&hasPdf=1

Southwest Airlines. (n.d.-a). Branded fares format overview. www.southwest.com/assets/pdfs/corporate-commitments/Southwest-Airlines-Branded-Fares-Format-Overview.pdf

Southwest Airlines. (n.d.-b). Low fare calendar. www.southwest.com/air/low-fare-calendar/index.html?clk=5736032

Spirit Airlines. (n.d.). Optional Services. www.spirit.com/optional-services

Toh, R. S. & Raven, P. (2003). Perishable asset revenue management: Integrated internet marketing strategies for the airlines. *Transportation Journal, 42*(2), 30–43.

US Department of Transportation, Bureau of Transportation Statistics. (2022). Passengers boarded and denied boarding by the largest US air carriers. www.bts.gov/content/passengers-boarded-and-denied-boarding-largest-us-air-carriersathousands-passengers

US Department of Transportation, Bureau of Transportation Statistics. (n.d.). Airline activity: National summary (US flights). www.transtats.bts.gov/

US Department of Transportation. (2015). *February 2015 Air Travel Consumer Report*. www.dot.gov/sites/dot.gov/files/docs/2015FebruaryATCR.pdf

US Department of Transportation. (n.d.) Bumping & oversales. www.transportation.gov/individuals/aviation-consumer-protection/bumping-oversales

Vinod, B. (2021). The evolution of yield management in the airline industry. Springer.

Walczak, D., Boyd, A., & Cramer, R. (2012). Revenue management. In C. Barnhart & B. Smith (Eds.) (pp. 101–160). Boston: Springer.

Weatherford, L. R., & Bodily, S. E. (1992, Sep/Oct). A taxonomy and research overview of perishable-asset revenue management: Yield management, overbooking and pricing. *Operation Research, 40*(5) 831–844.

Wensveen, J. (2011). *Air transportation: A management perspective* (7th ed.). Ashgate.

Wittman, M., & Belobaba, P. (2018). Dynamic pricing mechanisms for the airline industry: a definitional framework. *Journal of Revenue and Pricing Management* 18(2), 100–106. © 2018 Springer Nature Limited

Zizka, L., McGunable, D., Clark, P., & Essary, M. (2020). Airline overbooking: Customer (dis)satisfaction and communication challenges. SAGE Business Cases Originals. https://dx.doi.org/10.4135/9781529722420

Review Questions

1. What is the relationship between airline fixed costs and marginal costs and how do these relate to revenue management?
2. Why is a segmentable market important to a revenue manager?
3. What is meant by a perishable product, and why are airline seats perishable?
4. Explain why a perishable inventory is a very important characteristic of an airline's product.
5. Why do airlines overbook and why do some overbook more than others?
6. What is the difference between overbooking and overselling?
7. Oversold passengers can be categorized into two groups. What are these two groups and how does their compensation differ?
8. What is expected marginal seat revenue and why is it so important a factor in the seat allocation process?
9. What is fare nesting and how does it protect higher-paying seats?
10. How is network seat allocation different from leg seat allocation?
11. How does passenger revenue management differ from airfreight revenue management?

10 Distribution

Upon completion of this module, you should be able to:

- Identify key events in the history of airline computer rservation systems development.
- Describe what led to the development of the travel agency industry.
- Analyze the need for US government regulation of computer reservation systems and what led to its deregulation.
- Understand the impact of the internet on the travel agency industry.
- Describe the capabilities of the new distribution capability.

10.1 Introduction

Your company might manufacture a great product, but selling that product means running a successful marketing plan that culminates in getting your product into the hands of your customers. As discussed in Chapter 4, marketing a product requires a mix of activities. In business, it boils down to the *four Ps*:

- *Product*. The right product for the right market.
- *Price*. A price that provides value to the customer as well as adequate revenue to the seller.
- *Promotion*. Advertising that communicates information about the product to the customer.
- *Place*. A distribution channel that connects the product to the customer (Wensveen, 2011).

The last P, place, refers to the physical movement or distribution of the product from the manufacturer to the customer. This can be accomplished in several ways, but for many products, it is through a retail seller. A manufacturer creates a product and then transfers that product either directly, or through a distributor, to the retail seller. In some cases, the manufacturer might sell directly to the customer, possibly through internet sales or a factory outlet store.

DOI: 10.4324/9781003290308-10

The airline product is different than most retail products because it's not a physical product but rather a service. In service industries, three tasks must be accomplished to sell and distribute the product. The customer must be able to:

- Learn about the services offered. For airlines, this means finding information about flights and other services the airlines offers like baggage service and in-flight entertainment.
- Find out the prices. This would include the base fare and other additional fees that might apply.
- Obtain the desired service by making payment and receiving the product. In this case, the airline ticket.

Like other industry products, the airline product is sometimes sold through distributors, like travel agencies, and other times directly from the airline to the customer.

This chapter focuses on three primary distribution channels airlines use to disseminate their flight schedule, seat availability, and fare information to prospective customers and, if they are successful in making a sale, accepting payment and getting a ticket into the hands of the customer. The three primary distribution channels are:

- *Direct sales* from the airline to the customer through either airline reservation agents at call centers, or airline internet sites.
- Sales through the *global distribution system* (GDS) which include most "brick-and-mortar" travel agencies and many online travel agencies.
- Sales through the *new distribution capability* (NDC) that include some travel agencies and some airlines. (Gunther et al., 2012)

10.2 Airline Distribution History

In the US, the airline industry borrowed and built on many aspects of the railroad industry that they initially supplemented and eventually displaced for long-distance passenger transportation. Like the railroads, airlines themselves disseminated information on schedules and fares and sold tickets directly to passengers. As the airlines began flying early in the 1920s, airline sales distribution was decentralized consisting of airline representatives selling tickets and controlling inventory for flights at the point of the flight's departure (McKenney et al., 1995). As with railroads, a paper ticket was issued to the traveler that eliminated not only the need for the traveler to carry funds but also the need for the flight crew to collect fares.

To make a reservation, the traveler, or their agent, contacted the airline's reservation office at the departure city either in person or by telephone to check the flight schedule, seat availability (inventory), and price. Many times, these offices were located downtown to better accommodate business customers. The airline reservation agent would usually have to research the request and then

Figure 10.1 Early Reservation Center.
Source: Photo courtesy of American Airlines C.R. Smith Museum.

call the customer back with information about flight availability. If the schedule, availability, and price were acceptable, the customer would book the flight.

The airline reservation agent would note the transaction with a written reservation slip or notebook and post the reservation on large boards (maybe chalkboards) for all to see. This was known as the *Request and Reply system*.

Figure 10.1 is a depiction of an early reservation center. Reservation agents then communicated reservation information to the airline's centralized inventory control office. R. F. Meyer (as quoted in Copeland et al.,) descriptively portrays the scene at an early reservation office.

A large cross-hatched board dominates one wall, its spaces filled with cryptic notes. At rows of desks sit busy men and women who continually glance from thick reference books to the wall display while continuously talking on the telephone and filling out cards. One man sitting in the back of the room is using field glasses to examine a change that has just been made high on the display board. Clerks and messengers carrying cards and sheets of paper hurry from files to automatic machines. The chatter of teletype and the sound of card sorting equipment fills the air. As the departure date for a flight nears, inventory control reconciles the seat inventory with the card file of passenger name records.

(1995, p. 31)

10.2.1 The OAG

In 1929, an organization publishing railroad schedules began publishing a list of flights offered by airlines. Initially, it included 35 airlines, listing about 300 flights in a 24-page pamphlet called the Official Aviation Guide of the Airways. In 1948, it merged with a competitor and became the Official Airline Guide – OAG (Figure 10.2). The OAG included schedule information but no information about fares or seat availability. (The OAG is probably the "thick reference book" Meyer refers to in the quote above.) Although the OAG has gone through many changes over the years, it still exists today as a flight information database and is used as a source for flight schedule information. With an OAG subscription, travel agents (and maybe corporate travel offices) had information concerning flight schedules at their fingertips, but they still needed to contact the airline for fare information and seat availability (Gunther et al., 2012). Once the customer determined the availability of a flight they wanted, the customer (or their agent) still had to book the reservation with the airline, make payment, and receive a ticket.

Many times, manually constructing complicated linear-type itineraries required agents to make phone calls to multiple airline reservation offices. This made the process of determining availability and confirming reservations lengthy, – hours or maybe even days (McKenney et al., 1995). As airlines grew,

Figure 10.2 October 1954 OAG Cover and Sample Page.
Source: Images courtesy of Official Airline Guide.

the job performed by the reservation offices of manually tracking reservations, bookings, and available seats was quickly becoming unmanageable.

10.2.2 Payment

In 1938, one of the duties of the new Civil Aeronautics Board (CAB) was to promote and regulate air travel. And regulate they did – they controlled all scheduled airlines' fares, routes, and schedules. A few years later in 1945, a coalition of airlines called the Air Transport Association of America (later called the Air Transport Association and now Airlines for America) founded the Air Traffic Conference of America (ATC). The ATC controlled travel agency accreditation and the airline ticket settlement system. All agents selling tickets for airlines were required to be accredited through the ATC (Nicholls, 1985). The ATC also reconciled and distributed funds among airlines based on submitted ticket coupons. The ATC tariff department (later called the Air Tariff Publishing Company – ATPCO) published the scheduled airlines' fares.

10.2.3 Ticketing

Originally, tickets were paper, carbonized, multi-copy coupons. Each airline designed and issued its own tickets, so all airline tickets were different. Tickets issued by the airlines were transferable, negotiable documents and, as such, needed to be treated almost like cash. Ticket blanks (Figure 10.3) were kept in safes, accounted for, and tracked. Only the airline or a few airline-approved travel agents could issue the ticket needed to board the aircraft. Multiple leg flights needed several ticket coupons and the process involved several steps in which ticket data was transcribed.

After the 1929 Warsaw Convention spelled out the required contents of international airline tickets, the International Air Traffic Association (predecessor to the current International Air Transport Association – IATA) worked with airlines to develop the first standardized airline ticket in 1930. Handwritten tickets (Figure 10.4) persisted until 1971 when the IATA automated ticket printing with the Transitional Automated Ticket (TAT) as shown in Figure 10.5. In 1972, the IATA introduced the neutral TAT ticket ("The Paper Ticket," 2008).

In 1994, airlines began to use electronic tickets, and, by 1997, the IATA had instituted a global standard for electronic tickets. A few years later, in 2004, airlines began to phase out paper tickets that were virtually gone by 2008 ("The Paper Ticket," 2008). In fact, during 2013, the total number of paper airline tickets issued was only 28,400 out of a total of 143 million transactions (Rice, 2014). Twenty-six of the 206 airlines affiliated with the Airlines Reporting Corporation (ARC), a company that settles financial ticket transactions between airlines and travel agencies, no longer accept paper tickets. The ARC also reported that only three airlines do not support e-ticketing (Rice, 2014).

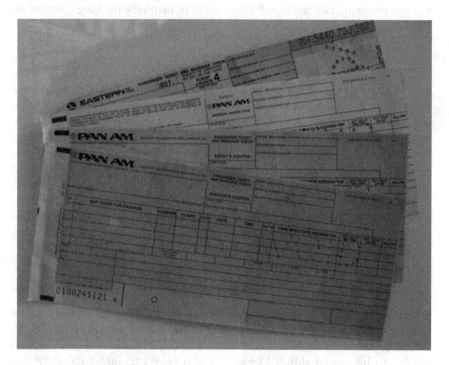

Figure 10.3 Ticket Blanks.
Source: Photo by B. Billig from "The Paper Ticket," 2008.

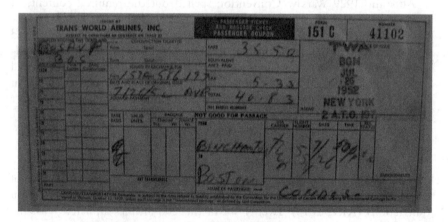

Figure 10.4 Handwritten Ticket From 1952.
Source: Photo by B. Billig from "The Paper Ticket," 2008.

Figure 10.5 Automated Ticket.
Source: Photo by B. Billig from "The Paper Ticket," 2008.

10.2.4 Growth of Travel Agencies

As the CAB allowed the growth of local service carriers, air travel was available between most of the large and midsized cities in the US. However, many itineraries involved traveling on several different airlines to reach your destination. Booking a long itinerary involving several legs on different airlines was a tedious job that required contacting several reservation offices and made the point of departure sale very cumbersome.

Even local customers often found the airline's airport office inconvenient. To improve this situation, airlines also operated distribution offices, or retail outlets, typically in the center of large cities. This made travel booking more convenient for business passengers who constituted most passengers in the early years of commercial aviation. But as airlines rapidly expanded after World War II, airline reservation offices at airports and downtown were not convenient for many potential passengers.

Further, accessing and understanding the connecting itineraries and fares was a complex task best suited to a trained specialist. Expanding their reservation office system to serve this rapidly growing customer base would have been prohibitively expensive for airlines, so they turned to independent travel agents to distribute their product information and sell tickets.

Airlines liked the idea of not having to burden their bottom lines with the extensive infrastructure and overhead needed to build and operate more airline offices. Travel agents were also attractive to the airline business because travel agents were compensated by airlines with a commission on tickets sold. By paying travel agents on commission, the airlines not only avoided the fixed costs incurred of having to pay employees even if no business was being conducted but also gave travel agents an incentive to sell the highest fares. Travel agents provided skilled and personalized service needed to help the customer find

Table 10.1 Travel Agency Locations, Sales, and Commission Rates

Year	Number of US travel agency locations	Sales (domestic and foreign carriers in billions)	Commission Rate (average domestic and international)
1977	15,053	$9.4	8.3%
1978	16,628	$11.4	8.3%
1979	18,121	$14.7	8.0%
1980	17,339	$18.1	8.5%
1981	19,203	$20.0	9.3%
1982	20,962	$21.8	9.5%
1983	23,059	$25.8	10.0%
1984	26,037	$25.9	10.1%
1985	26,297	$33.4	10.2%

Source: Data from Nicholls (1985).

information, make reservations, issue tickets, and deal with other travel issues like passports, vacation packages, hotels, and rental cars. Travel agents soon grew to be the predominant distribution method.

Before airline deregulation, travel agents' commissions were also regulated by the CAB at rates that allowed even smaller agencies to stay in business. The regulated airfares and regulated commissions made the travel agency business very stable (Levine, 1987).

Airline deregulation in 1978 also deregulated travel agencies. The ensuing explosive growth of airline traffic combined with the increased complexity of schedules and routes made skilled travel agents even more of a necessity. It's no surprise that travel agencies eventually grew to dominate the airline distribution system, selling 80% of all airline tickets by the mid-1980s. Table 10.1 shows the growth rate for US domestic travel agencies from deregulation through the mid-1980s.

10.2.5 The Reservisor

With little or no automation, managing an airline's inventory of seats available for sale was a paper nightmare. A trunk carrier in the late 1960s might have 50 airplanes of 100 seat capacity flying five flights per day. This created a requirement to track not only the availability of millions of seats per year but also the associated fare information. The paper inventory management system was complex, inflexible, unwieldy, and subject to error. The need for computer-based automation was critical.

In 1943, Charles Amman, the head of American Airlines' Systems and Methods Division, had an idea for electronically automating the reservations process. He was originally trained as a radio engineer and with that experience, he devised a three-step process that would lead to reservations automation. The first step was to automate seat inventory information, the second step was to automatically adjust the inventory, and the third step was to automate the transmission of passenger information (Eklund, 1994; McKenney et al., 1995). Amman's boss, American Airlines CEO C. R. Smith, envisioned substantial growth for American Airlines after the war and liked the idea of an automated reservation system. At the time, however, technology like this was not being actively developed, so, in 1944, Amman, using his engineering skills, built a working model of a simple electronic inventory control system. Seeing the functionality of such a system, Smith approved funding which led to a contract with Teleregister Corporation, a pioneer in the development of special-purpose equipment, to design and build a prototype (McKenney et al., 1995).

The project was completed in February 1946 and installed in American's Boston reservation center (Davidow & Malone, 1992). It was called the *Reservisor*. This first model resembled the stock trading board for which Teleregister was known.

This first-generation Reservisor automated activities in the reservation office, but, because it had no communication capability, it still required travel agents or travelers to telephone the reservation office. Another limitation was that it automated inventory control only; index cards with passengers' names were still carried from desk to desk. Although limited, the Reservisor allowed the Boston office to process an additional 200 passengers daily with 20 fewer employees (Edelman, 2009; Davidow & Malone, 1992).

10.2.6 The Magnetronic Reservisor

A few years later, in 1952, with Amman still leading the project for American, Teleregister built a second-generation Reservisor called the *Magnetronic Reservisor* (nicknamed Girlie) that included a new desk set (Figure 10.6) and a new-technology storage capability (Figure 10.7).

The central Magnetronic Reservisor was installed in the spring of 1952 at LaGuardia airport in New York (Petzinger, 1995; McKenney et al., 1995; "Teleregister," n.d.). Between 50 and 80 agent desk sets were tied into the LaGuardia Reservisor, most of which were in American's downtown ticket sales office just a few miles away on 42nd Street in New York City.

In 1956, another larger and faster version of the Magnetronic Reservisor was installed at American's West Side Terminal in New York that could handle 2,000 flights in a 31-day period. Although the Reservisor could remotely track inventory (it knew how many seats were left on a given flight), it could not attach a name to a seat, or track any other customer data (like a phone number or address). As a result, passenger manifests were still completed manually (Petzinger, 1995). As larger jets entered the airlines' inventories in the early 1950s, this became a huge limitation.

Teleregister also installed similar reservation systems for several other airlines in the mid-1950s, including United's "Unsil" system, and unnamed systems for National and Braniff airlines. Teleregister's inventory control systems were also used for railroads and industrial warehouse inventory automation ("Teleregister," n.d.).

Figure 10.6 Magnetronic Reservisor deskset on display at the American Airlines C.R. Smith Museum.

Source: Photo from Wikimedia Commons (CC BY-SA 4.0) (Z22, 2020).

10.2.7 Reserwriter

Amman's third step, to automate passenger information, began in 1954. Working with IBM, Amman developed a system called the *Reserwriter*. The Reserwriter system read punched cards containing passenger information and then converted them to paper tape that could be transmitted over the teletype. By 1956, the Reserwriter systems were linked into a nationwide network and by 1958, Reserwriter terminals were in key locations across the country in American Airlines' network (Davidow & Malone, 1992, p. 45; Santiago, 2015a, p. 5).

Although a vast improvement over the Request and Reply system, the Reservisor/Reserwriter systems were prone to error about 8% of the time and still required agents to fill in many steps not automated. For example, "a round trip reservation between Buffalo and New York LaGuardia required 12 people, 15 distinct steps, and could take as long as three hours" to complete (Santiago, 2015a, p. 5).

10.2.8 Sabre

Maybe only by chance, but once again, American Airlines led the industry to the next level of reservation automation. In the summer of 1953, a sales

Figure 10.7 The Magnetronic Reservisor.
Source: Photo courtesy of American Airlines C.R. Smith Museum.

representative from IBM named R. Blair Smith was boarding an American Airlines flight from Los Angeles to New York. He chose a window seat in the last row, and he was soon joined by Smith – the CEO of American Airlines. At the time, a flight from Los Angeles to New York was an extended journey in a propeller-driven aircraft, requiring 10 hours, including intermediate fuel stops. Blair Smith knew of Teleregister's work on American's Reservisor and during the multi-stop transcontinental flight, he explained to Smith that IBM was researching the possibility of building a system that would be much more capable than the Reservisor (Mapstone, 1980). IBM at the time was working on a Cold War project for the Air Force to computerize control of US air defense. This project was named the Semi-Automatic Ground Environment

(SAGE) and was one of the first real-time applications of computer technology (Petzinger, 1995; Eklund, 1994). IBM was looking for possible commercial applications for SAGE technology that included remote terminals and real-time teleprocessing. IBM named this commercial spin-off project Semi-Automatic Business Environment Research (SABER). Smith was very interested in what Blair Smith was saying and invited him to tour the Reservisor facility at LaGuardia Airport. After the tour, Blair Smith consulted with his boss and wrote a letter to Smith suggesting a joint research and development project between IBM and American Airlines (Mapstone, 1980). The project development with American went on for about five years and eventually led to a contract proposal from IBM in 1958. At first, the team at American did not have a name for their company-specific project, but in 1959, the project manager decided to call it Sabre[1] which stood for Semi-Automatic Business Research Environment. In 1959, American accepted IBM's proposal and in 1960, the first Sabre system was installed in Westchester County, north of New York City (Santiago, 2015b).

Sabre's mainframe computer became the largest private real-time data processing system in the country. Because of the ground-breaking nature of the project, there were a few setbacks, and it wasn't until 1964 that Sabre (Figure 10.8 top) became fully operational, initially with 1,500 remote terminals, mostly in airline reservation offices and selected travel agencies (Mapstone, 1980). Sabre turned out to be a 400-man-year effort costing American Airlines $40 million; however, Sabre saved American 30% on reservation personnel staffing ("Sabre History," n.d.).

Sabre's major technological breakthrough was the inclusion of *passenger-name record* (PNR) information which tied a passenger's name to a seat reservation (Copeland & McKenney, 1988). The lack of PNR was one of the main limitations of the Reservisor. Sabre allowed agents to book flights directly, bypassing telephone calls to the airline to check flight availability and confirm bookings.

While developing Sabre, IBM was concurrently working on projects for Delta (DELTAMATIC) and Pan Am (PANAMAC). Figure 10.8 (bottom) depicts the IBM PANAMAC workstation. These systems became operational shortly after Sabre. They were functionally similar to Sabre but used different IBM processors (Boyd, 2007; Copeland & McKenney, 1988). At about that time, IBM came out with a new technology computer architecture called System 360. Delta updated their system in 1968 which they then called the Delta Automated Travel Account System II (DATAS II). The race was on for other airlines to catch up. Eastern, TWA, and United needed computer-based PNR systems to keep pace.

IBM's experience gained through the Sabre project and their new System 360 computers helped them develop their own Programmed Airline Reservation System (PARS) which they developed to market to airlines. The only major airline interested, however, was Eastern, calling it System One. By the end of 1971, both United and TWA were using modified PARS systems. TWA's PARS

Figure 10.8 (Top) Sabre Advertisement. Note. Reprint Courtesy of IBM Corporation©. (Bottom) PANAMAC Workstation.

Source: Photo Courtesy of the Pan Am Historical Foundation/www.panam.org.

system was like Eastern's, and United's was an expanded PARS system they named Apollo. In 1972, American also completed upgrading Sabre with the newer PARS technology. These new systems were generically named *computer reservation systems* (CRS). By 1972, Northwest Airlines had a Univac system and was the single airline (of the 10 trunk airlines) that was not running an IBM PARS-based CRS (Copeland & McKenney, 1988).

10.2.9 Travel Agents Get Involved

Although several airlines had automated reservation technology by the early 1970s, most of that technology was used internally within the airline. Either the traveler or the travel agent still had to call or visit an airline reservation agent. As late as 1968, airlines had installed only a few airline-specific CRS machines in a select few travel agencies to support their operations, but there was no overall use of airline CRSs outside the airline. Travel agents needed a system like the airlines had, but they wanted a single system that could process reservations for all airlines (Boyd, 2007). In 1972, the president of the American Society of Travel Agents (ASTA) joined American Airlines to develop a joint computer reservation system for travel agents that would be common to all airlines. As ASTA and American collaborated on the project, named the Joint Industry Computerized Reservation System (JICRS), they tried to build interest from other airlines. Other airlines were quietly avoiding commitment to the project, and in January 1976, United officially announced it would not support JICRS and instead outlined a plan to begin installing their Apollo systems in travel agencies the following September. Without United's support, the idea of a single travel agent CRS was going nowhere and JICRS was eventually dropped.

Bob Crandall, then CEO of American, quickly found himself in a battle with United's Dick Ferris to hard-wire the most CRS terminals (Sabre and Apollo, respectively) in travel agencies (Petzinger, 1995). Although United took the first step in announcing their plans, their September 1976 target date gave American time to get the jump on them. Crandall, at American, had already been working on a contingency plan and they jumped into action. Crandall, who felt he was fighting for the life of American Airlines, decided that the added cost to accelerate the effort was worth it, especially after being told that a travel agent using a Sabre terminal instead of an OAG book could more than double their annual bookings (Petzinger, 1995). American began installing Sabre terminals in April 1976 and, by the end of the same year, American had about 130 travel agencies equipped with Sabre. United met its September goal but with only four pilot locations (Copeland & McKenney, 1988).

Over the next few years, both American and United invested millions of dollars installing Sabre and Apollo CRS systems in travel agencies. TWA also opted to install their PARS systems, but to a smaller extent than American and United. Eastern and Delta, who competed mostly with each other, were also less aggressive with their installations of System One and DATAS II, respectively.

"Although multiple computer reservation systems were available to travel agencies, most relied on only one system" (Edelman, 2009; "Airline Ticketing: Impact of Changes," 2003). Using more than one system was very inefficient due to additional employee training and customer record-keeping requirements.

With most travel agencies having only one airline's CRS system, airlines found that in order to have their tickets available for sale at all travel agencies, they had to be able to list their flights on competitors' systems. For example, most travel agencies in the Dallas area used American's Sabre system, while most agencies in Chicago used United's Apollo system. To sell tickets in the Dallas area, United had to make a deal with American to list their flights on Sabre. Likewise, American had to get their flights listed on Apollo to sell tickets in the Chicago area ("Airline Ticketing: Impact of Changes," 2003). "Airlines had little choice except to participate in each CRS" and before long, most airlines' flight information was available on all CRS systems ("Airline Ticketing: Impact of Changes," 2003, p. 9). This led to a co-host system, where airlines with their own CRS systems agreed to co-host each other's flights. Airlines who did not own their CRS system, however, had little choice but to pay the CRS-owner airlines to have their schedules displayed on the system owner's CRS.

With their CRS systems in travel agencies, American and United, and, to a lesser extent, TWA, Northwest, Delta, and Eastern Airlines, enjoyed a significant competitive advantage over other airlines who did not own their own CRS. These advantages fell into two principal categories: market intelligence and the ability to monitor its competitors. Airlines with CRSs in travel agencies were able to gather real-time information concerning the market preferences of travelers and the success of advertising, promotions, or other sales initiatives. A CRS owner airline could track the sales of competitors' inventory, measure their full-fare business traffic versus discounted leisure traffic, and even see the utility of their competitors' frequent flier program.

10.3 Birth of the Global Distribution System

Airline CRSs expand by adding travel services, including hotels and rental cars to airline flight information. This was the beginning of a new industry term – the systems that were called airline "computer reservation systems" had now turned into *global distribution systems* (GDS), reflecting the increasingly international and diverse nature of travel they encompassed ("Airline Ticketing: Impact of Changes," 2003).

Figure 10.9 shows a typical GDS display a travel agent might see. This example is from a Sabre system, but the other GDS displays are similar. The top line shows the travel date and origination-destination; in this case, the flight date was Sunday, 20 July, with the origination in San Antonio (SAT) in Central Daylight Time and the destination Seattle (SEA) in Pacific Daylight Time. This is a relatively simple example, showing one non-stop flight (an Alaska flight on line 1) and two connecting flights on the first screen (a Delta connecting flight

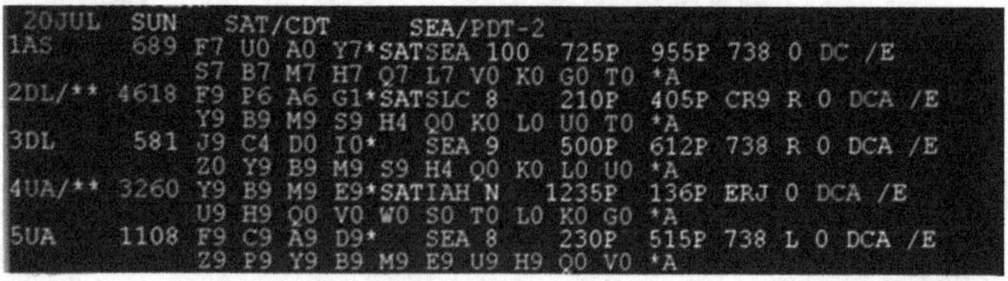

```
20JUL    SUN      SAT/CDT        SEA/PDT-2
1AS       689  F7 U0 A0 Y7*SATSEA 100   725P   955P 738 0 DC /E
              S7 B7 M7 H7 Q7 L7 V0 K0 G0 T0 *A
2DL/**   4618  F9 P6 A6 G1*SATSLC 8     210P   405P CR9 R 0 DCA /E
              Y9 B9 M9 S9 H4 Q0 K0 L0 U0 T0 *A
3DL       581  J9 C4 D0 I0*    SEA 9     500P   612P 738 R 0 DCA /E
              Z0 Y9 B9 M9 S9 H4 Q0 K0 L0 U0 *A
4UA/**   3260  Y9 B9 M9 E9*SATIAH N    1235P   136P ERJ 0 DCA /E
              U9 H9 Q0 V0 W0 S0 T0 L0 K0 G0 *A
5UA      1108  F9 C9 A9 D9*    SEA 8    230P   515P 738 L 0 DCA /E
              Z9 P9 Y9 B9 M9 E9 U9 H9 Q0 V0 *A
```

Figure 10.9 Typical Sabre GDS Screen Display.
Source: Photo by B. Billig.

on lines 2 & 3 and a United connecting flight on lines 4 & 5). More options
would be available on subsequent screen pages). Looking at the non-stop flight
on line 1, moving left to right:

AS is the two-letter abbreviation for the airline – in this case, Alaska Airlines.
Each airline, worldwide, is assigned a two-letter alpha-numeric identifier. Also
shown in this example is DL, the code for Delta Air Lines, and UA for United
Airlines.

689 is the flight number.

The next entries are the available seats, listed by fare categories in descending
price order. The top line shows four fare categories (F, U, A, and Y) and the
corresponding number of seats available. The second line shows 10 more fare
categories (S, B, M, H, Q, L, V, K, G, and O) with their available seats. To save
space on the page, and not disclose proprietary seat allocation figures, airlines
show only a single digit for the number of seats available in each fare category. In
this example, Alaska's numbers only go up to 7, but the actual number is either
7 *or higher.* If the number of seats drops below 7 the single-digit will reflect the
actual number of seats. Other airlines may use different numbers: note that DL
and UA on the lines below stop at 9 seats.

Before continuing with the other data, this is a good opportunity to review
seat allocation and fare categories from the previous chapter. First, look at the
first five fare categories (J – Z) for Delta flight 581 on lines 3 & 4. These are
all business-class seats. Note how the lower fare categories (D, I, and Z) sell
out first. You might recall that with fare nesting, a higher fare will never sell
out with a lower fare still available. Now, look at United flight 3260 on line
4. It looks like many of the most discounted seats on line 2 are sold out. This is
explained by looking at the aircraft. It shows an ERJ which is a smaller regional
jet that seats only 35 to 50 people.

Now getting back to the line 1 Alaska flight 689 example:

The flight operates non-stop from SAT to SEA. Note that line 2 indicates
a connecting flight; line 2, flight 4618 is from San Antonio to Salt Lake City
(SLC) with a 55-minute layover, then continues with line 3, flight 581 from
SLC to SEA.

The line 1 departure is at 7:25 pm CDT and arrival is at 9:55 pm PDT.

The aircraft code is 738 indicating a Boeing 737–800.

The following 0 indicates non-stop.

The remaining alpha–numeric entries are for system administration.

Once the agent reviews these flights, he or she can select the next screen to review more flight options. With another keyboard entry, the agent can view the fare information.

10.3.1 CRS Favoritism

Although many other airlines listed flights on host airlines' computer systems, the host still controlled how the system worked and that meant that whenever possible, the system favored the host airline. There were several ways this favoritism worked, but perhaps more than any, CRS display bias or *screen bias* drew the ire of smaller airlines and the attention of Congress and the Department of Transportation.

10.3.1.1 Screen Bias

When looking for a flight for a customer, a travel agent typed in the traveler's needs and the screen would light up with available flights like those shown in Figure 10.9. It didn't take the host airlines long to figure out that agents tended to book the first flight they saw on the screen (like Alaska flight 689). Statistics showed agents booked the top flight on the screen 45% to 50% of the time. When available flights filled multiple screens, the agent would book off the first screen 80% to 90% of the time (Shaw, 2011). To favor their airline, the host airline would manipulate the programming so that the screen always showed their flights first (or at least on the first screen if there were multiple screens). This screen biasing led to millions of dollars of extra revenue for the host computer owners. The two airlines with the most CRSs in travel agencies, American and United, were the biggest perpetrators of screen bias, but other airlines did it as well.

10.3.1.2 Halo Effect

Another way host airlines benefited from agents using their system was through the *halo effect*. Information on seat availability and reservation confirmations tended to be more timely, accurate, and reliable for host airlines because the host's internal reservations were managed by the same computer as the CRS while non-host airline information needed to be transmitted between systems. As a result, travel agents were more confident booking flights on the host airline than on non-host airlines and tended to disproportionately book on the host airline (US Department of Justice, 2003). Additionally, travel agents favored the locally dominant airline's CRS which was probably the one installed in their office. Override commissions paid to travel agents for surpassing a target sales goal added to this dominance of the local travel market. This broad effect became known as the halo effect (Borenstein, 1992).

10.3.1.3 Co-Hosting

The host airlines added another revenue source by selling *co-host* status to airlines. For a fee, an airline could boost their flights to a better position on the agents' screen (but not ahead of the host airline). This was accomplished by modifying the programming, or by simply assigning penalty minutes to the flight times of the non-host airlines. As an example of how important co-hosting was, in 1981, American had nine co-hosts on the Sabre system, making them an additional $6.9 million just in co-host fees (Copeland & McKenney, 1988).

10.3.1.4 Screen Padding

Airlines who code-shared with other airlines could *screen pad* by listing flights under two different flight numbers. For example, Delta Air Lines and Western Airlines were code-sharing before their merger in 1987. A single flight flown by either company might be listed as both a Delta flight number and a different Western flight number. Listing the flight under both numbers takes twice the screen space and, therefore, twice the travel agent's attention.

10.3.2 GDS Regulation

After airline deregulation in 1978, airlines were not shielded from competition – they now had to compete to stay in business. This new competition stimulated the use of computer reservation systems. By 1984, travel agents were responsible for 60% of all airline revenue, and 90% of those sales were made using a CRS. This reliance on CRSs increased the effect of screen bias on non-host airlines. Some non-host airlines complained that the host airlines were using their CRSs as a competitive weapon to hinder competition At least one airline, People Express, met its demise at least partly because of the lack of a competitive computer reservation system (Ravich, 2004). It was very clear those airlines that hosted CRS systems not only exercised a great degree of leverage in the airline business but also created incremental revenues not available to non-host airlines. In 1984, there were a total of six GDS systems, and five of those six were owned by airlines (US Department of Justice, 2003):

Sabre (owned by American)
Apollo (owned by United)
System One (owned by Eastern)
DATAS II (owned by Delta)
PARS (owned by TWA)
MARS PLUS (independent)

Host airlines, like American, justified the benefit they gained from screen bias as merely a return on the $467 million investment they made between 1976 and 1983 in CRS development (Copeland et al., 1995). The debate over the use of screen bias grew to allegations of antitrust violations as non-host airlines claimed the anti-competitiveness went so far as to harm the consumer. A CAB study concluded that airlines' abuse of the CRS system was indeed hurting the consumer. In November 1984, one of the last official actions of the CAB came just six weeks before its closing. The outgoing CAB put into effect regulations governing airline-owned computer reservation systems. These rules include:

- *Prohibited screen bias.* One of the most distinguishing features of CRS regulation was the prohibition against ranking flights based on carrier identity (display bias). The rules did not prescribe specific prohibitions, but only that CRSs contain at least one integrated display that lists the schedules, fares, and availability of all participating carriers. Additionally, the criteria used to rank flight displays had to be disclosed to participating carriers.
- *Required equal functionality.* The CRSs must allow equal reliability, accuracy, and accessibility of information for any participating airline. All participating carriers must have equal access to software enhancements and speed of data transmissions.
- *Prohibited discriminatory booking fees.* Host airlines could not charge competitors higher booking fees. This rule also helped smaller carriers with less bargaining power benefit from fees negotiated by larger airlines. (US Department of Justice, 2003)

The feeling in Europe was much like that in the US. The CRSs predominantly used in Europe, Amadeus and Galileo, both Sabre/PARS derivatives, were owned by airline consortiums, and it was thought that the airline ownership fostered a "market situation … incompatible with the principles of free competition" (Cavani, 1993, p. 446). As a result, the European Community passed three regulations designed to guarantee "complete and nondiscriminatory information, fair access to the computer systems for all air carriers, and a policy to protect free market equilibrium" (Cavani, 1993, p. 447). The primary regulation, passed in July 1989, instituted a Code of Conduct for Computerized Reservation Systems (European Union, 1989). It was later amended requiring CRS parent airlines to participate equally in all other CRSs.

Although Canadian airlines did not own any CRSs in 1995, the Canadian Aeronautics Act instituted CRS regulations to ensure, among other things, that displays were "comprehensive, neutral and non-discriminatory" (Canadian Minister of Justice, 2022, p.4).

Figure 10.10 shows travel agent market shares of the top five US systems in 1988. By 1990, 93% of the 35,000 travel agents in the US were using a CRS (Davidow & Malone, 1992).

In 1992, the DOT reexamined the CAB rules and elected to continue them, noting that GDSs were even more powerful than in 1984. Travel agents

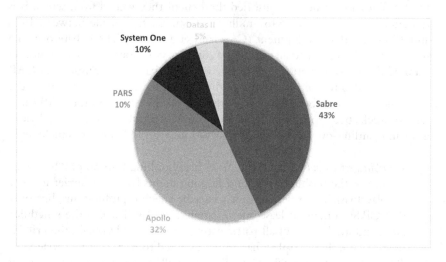

Figure 10.10 US CRS Market Share in 1988.

accounted for 75% of airline revenue and over 95% of travel agents relied on a GDS (US Department of Justice, 2003). The DOT also included a sunset date for the rules in December 1997, later revised to March 2003, and again revised to January 2004 ("CRS regulations," 2002).

The non-hosting airlines were not satisfied, however, and filed a $400 million lawsuit seeking $250 million from American and $150 million from United charging that they "possessed a monopoly in electronic booking of airline seats and were using that status to limit competition" (Copeland et al., 1995, p. 51). The Government Accounting Office studied the issue and determined there was insufficient data to prove booking fees were substantially above costs and harmful to competition. Some of the claims were rejected and others were settled out of court.

10.4 GDS Divesture and Worldwide Growth

Beginning with American Airlines' sale of Sabre in 1996 and United Airlines' sale of Apollo in 1997, airlines began selling off their GDS systems to raise cash. Finally, by 2003, when Worldspan, a combination of PARS II and DATAS II, was sold to private investors ("Airline Ticketing," 2003), all airline-owned CRS systems were independently owned and operated. Once divested from airline ownership, the already unpopular GDS practices became even more objectionable to airlines. One practice that airlines found particularly loathsome was when GDSs began charging airlines a booking fee averaging about $12.50 per flight but then rebated $3.00 to $5.00 of that fee to the travel agent as an incentive to use the GDS. This incentivized agents to book travelers on circuitous routings, maximizing the number of segments and fees.

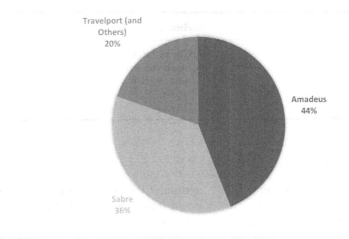

Figure 10.11 Worldwide GDS Market Share in 2017.
Source: Data from Business Travel iQ. (2017).

The first non–North American CRS, Amadeus, was developed in 1987 by a consortium of four European airlines (Air France, Iberia, Lufthansa, and SAS) to serve as an alternative to Sabre. The Amadeus software was an adaptation of System One software (which was an offspring of Sabre) and became fully operational in 1992. Later, in 1995, Amadeus consolidated with System One.

Another European system, Galileo, was also formed in 1987. Like Amadeus, Galileo was formed by a collection of nine European airlines and modeled after United's Apollo which was absorbed by Galileo. Later, Travelport bought Galileo, which continued to operate independently, and added Worldspan in 2007.

These mergers resulted in the three main GDS systems we see today: Sabre, Amadeus, and Travelport. As shown in Figure 10.11, Amadeus has the highest worldwide market share and is the most popular in Europe, the Middle East, and Africa. Sabre is most popular in North America and Sabre subsidiaries are widely used in Asia. Worldspan is the least used overall but has an evenly spread market share in most regions of the world.

Although these GDSs are considered the major systems for worldwide coverage, there are several smaller, regional distribution systems. Table 10.2 lists regional GDS systems and their area of most use.

Since 2000, TravelSky, a Chinese state-backed GDS, has been the main GDS platform in China, handling all Chinese airlines. With the potential growth in the Chinese aviation market in the future, TravelSky may soon be one of the largest GDS companies.

Table 10.2 Regional GDS Systems

GDS	*Area of Most Use*
TravelSky	China (state-owned)
KIU	Latin America
Abacus	Asia-Pacific
INFINI	Japan
Sirena	Russia

10.5 Deregulation

The GDS regulations were mostly continued for almost 20 years. In 2002, the DOT published a Notice of Proposed Rulemaking (NPRM) asking for comments from the industry concerning the then-upcoming (January 31, 2004) sunset of the GDS regulations. On January 7, 2004, the DOT published its final ruling stating that most of the GDS regulations would be allowed to sunset as planned ("CRS Regulations," 2004). There was little immediate effect with CRS deregulation as most airlines already had extended contracts with GDSs through 2006. However, airlines were under no obligation to participate with any GDS and airlines were free to negotiate with any or all systems once their contracts expired. The DOT cited two primary reasons for its decision:

- First, "all of the US airlines that had controlled a [CRS] system have divested their CRS ownership interests" ("CRS Regulations," 2004, p. 977). Without airline ownership in GDSs, the GDSs had no incentive to use screen bias to favor one airline over another. As a result, regulators felt that screen bias was no longer an issue.
- Second, "airlines are selling an increasingly large share of their tickets through their internet websites and a diminishing share through travel agencies using a [CRS] system" ("CRS Regulations," 2004, p. 977). The emergence of internet availability took away much of the influence of CRS systems by allowing customers easy opportunities to search schedules, compare fares, and purchase tickets.

In Europe, GDS deregulation was not as forthcoming. At the time of US GDS deregulation, Amadeus, the primary system used in Europe, was still owned by its founding airlines. In 2005, after the airlines sold off Amadeus, the EU simplified but did not end CRS regulation. As of 2022, the EU is reviewing the CRS regulations to determine if there is still a need. Canadian CRS regulations still exist to ensure "consumers have access to comprehensive, neutral information about passenger air services that are displayed or sold in Canada" (Canadian Minister of Justice, 2022, introduction).

10.5.1 New Entrants

Deregulation also sparked some new entrant alternatives to the established GDSs. These companies were called GDS New Entrants or GNEs (pronounced "genies"), with the idea that they could lead to the end of the dependence of airlines on traditional GDSs. Two major GNEs were ITA Software and G2 SwitchWorks (Field, 2007).

ITA Software, best known for a fare-shopping metasearch engine, promised an alternative with all the capabilities of a traditional GDS – display of flight availability, schedules, and passenger information; making reservations; and issuing tickets, refunds, and exchanges. Google acquired ITA in 2011, and now called ITA Matrix is the leading search engine for fares. However, seats cannot be booked directly from ITS Matrix – the customer has to go to the airline website or a travel agent.

The other major new entrant, G2 Switchworks, attempted to provide a more supplier-friendly GDS and took a travel agency viewpoint. G2 SwitchWorks failed in this attempt as a new entrant and was purchased out of bankruptcy by Travelport. Of course, this is yet another example of the high failure rate of new business ventures trying to break into an oligopolistic market.

In the end, about all the GNEs accomplished was to provide airlines leverage for negotiations with the traditional GDSs resulting in some savings for the airlines (Field, 2007).

10.6 The Rise of the Internet Changes the GDS Environment

Before the mid-to-late-1990s, travel agents, long the dominant airline distribution channel, traditionally provided two services to the passenger. The first was information. Travel agents had easy access and the skills necessary to use their GDSs to gain information on schedules, fares, and seat availability. The second was the need for a paper ticket. Paper tickets were required for all airline travel and because only a travel agent or an airline could issue tickets; passengers were further tied to travel agency distribution.

Two developments rapidly diminished this traditional travel agency's hold on airline distribution. The first development was the rise of the internet in the 1990s. Southwest Airlines launched the first airline website in 1995 (Gunther et al., 2012) and it wasn't long before most airlines had their own websites to serve as online portals to their flight schedules and seat inventories. As the websites matured, passengers could check flight availability, review fare information, and make reservations – all of the services passengers previously needed the travel agent and GDS system to access. The money the airlines could save in distribution costs if passengers used their website versus travel agencies was significant – $ 8.50 for booking through a travel agent versus as little as $0.25 if booked on the airline's website.

The second development was ticketless travel. Ticketless travel, or e-tickets, originated with Morris Air in the early 1990s. Morris was owned and operated by a travel and tour agency out of Salt Lake City. Morris was subsequently purchased by Southwest which used Morris' developed technology to initiate ticketless travel in 1994, largely in response to a dispute with Sabre and other GDSs. The cost of issuing a paper airline ticket was about $10 compared to about $1 for an e-ticket (Belobaba et al., 2009). Passengers with e-tickets are spared the hassle of having to take possession of tickets, storing them in a secure place until their flight, and then finding them to present to the agent. E-ticketing also saves airlines money. For example, the cost to process an e-ticketed passenger who prints the boarding pass at home or an airport kiosk and uses the internet for flight check-in is about 16¢ compared to about $3.62 for check-in with a passenger service agent. The International Air Transportation Association (IATA), which mandated that all 230 of its airline members utilize e-ticketing by 2008, reports that the 100% use of e-ticketing saves the airlines a total of $3 billion a year worldwide ("IATA Fact Sheet," n.d.).

E-ticketing also works very well in conjunction with internet reservations. Instead of having to physically visit a travel agent or airline office to print a ticket (and pay the additional fee), e-ticketed passengers get only a confirmation number from either the airline's website or from an agent over the phone. When they get to the airport, they provide the airline agent (or kiosk) with the confirmation number and their identification and they are issued a boarding pass for their flight. The two main services the traveler needed from travel agents and GDS systems were now available directly from the airline via the internet.

Using the internet to sell reservations could save the airlines each tens of millions of dollars in travel agent commissions and GDS booking fees annually. However, it would take time for travelers to transition to the internet process. In 2004, America West estimated it costs about $2 to process a booking on its own site versus $15 through a GDS. This could be significant – at the time, of the $58 paid for a ticket between Phoenix and Southern California booked on America West through a GDS, $15 went to the GDS.

Some airlines drive potential customers to their internet site by charging an additional fee for using a GDS. As an example, in 2015, Lufthansa began adding a *distribution cost charge* (DCC) of €16 ($18) to their tickets processed through a GDS transaction (Flottau, 2015, p. 1). Tickets purchased through their website were not charged the DCC. Lufthansa even suggested travel agents and corporate customers use the online portal instead of the GDS (Lufthansa, 2016). Amadeus, Europe's largest GDS, naturally opposed the charge, saying it "will penalize travelers based on the shopping channel they use" (Flottau et al., 2015, p. 1). Lufthansa's DCC increased to €19 ($21) in 2020 (Hoskins, 2020).

Limiting the fares available on the GDS was another airline tactic. To promote their websites, the airlines came up with the idea of holding back some fares, mostly the highest discounted fares, from the GDSs and making them available only on the airline's site. The airlines began to advertise these "internet-only" fares, hoping customers would give their websites a try. With airlines withholding some fares, travel agents using GDSs could no longer promise their customers they were getting the best fares. In response, GDSs negotiated with airlines for all fares (full content), and in return gave discounts to the airline. GDSs now have Full Content Agreements with almost all major carriers.

10.6.1 Airlines' Internal Reservation Systems

An airline's internal reservation system (RES), sometimes called *Passenger Service System* (PSS), contains its flight schedules, prices, seat inventories, and real-time operational information. In many respects, the internal reservation system is "the heart of the airline's operation" (Belobaba et al., 2009, p.447). When the airlines owned the GDSs, most of them used their GDS as their internal reservation system as well. In 1992, United used Apollo, American used Sabre, and Northwest and TWA used Worldspan as their internal reservation system. Only Continental, who owned System One, did not use it internally ("Computer Reservation Systems: Action Needed," 1992). The divesture of GDSs from airline ownership required the establishment of new business relationships between RES and GDS systems.

Airline planning and operations functions – like revenue management, crew scheduling, schedule optimization, and pricing – feed solutions into the airline's RES. The internal system then processes information and releases non-proprietary information to the GDSs. The GDS distributes the schedule, pricing, and availability information contained in the RES system to traditional travel and online travel agents. The internal reservation system used by an airline may be purchased from a vendor or developed internally. Similarly, it may be housed in the airline's internal IT system or hosted by a vendor, usually one of the GDSs (Belobaba et al., 2009).

10.6.2 Shift in Travel Agency Approach

Initially, travel agencies were agents for the airlines. They were paid a base commission by the airline based on the value of the tickets they sold – the higher-priced ticket they could sell, the more money they made. Some airlines also paid incentive payments to encourage travel agents to meet specific sales quotas. These payments, called TACOs (Travel Agent Commission Overrides), were in addition to the base commissions. Because TACO was paid only above a negotiated level of sales, agents typically favored the largest airlines in their region.

As airlines sought to decrease their distribution costs, they reduced the base commissions. Before these reductions, the standard commission was 10%, but during the period from 1995 through 1997, most airlines decreased their base commissions to 8%. Then in 1995, Delta was the first airline to cap travel agency commissions at $50 for a round-trip fare. A few years later, some airlines reduced base commissions to 5%, with a $10 cap. Other airlines followed and the commission cap practice stuck ("Report on Travel Agent Commission Overrides," 1999).

In 2002, Delta went further to eliminate base commissions to travel agents for tickets issued in the US and Canada. This action was quickly matched by the rest of the industry with Southwest the last major carrier in 2003 to eliminate base commissions. Despite airlines' ties to travel agents, with total distribution costs of more than 12% of industry operating expenses, the potential savings were too high to ignore, especially with the growing competition from low-cost carriers (LCCs).

The elimination of base commissions placed travel agencies under great pressure. Combined with the emergence of LCCs, passengers' interest in low fares, and competition from internet distribution, most airlines now pay nothing to travel agents. The traditional travel agency business model was no longer profitable, and many agencies went out of business. The US Bureau of Labor Statistics reports that from a high of about 124,000 travel agents in 2000, only 74,000 remained in 2014 (Elliott, 2018). COVID-19 pandemic losses were high with only 37,000 agents remaining in 2021 (US Bureau of Labor Statistics [BLS], 2022b). Although some post-pandemic rebound is expected, the growth of the travel agent career field from 2020 to 2030 is projected at only 5% for the entire decade (BLS, 2022a).

With the decrease or elimination in airline commissions, travel agencies began shifting their fees to the passenger. As a result, their allegiance also shifted to the passenger. Instead of trying to sell the highest fare to improve their commission, travel agents were more likely to provide the best service to the traveler, possibly finding cheaper seats or promotions to save the traveler money and hope for repeat business. If anything, the airlines paid only minor transaction fees to travel agencies for specific services, and instead, the traveler now paid a surcharge to the travel agent for their time and service, possibly as much as $25 for a one-way booking or $50 for a round-trip booking.

Some travel agencies shifted their business model to favor business travel, where they provide specialized travel services for corporations. These companies provide not only corporate travel management, consulting, and executive travel services, but often meeting or conference support logistics as well. Agencies taking this approach are often called *travel management companies* (TMC).

10.7 Rise of Online Travel Agencies

As the internet began to take shape in the early 1980s, American's Sabre was the first to offer limited online availability to the traveler when they introduced EAAsySabre in 1985 ("Sabre History," n.d.; Gunther et al., 2012). Customers with internet capability could access the Sabre system through CompuServe from their personal computers. "For the first time, travelers were able to check schedules (via [electronic access to the] OAG) and book travel [via EAAsySabre]." (Gunther et al., 2012, p. 175) As the internet became more available in the mid-1990s, online travel agencies (OTAs) appeared. Using an OTA (also sometimes referred to as an ITA – internet travel agency) travelers pay a small service fee in addition to the ticket price and the airline pays the standard segment fee to the GDS system if one is used.

10.7.1 Standard Online Travel Services

Online travel agencies, sometimes owned by the GDSs themselves, are very much like traditional brick-and-mortar travel agencies except they operate via a user-friendly computer interface directly between the traveler and the GDS. Most still use the GDS to book flights but without the overhead, they not only offer the service at a lower cost but also provide better transparency to the customer (the ability to compare fares side-by-side). Online travel agencies are generally favored by price-sensitive leisure passengers. Initially, the big three online travel agencies were Travelocity, Expedia, and Orbitz.

10.7.1.1 Travelocity

The first real online travel agency and successor to EAAsySabre was Travelocity (Gunther et al., 2012). Travelocity started in 1996 as a joint venture between Sabre and Worldview Systems Corp, with Sabre providing the GDS capability and Worldview providing the travel-related content.

10.7.1.2 Expedia

Expedia is an online travel booking site catering to both leisure and business travelers. Expedia began in 1996 when a small division within Microsoft launched Expedia.com to give consumers a revolutionary new way to research and book travel. In 1999, Expedia Inc. spun off into its own company. By 2001, Expedia took over Travelocity's number one online travel agency position. Expedia utilizes Sabre, Amadeus, and Travelport GDSs (Expedia, 2022).

10.7.1.3 Orbitz

In 2000, at a cost of about $100 million, five major US airlines (United, Delta, Continental, Northwest, and later American) teamed up to start Orbitz. Initially, Orbitz used Worldspan GDS for flight information and booking, but their goal was to create a new technology that could book tickets by direct access to the airlines' internal reservation systems.

10.7.1.4 OTA Industry Leaders

In January 2015, Expedia Group bought Travelocity, then a month later, in February, Expedia Group proposed a merger deal with Orbitz. The US Department of Justice approved the merger in September 2015, and Orbitz is now a wholly owned subsidiary of Expedia Group (Orbitz Worldwide, Inc., 2014). In 2018, Priceline Group became Booking Holding (Booking Holding,

2022). As a result, there are currently only two major OTAs towering above the other competition – Expedia Group and Booking Holding. In 2021, these two OTA giants now control about 93% of the OTA leisure and unmanaged business travel market (PhocusWright, 2022).

10.7.2 Opaque Travel Services

So-called opaque online travel agents are OTAs that do not fully disclose certain information about flights before booking. As discussed in the previous chapter, the marginal cost of producing an airline seat is very low. Because of this, airlines can sell seats remaining in inventory that would probably end up as spoil (empty seats at departure time) at deeply discounted prices. If the airline offers super-low fares to fill these seats, however, they risk diluting the value of other seats they have for sale. If the airline offers those seats through an opaque seller, however, they can remain nameless until after the sale. Although the uncertainty of the opaque transaction is uncomfortable for some customers, the savings appeals to others. Two OTAs that sell opaque tickets as well as standard tickets are Expedia Group subsidiary Hotwire and Booking Holdings subsidiary Priceline.

Hotwire. Established in 1999 by the Texas Pacific Group and six major airlines (American Airlines, America West Airlines, Continental Airlines, Northwest Airlines, United Airlines, and US Airways), Hotwire launched operations in 2000. Hotwire is considered an opaque seller because their "Hot Rate" flights do not reveal the name of the airline or the departure time until after the sale. Hotwire is a subsidiary of Expedia Group.

Priceline. Priceline was created in 1997 by Walker Digital and launched in 1998. Priceline offers standard OTA (price-disclosed) sales as well as opaque "Express Deals." Priceline primarily serves customers in the US and includes deals on airline tickets, hotels, rental cars, vacation packages, and destination services.

An additional method considered opaque is an airline ticket sale bundled with another travel service, like a hotel booking or rental car. When purchasing a bundled product, a traveler is often unable to discern specific prices for the airfare, so fare-only comparison is impossible.

10.7.3 Fare Aggregators and Metasearch Engines

The highly profitable GDSs are a good target for competitive attacks. As airlines were doing their best to bypass the GDSs by promoting their own websites, search engine technology was being developed that could search airline websites for flight information. These so-called metasearch engines or vertical search engines do not sell travel services at all – they search airline websites and then redirect the customer to the airline or online travel agent for the final purchase of a ticket. The metasearch engines make their revenue from advertising and some referral fees if a customer uses an online travel agency (GDS) to book their travel. The biggest metasearch travel site is Booking Holding subsidiary Kayak, but there are several others, including Skyscanner and Google Flights.

10.8 The Future of Airline Distribution

With GDS systems developed before the rise of the internet, they use an inflexible pre-internet message protocol called EDIFACT (US Department of Transportation [US DOT], 2014; Popovich, n.d.). This text-only method of data display, sometimes called "green screen," was state-of-the-art in the 1980s, providing the basic information travel agents needed – fares and availability. This system worked well when airlines were selling only seats (Figure 10.12). However, now that airlines have unbundled their products, the sales process has become more complex and GDS processing using EDIFACT has become outdated and impractical.

Conversely, airlines, using modern-day XML (Extensible Markup Language) internet language, build very user-friendly sites to sell their multitude of unbundled products like seat upgrades, baggage fees, and other ancillary fees. Airlines can also customize travel packages based on their customers' travel history and stated desires. These services are either totally unavailable or at best not easily available to travel agents using EDIFACT GDS.

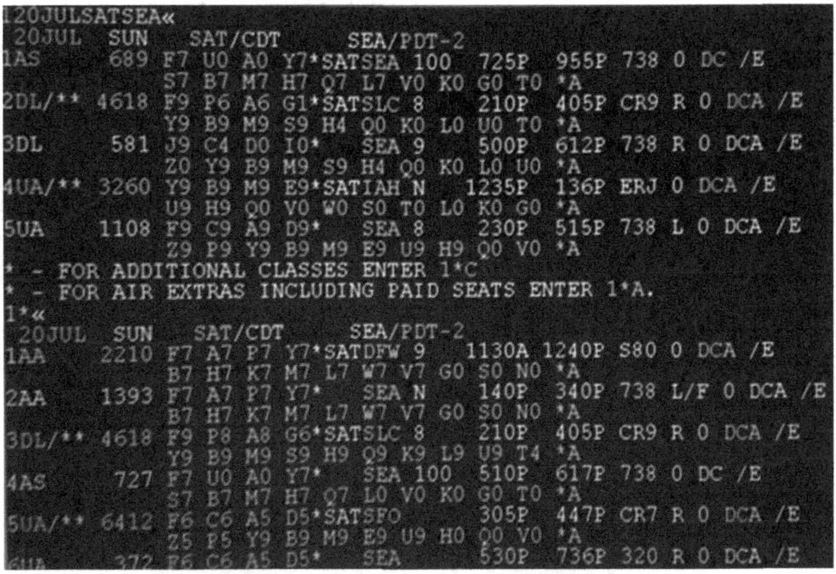

Figure 10.12 EDIFACT Screen Presentation.
Source: Photo by B. Billig.

10.8.1 New Distribution Capability

Beginning in 2012, The International Air Transport Association (IATA), an airline consortium, led an effort to update the airline distribution process. This new system, called *New Distribution Capability* (NDC), was designed to establish a new technical standard for data exchange using XML. Using XML, the NDC allows an enhanced and simplified distribution capability. Figures 10.12 and 10.13 compare the original EDIFACT "green screen" presentation and the NDC's new XML presentation.

The US DOT approved the NDC standard in 2014.

> [According to the DOT, the NDC] would help modernize airline product distribution by generating common industry-wide, real-time communications standards and protocols so that all the participants in the distribution chain – airlines, travel agents, GDSs, and consumers – could speak the same electronic language in their communications with each other.
>
> (US DOT, 2014, p. 10)

The DOT's approval stimulated industry interest and by 2021, over 75 airlines have been certified at some level of NDC capability (Duffel, 2022). Many believe the industry's recovery from the pandemic will speed up the shift to NDC as those upgrading will have a distribution advantage (Hoffman & Walker, 2021).

Aside from simplified operation with XML, NDC greatly enhances the capabilities of the travel services seller. Not only are ancillary sales possible, but also special fares and improved pricing capabilities. Table 10.3 lists six airline examples with their respective added capabilities.

Most airlines are very excited about the prospect of NDC, feeling that the new XML-based standard is a new "vibrant marketplace that is not possible with today's closed proprietary systems" (Popovich, n.d., p. 1). Airlines claim NDC will help travel agents better serve their customers by making more airline ancillary services available and by allowing customers to make fully informed choices.

On the downside, implementing NDC will be expensive and time-consuming. Until NDC matures and replaces the time-tested GDS, they will have to coexist. Travel agents adopting direct NDC connections to airlines have found that to maintain the same level of flight inventory they need NDC connections to many airlines. This has led to a new middleman, the *NDC aggregator*. The aggregators pull together multiple NDC connections and offer that access as a service to travel agents (Anderson, 2020).

Not surprisingly, the major GDSs, Sabre, Amadeus, and Travelport, were initially opposed to NDC. They felt that NDC would be used to bypass their services and destroy their businesses. However, after the IATA reasserted that NDC was not mandatory and GDSs (or something much like them) would still be needed to supplement NDC, opposition lessened. The major GDS companies

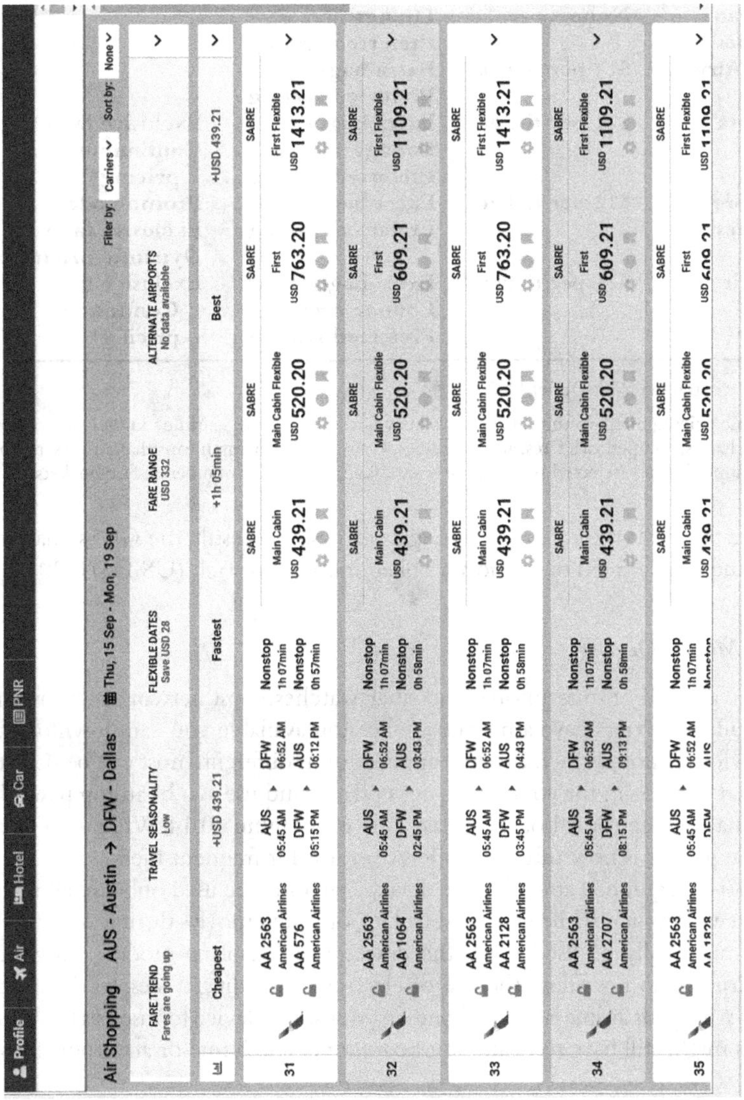

Figure 10.13 NDC XML Screen Presentation.
Source: Photo by B. Billig.

Table 10.3 Airline NDC Enhancements

Airline	GDS Fee Savings	Auxiliaries Available	Other
Air Canada	None	**Lounge access** **Preferred seating** **Special meals**	**Promo codes** **Fare bundles**
American Airlines	None	**Lounge access** **Preferred seating**	
British Airways	**$14 per ticket**	**Extra bags** **Preferred seating**	
Lufthansa	**$21 per ticket**	**Extra bags** **Lounge access** **Preferred seating**	**Exclusive fares** **Continuous pricing★**
Singapore Airlines	**$12 per ticket**	**Extra bags** **Preferred seating**	**Promo codes** **Exclusive fares** **Dynamic pricing★**
Swiss Air	**$21 per ticket**	**Extra bags** **Lounge access** **Preferred seating**	**Exclusive fares** **Continuous pricing★**

Note: Date extracted from Duffel (2022).

★ Dynamic pricing refers to the capability to vary fares based on the airline's knowledge of your buying habits, competitor's fares, and schedules. Continuous pricing means that airlines' pricing is not limited to pre-selected price points – essentially, a limitless number of fare buckets.

also have taken on the role of NDC aggregators. As a result, the GDSs changed their minds and agreed to support a collaborative approach (US DOT, 2014a).

10.8.2 Mobile Devices

With the growth of smartphones and smartwatches, most network carriers and many budget carriers have a mobile application available you can download to your device. Although each carrier's application is different, most can be used to buy tickets, check-in for your flight, get updates, and use as a boarding pass, like the China Air electronic boarding pass shown in Figure 10.14. What used to be just a fun gadget is now often considered a must for frequent fliers.

On Wi-Fi-equipped aircraft, these devices can often be used onboard allowing you to keep in touch while you are traveling, or watch movies during your flight. (Wi-Fi can usually be used while the device is in "airplane mode.") However, a few airports do not allow the use of electronic boarding passes, so do a little research with your airline before relying on your smart device for use as a boarding pass. You might still have to print your boarding pass at home or an airport kiosk.

10.9 Summary

The goal of the airline's distribution system is to provide the *place* or the distribution channel that connects the product to the customer. Figure 10.16 shows this graphically. A customer has three places to go to for their airline booking:

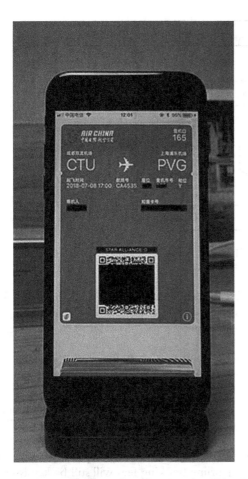

Figure 10.14 Smartphone Boarding Pass for China Air.
Source: Wikimedia Commons (Ziaho, 2022) (CC BY-SA 4.0).

- Make a phone call to the airline's call center. The airlines like to avoid this – call center staffing is expensive.
- Go online to the airline's website (the best option for the airline). This can be done after consulting a metasearch engine (like Kayak) to compare fares.
- Visit a brick-and-mortar or online travel agency. The travel agency will probably use either the GDS, NDC, or an NDC aggregator to book the flight.

It's no surprise that the airlines attempt to distribute their product for the least cost possible. The least cost distribution from the customer to the airline, as shown by the dashed line in Figure 10.15, is through the airline's website. This

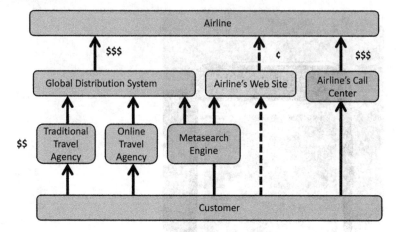

Figure 10.15 Airline Distribution Diagram.

method costs the airline just a few cents to complete. Alternatively, any booking that goes through the GDS system costs the airline a substantial booking fee. Even using the airline's call center requires expensive additional overhead and labor costs.

The future of airline distribution will find traditional travel agencies continuing to transition more to corporate travel functions and calling themselves travel management companies (TMCs), as they now see themselves as consumer travel consultants rather than airline agents. Airlines will attempt to increase direct booking via the internet and re-kindle relationships with TMCs to develop global business strategies for corporate travel. Airlines will also work with their GDSs to transition to the New Distribution Capability (NDC) and shift away from EDIFACT. High airline booking fees will still be key issues in the airlines' continuing shift away from GDSs to direct booking and GDSs' airline revenue will continue to decrease. GDSs feel they will still maintain airlines as a major part of their revenue, but they also understand they will need to diversify to reduce their dependence on airline revenue. That will require changing their business model and developing new pricing mechanisms for airlines (Alamdari & Mason, 2006).

10.10 Case Study: The Rise and Fall of a Travel Agency Legend

Earlier, this chapter describes the use of travel agents in the airline industry and the rise and fall of the brick-and-mortar travel agency business. This case study highlights one brick-and-mortar travel agency that expanded into airline operations, only to find itself unable to adapt to new competition in both endeavors, leading to failure and eventual bankruptcy.

Figure 10.16 Thomas Cook (date unknown).
Source: From Wikimedia Commons, author unknown (2021) (GFDL).

The travel agent business started probably farther back than you might think. Some say Englishman Richard Cox, who, as a clerk for an English general in 1758, learned a trade he turned into a travel business called Cox and Co. Flourishing through the late 18th century in Great Britain, Cox and Co. still exists as Cox & Kings, a very respected luxury travel agency well known in the Middle East, Africa, the Far East, and Europe (Cox & Kings, 2021). However, many contend that Cox & Kings was more of a banking business that helped rich clients set up lavish tours, and that Thomas Cook (Figure 10.16), also an Englishman, was the father of the modern travel agency industry.

Starting small in 1841, Thomas Cook organized tours and rallies for the temperance movement in England. Although initially small, Thomas Cook Travel expanded rapidly and two decades later, Cook was providing international tours to Europe. In 1865 he expanded to North America with sea and rail holiday packages, that, by 1872, were continuing across the US and on to the Far East.

Figure 10.17 Thomas Cook Airlines A-330 in Las Vegas.
Source: From Wikimedia Commons (Del Coro, 2021) (CC BY-SA 2.0).

After his death in 1892, Cook's company was operated by his son and then, later, his grandsons. Expansion of the company continued into the Middle East, Egypt, Australia, and New Zealand. Cook's descendants sold the prospering company in 1928. During WWII, the company struggled and was nationalized in 1948 to become part of the British Transport Commission. The company began to profit in the 1960s and was re-privatized in 1972.

Thomas Cook Travel's expansion into the US travel market was the subject of a merger with Crimson Travel Service and Heritage Travel Inc. in 1989. The combined sales of the new company, retaining the Thomas Cook name, exceeded $1.3 billion making it the third-largest travel company in the US. By 1994, Thomas Cook (in the US) was very successful, serving mostly corporate clients. In 1994, the owner sold the company, consisting of 375 brick-and-mortar offices, to American Express Travel.

The original company, now an Anglo-German company, Thomas Cook AG, merged with UK-based MyTravel Group in 2007. MyTravel also operated MyTravel Airways, which turned into Thomas Cook Airlines.

Thomas Cook Airlines flew to 82 destinations in Europe, Africa, and the Americas with Boeing 757–200/300 and Airbus A-320, A-321, and A-330 aircraft. Figure 10.17 shows a Thomas Cook A-330 in Las Vegas in 2019.

Good times didn't last long though. By 2009, Thomas Cook Airlines was in financial trouble as debt was growing rapidly. By 2017, debt had reached over $2 billion. The rise of low-cost carriers in Europe as well as the ease of booking flights and vacations directly online or through an online travel agency was behind the worsening financial conditions. Two years later, in 2019, the company declared bankruptcy, ceasing operations suddenly on September 23.

With its surprise collapse, many travelers were left stranded. To expedite their return, the UK government launched a repatriation effort returning more than 140,000 people to the UK. Called "Operation Matterhorn," the cost of the effort, put at £100 million, was shared by the UK taxpayers and the Air Travel Organisers' Licensing (ATOL), a UK CAA scheme that provides financial protection to travelers.

Like several bankrupt airlines whose names seem to reappear, the name Thomas Cook continues. Chinese travel firm Fosun purchased rights to Thomas Cook and reopened the company as an online travel agency in the UK and China in July 2020.

Concepts: Travel agency history; online travel agency; global distribution systems; e-tickets.

Sources: Del Coro, 2021; Cox & Kings, 2021; Operation Matterhorn, 2019; Hayward,2021; United Press International, 1989; BBC News, 2020; FundingUniverse, n.d.

Note

1 Accounts vary, but apparently IBM's umbrella SABER project was also being looked at by Delta and Pan Am, so American needed a unique name for their system, presumably in order to copyright it (Mapstone, 1980). Most say Sabre stands for Semi-Automatic Business Research Environment (simply swapping the last two words from the original SABER), but others say the name came into being after an executive saw an advertisement for a Buick LaSabre in a magazine (Boyd, 2007; Petzinger, 1995; Copeland & McKenney, 1988; McKenney et al., 1995). The *Sabre*™ company today states that historically the name was an acronym for Semi-Automatic Business Research Environment, but now treats the name as if was not an acronym at all ("Sabre History," 2013).

References

Airline ticketing: Impact of changes in the airline ticket distribution industry. (2003). US General Accounting Office Report to Congressional Requesters. www.gao.gov/products/GAO-03-749

Alamdari, F. & Mason, K. (2006). The future of airline distribution. *Journal of Air Transport Management, 12*, 122–134.

Anderson, P. (2020). NDC in 2020: So close but so far. PhocusWire. www.phocuswire.com/NDC-in-2020-so-close-but-too-far

BBC News. (2020). Thomas Cook to be revived as online travel firm. *BBC News*. www.bbc.com/news/business-54070587

Belobaba, P. P., Swelbar, W., & Barnhart, C. (2009). Information technology in airline operations, distribution and passenger processing. In P. Belobaba, A. Odoni & C. Barnhart (Eds.), *The global airline industry* (pp. 73–111). Chichester, UK: Wiley.

Booking Holding. (2022). History. www.bookingholdings.com/about/history/

Borenstein, S. (1992). The evolution of US airline competition. *Journal of Economic Perspectives, 6*(2), 45–73.

Boyd, A. (2007). *The future of pricing: How airline ticket pricing has inspired a revolution.* New York: Palgrave Macmillian.

Business Travel iQ. (2017). GDS market shares and more. *Business Travel News.* www.businesstravelnewseurope.com/Analysis/GDS-market-shares-and-more

Canadian Minister of Justice. (2022). Canadian computer reservation systems (CRS) regulations.   https://laws-lois.justice.gc.ca/eng/regulations/SOR-95-275/page-1.html#h-973226

Cavani, R. (1993). Computerized reservation systems for air transport: Remarks on the European Community legislation. *Fordham International Law Journal, 17/2,* pp. 441–457.   http://ir.lawnet.fordham.edu/cgi/viewcontent.cgi?article=1377&context=ilj

Computer reservations system (CRS) regulations. (2004). *Federal Register, 69*(4), 976–1033.

Computer reservations systems: Action needed to better monitor the CRS industry and eliminate CRS bias. (1992). United States General Accounting Office, GAO/RCED-92-130 Computer Reservations Systems

Copeland, D. G., & McKenney, J. L. (1988). Airline reservations systems: Lessons from history. *MIS Quarterly, 12*(3), 353–370.

Copeland, D. G., Mason, R. O., & McKenney, J. L. (1995). Sabre: The development of information-based competence and execution of information-based competition. *IEEE Annals of the History of Computing, 17*(3), 30–56.

Cox & Kings. (2021). The history of Cox & Kings: Over 260 years of discovery. www.coxandkings.co.uk/about-us/history

Davidow, W. H. & Malone, M. S. (1992). *The virtual corporation: Structuring and revitalizing the corporation for the 21st century.* New York: HarperCollins.

Del Coro, T. (2021, May 27). Thomas Cook Airlines A330–200 (G-TCXB) at LAS. *Wikimedia Commons, the free media repository.* Retrieved 21:21, July 30, 2022 from https://commons.wikimedia.org/w/index.php?title=File:Thomas_Cook_Airlines_A330-200_(G-TCXB)_@_LAS,_April_2019.jpg&oldid=565218805.

Department of Justice. (2003). Statement by Assistant Attorney General R. Hewitt Pate regarding the closing of the Orbitz investigation. www.justice.gov/atr/public/

Duffel. (2022). The complete NDC guide for travel sellers https://duffel.com/ndc

Edelman, B. (2009). Distribution at American Airlines. In J. Wirtz, P. Chew, & C. Lovelock *Essentials of services marketing* (2nd ed.). Singaopre,: Pearson.

Eklund, J. (1994). The reservisor automated airline reservation system: Combining communications and computing. *IEEE Annals of the History of Computing, 16*(1), 62–69.

Elliott, C. (2018). This is why travel agents want to be called travel advisors. *Forbes.* www.forbes.com/sites/christopherelliott/2018/11/11/this-is-why-travel-agents-want-to-be-called-travel-advisors/?sh=10e957121ca3

European Union. (1989). Council Regulation (EEC) No 2299/89. https://eur-lex.europa.eu/legal-content/EN/TXT/PDF/?uri=CELEX:31989R2299&from=EN

Expedia. (2022). 2021 Form 10K report. https://s27.q4cdn.com/708721433/files/doc_financials/2021/q4/936c054e-9c59-465f-a87e-c9144fb28bb1.pdf

Field, D. (2007). Back in the bottle: What happened to GDS new entrants? www.flightglobal.com/news/articles/back-in-the-bottle-what-happened-to-gds-new-entrants-214686/

Flottau, J. (2015). Lufthansa introduces surcharge for GDS bookings. *Aviation Daily, 2* June, 2015. P.1.

Flottau, J., Buyck, C., & Sumers, B. (2015). Tsunami ahead: Lusthansa's imposition of fees for GDS bookings is making waves. *Aviation Week & Space Technology*, June 22-July 5, 2015.

FundingUniverse. (n.d.). History of Thomas Cook Travel Inc. www.fundinguniverse. com/company-histories/thomas-cook-travel-inc-history/

Gunther, D., Ratliff, R., & Sylla, A. (2012). Airline distribution. In C. Barnhart & B. C. Smith (Eds.), *International Series in Operations Research and Management Science: Quantitative Problem Solving Methods in the Airline Industry: A Modeling Methodology Handbook*. (pp. 163–237). Boston: Springer.

Hayward, J. (2021). The rise and fall of Thomas Cook Airlines. *Simple Flying*. https://simpleflying.com/the-rise-and-fall-of-thomas-cook-airlines/

Hoffman, K. & Walker, K. (2021). Why airlines should use technology to stay ahead with distribution. *Aviation Week Intelligence Network*. https://aviationweek-com.ezproxy. libproxy.db.erau.edu/air-transport/airlines-lessors/why-airlines-should-use-technol ogy-stay-ahead-distribution

Hoskins, A. (2020). Lufthansa Group to increase DCC fee from October. www.busin esstravelnewseurope.com/TMC-Distribution/Lufthansa-Group-to-increase-DCC-fee-from-October

IATA Fact Sheet. (n.d.). www.iata.org/pressroom/facts_figures/fact_sheets/Pages/stb-concluded.aspx

Levine, M. E. (1987). Airline competition in deregulated markets: Theory, firm strategy, and public policy. *Yale Journal on Regulation*, 4 (393–503).

Lufthansa. (2016). 2015 annual report. https://investor-relations.lufthansagroup.com/ fileadmin/downloads/en/financial-reports/annual-reports/LH-AR-2015-e.pdf

Mapstone, R. (1980). *An interview with R. Blair Smith/Interviewer: Robina Mapstone 29 May 1980*. Charles Babbage Institute, The Center for the History of Information Processing, University of Minnesota, Minneapolis. http://conservancy.umn.edu/ bitstream/107637/1/oh034rbs.pdf

McKenney, J. L., Copeland, D. C., & Mason, R. O. (1995). *Waves of change: Business evolution through information technology*. Boston: Harvard Business School Press.

Nicholls, J. A. F. (1985). Airline deregulation, computer reservation systems, and travel agents. FIU Hospitality Review, 3(2), 61–63.

Operation Matterhorn Carries Final 4,800 Thomas Cook Passengers Home. (2019). *Simple Flying*. https://simpleflying.com/operation-matterhorn-final-day/

Orbitz Worldwide, Inc. (2014). http://phx.corporate-ir.net/phoenix.zhtml?c=212 312&p=irol-SECText&TEXT=aHR0cDovL2FwaS50ZW5rd2l6YXJkLmNvbS9m aWxpbmcueG1sP2lwYWdlPTEwMTMzNjg1JkRTRVE9MCZTRVE9MCZTU URFU0M9U0VDVGVlPTl9FTlRJUkUmc3Vic2lkPTU3

Petzinger, T. (1995). *Hard landing: The epic contest for power and profits that plunged the airlines into chaos*. New York: Times Books.

PhocusWright. (2022). The state of online travel agencies. *Hospitalitynet*. www.hospita litynet.org/news/4109835.html

Popovich, A. (n.d.). Understanding IATA's new distribution capability. www.ttgasia. com/article_print.php?article_id=20740

Ravich, T. M. (2004). Deregulation of the airline computer reservation system (CRS) industry. *Journal of Air Law and Commerce*, 69(2), 387–412.

Report on travel agent commission overrides. (1999). US Department of Transportation Report Number: CE-1999-060. www.oig.dot.gov/sites/dot/files/pdfdocs/ce1999 060.pdf

Rice, K. (2014). Though e-tickets by far the norm for air travel, paper persists. Travel Weekly, March 4, 2014. www.travelweekly.com/Travel-News/Travel-Agent-Issues/Paper-persists-although-e-tickets-by-far-the-norm-for-air-travel/

Sabre history. (n.d.). www.sabre.com/home/about/sabre_history

Santiago, J. (2015a). Tails thru time: Short trips on the long road of aviation history. American automates reservations and ticketing, part one: Reservisor System. www.tailsthroughtime.com/2015/06/american-automates-reservations-and.html

Santiago, J. (2015b). Tails thru time: Short trips on the long road of aviation history. American automates reservations and ticketing, part two: Sabre. www.tailsthroughtime.com/2015/06/american-automates-reservations-and_16.html

Shaw, S. (2011). *Airline marketing and management* (7th ed.). Burlington, VT: Ashgate.

Teleregister: Special purpose electronic engineering...that sets the pace. (n.d.). http://archive.computerhistory.org/resources/text/Teleregister/Teleregister.SpecialPurposeSystems.1956.102646324.pdf

The paper ticket. (2008). Boeing Future of Flight Aviation Center display, Everett, Washington.

Thomas Cook (photo). (2021, September 17). *Wikimedia Commons, the free media repository*. Retrieved 21:15, July 30, 2022 from https://commons.wikimedia.org/w/index.php?title=File:Thomas.Cook.jpg&oldid=591734518

United Press International (UPI). (1989). Thomas Cook, Crimson and Heritage announce merger. www.upi.com/Archives/1989/12/11/Thomas-Cook-Crimson-and-Heritage-announce-merger/7127629355600/

US Bureau of Labor Statistics. (2022a). Travel agents: Summary. www.bls.gov/ooh/sales/travel-agents.htm

US Bureau of Labor Statistics. (2022b). Occupational employment and wage statistics: Travel agents. www.bls.gov/oes/current/oes413041.htm

US Department of Justice. (2003). Reply comments of the Department of Justice. www.justice.gov/atr/public/comments/201081.htm

US Department of Transportation. (2014). Action on IATA agreement issued by the Department of Transportation on the 21st day of May, 2014. www.regulations.gov/contentStreamer?objectId=09000064817054ea&disposition=attachment&contentType=pdf

Wensveen, J. G. (2011). *Air transportation: A management perspective* (7th ed.). Burlington, VT: Ashgate.

Z22. (2020). Magnetronic reservisor. *Wikimedia Commons, the free media repository*. Retrieved 20:09, July 29, 2022 from https://commons.wikimedia.org/w/index.php?title=File:Magnetronic_Reservisor.jpg&oldid=474094648

Ziaho, D. (2022). Electronic boarding pass of Air China flight on iPhone 7. *Wikimedia Commons, the free media repository*. Retrieved 17:58, July 31, 2022 from https://commons.wikimedia.org/w/index.php?title=File:Electronic_boarding_pass_of_Air_China_flight_on_iPhone_7.jpg&oldid=635330137

Review Questions

1. What was the Request and Reply system?
2. Why was the Official Airline Guide produced and what is its use today?
3. What were the factors that led to the government's regulation of the computer reservations systems? What led to its deregulation?

4. Why did the travel agent become the predominant distribution channel for airlines and what two services did travel agents provide that were needed by passengers?
5. What happened to transform computer reservation systems into global distribution systems?
6. How has internet distribution affected traditional travel agents?
7. Why did airlines offer internet-only fares?
8. What is the NDC and how will its use improve airline sales?
9. What is an NDC aggregator?
10. Which reservation process is the least costly for an airline?

11 International Air Transportation, Policy

Upon completion of this chapter, you should be able to:

* Describe the common elements in post–World War II air service agreements.
* Explain the rationale for Open Skies and liberalization of international airline markets.
* Evaluate the competitiveness of state-owned airlines.
* List the steps required to privatize a state-owned airline.
* Compare and contrast the costs and benefits of an airline joining a global alliance.
* Define an equity alliance and provide an example.
* List the forces that have driven consolidation in the global airline industry.

11.1 Introduction

In this chapter, our attention switches from an emphasis on domestic operations to international air transportation. It begins with the history of international air transportation economic regulation instituted after World War II. Next, we consider the reasons for expanding liberalization of world airline markets and the impact on airline competition, efficiencies, and management. From the earliest days of the industry, most airlines were state-owned and controlled. The systemic problems afflicting government-owned airlines have led governments to either partial or full privatization. In response to liberalization and increased competition, the largest full-service network carriers formed global alliances to increase market reach and offer customers seamless travel around the world. We consider the nature of these alliances, the benefits, and the costs of membership to both the airlines and the traveling public. Finally, mergers and acquisitions have accelerated in the 21st century bringing some promise of long-term stability. The COVID-19 pandemic upended a decade of relative profitability. Still, consolidation and the eventual emergence of true multinational airlines, as has occurred in other large industries, might be the future of the airline industry as well.

DOI: 10.4324/9781003290308-11

11.2 Air Service Agreements

In 1944, with World War II still raging, 52 nations gathered in Chicago to draft international aviation protocols. The Chicago Convention established the International Civil Aviation Organization (ICAO) that later became an agency of the United Nations. The ICAO is charged with developing and promulgating international aviation standards, including licensing of airmen and aircraft, rules of the air, meteorology, aviation charts, telecommunications, air traffic services, and noise and emissions standards. This worldwide standardization has proven very effective in promoting safe and efficient international air transportation.

The Convention also considered the protocols and restrictions for establishing international air transport among the various nations. The US pushed for minimal restrictions allowing airlines to operate routes without government approval or economic regulation. Most other countries, however, feared US hegemony because of its dominant position in aviation near the end of WWII. However, instead of a relatively free market in international aviation, the Convention settled on a system of negotiated government-to-government agreements regulating flights between two countries. Formally titled air service agreements (ASA), the common term is bilateral agreements, or just *bilaterals*, because each agreement is the result of negotiations between two countries.

Although not the first bilateral, the United States-United Kingdom ASA completed in 1946 set the pattern for many agreements that followed. Known as Bermuda I for the site of the negotiations, this ASA provided for restricted air service between the nations and to some points in Latin American and the British Colonies of Hong Kong and Singapore ("Bermuda Agreement," 2021). The Agreement limited the number of airlines, gateway cities, the capacity or number of seats offered by each carrier, flight frequency, and tariffs or ticket prices. Tariffs were based on a formula developed by the International Air Transport Association, a body formed by the airlines. Since the Chicago Convention and Bermuda I ASA, some 1,500 air service agreements have been negotiated worldwide between states (Doganis, 2019). Many, although often revised and updated, are still in place.

In 1977, the US and UK revised their first agreement replacing it with the Bermuda II ASA that remained in place until 2007. This second agreement was also highly restrictive. Access to London's Heathrow Airport, the most advantageous gateway to continental Europe, was a major point of contention as it has been since. Bermuda II allowed only two airlines from each country – American and United for the US and British Airways and Virgin Atlantic for the UK – to fly between London and a limited number of US cities. Flight frequency, code sharing, and pricing were also restricted.

Air service agreements provide for economic control of flights between countries not unlike the way the CAB once controlled US domestic flights. These agreements generally seek to split the traffic between the two countries' airlines and ensure some level of profitability. ASAs vary significantly in detail and restrictions, but all address the following points:

- Number of routes and gateways served.
- The airlines designated by each country to fly the approved routes.
- Capacity, including the number of seats and frequency of flights permitted. Competition was often further restricted by specifying the legroom (seat pitch), type of meals, and other amenities.
- Tariffs or prices charged by each carrier. These have traditionally been set according to formulae developed by the International Air Transport Association and subject to the approval of both governments.
- A requirement that designated airlines are owned and controlled by citizens of states who are party to the bilateral. The United States, for example, requires that US citizens own at least 75% of the voting stock in a designated publically-owned carrier.

Some ASAs allowed air service to continue beyond the signatory countries, known as fifth freedom rights. *Cabotage*, the carriage of passengers wholly within the other party's country, is rarely permitted.

11.3 US Open Skies

In 1978, the United States formally renewed its goal of less restricted air commerce between nations when President Jimmy Carter declared "*Open Skies*" as US international aviation policy. The announcement coincided with domestic airline deregulation. The policy, which remains in effect, seeks to promote international airline competition with a minimum of government interference. (Confusingly, the term Open Skies also applies to agreements between nations allowing reciprocal surveillance overflights.) Open Skies, however, stops well short of complete deregulation. It restricts foreign ownership of US airlines to 25% of the voting stock and prohibits effective control by foreign citizens. Second, it does not allow for cabotage, so although foreign airlines may operate between two points in the US as part of continuing or thru-flight service, they may not carry passengers solely between the two points.

Following the declaration of the Open Skies policy, the US sought nations willing to negotiate much more liberal air service agreements. The US stated its goals: (a) innovative and competitive pricing; (b) no restriction on capacity, frequency, and routes; (c) elimination of discrimination against US carriers in foreign services provided; (d) multiple destinations; and (e) liberal charter and cargo rights. To provide maximize pricing flexibility, the US proposed that airlines be allowed to set fares unless both governments disapproved. This provision made it difficult for a government to protect its home carrier.

The Open Skies policy enjoyed the support of US airlines because the domestic market had matured such that international flights offered a better opportunity for growth. US network carriers, with their expansive domestic

Table 11.1 Elements of US/Netherlands Open Skies ASA

Full route and destination access	*No frequency or capacity control*
Unlimited 5th freedom (the right to fly between the two countries and then continue to a 3rd country)	Change of aircraft for beyond flights permitted (break of gauge)
Full charter access	No pricing or tariff control
Multiple airline designation	Codesharing permitted

route systems, could tap a large US market to feed international routes, an advantage that most foreign carriers could not match.

11.3.1 US/Netherlands Open Skies

Open Skies agreements were not immediately forthcoming. The first true Open Skies agreement was not reached until 1992 when the US and the Netherlands penned a new agreement. The impetus behind this innovative ASA was a proposed investment in financially struggling Northwest Airlines by the Netherlands-based airline KLM. As shown in Table 11.1, the new ASA met the US policy objectives.

The United States subsequently granted antitrust immunity to Northwest and KLM allowing the airlines to closely coordinate their businesses in an international alliance. Northwest and KLM were allowed to share airline codes for their flights thus allowing each to seamlessly market the other airline's flights as their own. Further, the immunity allowed for the coordination of schedules, fares, and other commercial aspects that, in the absence of antitrust immunity, would be illegal.

The agreement became a US precedent. In 1995, the Department of Transportation adopted a policy of making code-sharing and antitrust immunity with foreign airlines contingent upon an Open Skies agreement between the two countries. The promise of Open Skies was largely fulfilled in 2002 when the US and France signed an agreement bringing the number of US open-skies partners to 56. At the time, only four European Union nations – the UK, Ireland, Spain, and Greece – were without fully liberalized aviation markets with the US. The US-France accord granted antitrust immunity to the global SkyTeam alliance members Delta, Air France, and several smaller partners.

11.3.2 US/EU Open Skies

Coincidently in 2002, as the US-France Open Skies agreement was concluded, the European Union court ruled that bilateral agreements between the US and individual EU countries violated the EU's single market principles. The EU and US then began negotiating an agreement to replace all US/European bilaterals with a single agreement covering all EU countries and the US. After years of difficult, on-again, off-again negotiations, an agreement was reached in March 2006. The key points of the agreement, which took effect in March 2008, are:

- US and EU member states are allowed open access to each other's markets with freedom of pricing and unlimited rights to fly beyond the EU and US to points in third countries.
- US carriers may fly between any EU cities but a reciprocal right was not granted to EU airlines.
- London's Heathrow airport was opened to all US international airlines as well as to all European carriers wishing to fly between the US and London subject to acquiring takeoff and landing slots. This last restriction is significant because Heathrow operates at capacity and is slot-controlled. Slots are occasionally available albeit at a high price. In 2016, Oman Air set a record by paying $75 million for a pair of slots (one takeoff and one landing) at Heathrow, beating the previous year's record of $60 million.
- The EU's demand for the elimination of US restrictions on foreign ownership of US airlines (limited to 25% voting stock) was left for later negotiations. Dropping citizenship rules would allow for trans-Atlantic airline mergers.
- Cabotage is still not allowed.

In 2011, the US and EU essentially agreed to continue the Open Skies agreement without further liberalization sought by the EU. US carrier Virgin America, purchased by Alaska Airlines in 2018, illustrates how important and contentious the citizenship restriction can be. Virgin Chairman Richard Branson was the driving force behind Virgin America and the many other Virgin airlines in other countries. Although Branson held only a 25% voting share in the new carrier, Continental and other US carriers succeeded in delaying Virgin America's new service for nearly a year by convincing the Department of Transportation that Branson would control the airline in violation of the US citizenship rules. Virgin was forced to make substantial changes in its financing and governance before finally winning DOT approval and starting flights in the fall of 2007.

Liberalization of the transatlantic market was expected to lead to more flights and lower fares across the Atlantic. Ryanair proposed entering the market but never did. Instead, the Norwegian entered the market with low fares only having to withdraw because of the pandemic.

With its exit from the EU in 2020, the United Kingdom was obliged to negotiate new bilateral agreements. A new US-UK agreement was penned shortly thereafter.

11.3.3 US/China Free Skies

The early bilateral agreement between the US and China was highly restrictive. But in 2004 the US and China inked a "landmark" air services agreement that would more than double the number of US airlines that may serve China, permitting a nearly five-fold boost in weekly flights between the countries over the following six years. US officials called the deal "free skies," indicating it fell short of an Open Skies agreement.

Continuing negotiations resulted in annual expansions of the agreement. By 2018, there are 61 non-stop routes between mainland China and the US operated by four US carriers – United, American, Delta, and Hawaiian – and six Chinese carriers – Air China, China Eastern, China Southern, Hainan, Xiamen, and Sichuan Airlines. The pandemic upended the progress. Both countries canceled flight authorizations so that by early 2022, flights were irregular and rare. With diplomatic relations strained, the resumption of regular service may take years.

11.3.4 US/Japan Open Skies

The US and Japan reached an open skies agreement in late 2009 following three decades of sporadic negotiations. The agreement lifted restrictions on carrier designation, cities served, capacity, frequency, pricing, and cooperative marketing arrangements, including code-sharing. The previous, restrictive bilateral air service agreement dated from 1952. It had granted highly privileged positions to Pan American World Airways and Northwest Airlines (formerly Northwest-Orient, now merged into Delta Air Lines) that were later obtained by United and Delta through a merger. This allowed Delta to control 22% of the capacity at Tokyo's Narita International Airport, the main international gateway. United held another 12% of Narita's capacity. The agreement was contingent on Japan's two major airlines, Japan Airlines (JAL) and All Nippon Airways (ANA), being granted antitrust immunity with US airline partners, a process concluded in 2011.

The agreement was expanded in 2019 to increase daytime service between Tokyo's more convenient and popular Haneda Airport and several US destinations.

11.4 Expansion of Open Skies

The US has continued to expand Open Skies agreements that include more than 130 countries as of late 2021 ("Open Skies," 2021). Liberalization is not limited to agreements between the US and other countries as other world regions have reduced economic restrictions on the airline industry. In 2003, the European Union began aggressively pursuing open skies and subsequently concluded agreements with the US, Canada, and Brazil, and created a common aviation area with several neighboring states, including the states of the West African Economic and Monetary Union. In 2015, the EU announced an "ambitious package of proposals" to negotiate comprehensive aviation agreements with the Association of Southeast Asian Nations (ASEAN) states, the Gulf Cooperation Council states plus Turkey, China, Mexico, and Armenia (European Commission, 2015).

All 10 ASEAN countries ratified an open skies agreement in 2016 intended to increase connectivity and enhance regional trade. Several of the members of the Latin American Civil Aviation Commission joined an open skies agreement beginning in 2019. For more than three decades, The African Union has pursued a pan-African Open Skies aviation agreement. Progress has been slow; however, by 2018, 34 of Africa's 54 countries have joined the Single African Air Transport Market (SAATM).

11.5 State-Owned Airlines

Coincident with the trend toward liberalization of international air transport markets, governments have moved to privatize state-owned airlines. There are several reasons that governments continue to own and control their national airlines.

- Early in the industry, airlines were not financially viable, but governments recognized the benefits of developing an air transportation system. The common solution was a government-owned national airline. The US was an exception choosing instead to subsidize airmail service and later opting for economic control under the Civil Aeronautics Board.
- National pride often demands and is reflected in a flag carrier even when the market is too small for financial success.
- A state-owned airline can provide low airfares with government subsidies. Railroads are subsidized and/or state-owned in most of the world for the same reason.
- The national carrier provides a link to dispersed ethnic groups worldwide, again often with a subsidy.
- A national airline earns hard currency for foreign exchange, usually US dollars.
- The airline provides employment, technical skills, and training, but also patronage jobs for political advantage.
- The flag carrier helps domestic aerospace industries develop by providing a market for products.
- The airline is critical for national defense. Transport aircraft developed in the former Soviet Union, for example, included design features for military use.

11.5.1 State versus Private Ownership

As Figure 11.1 shows, the majority of airlines in North America, Europe, Asia-Pacific, and Latin America are privately owned – the government may hold a minority stake – in contrast to Africa and the Middle East.

Although it exercised broad economic regulation of the domestic airline industry for 40 years, the US government has not owned an airline. As part of Canada's airline deregulation, Air Canada, the flag carrier, was fully privatized

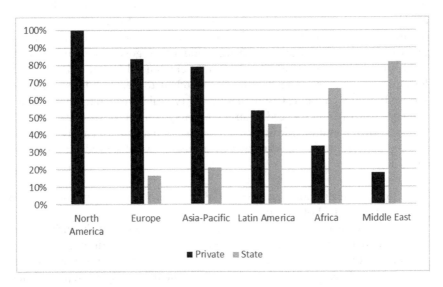

Figure 11.1 Private versus State-Owned Passenger Airlines.
Data source: Seat Guru and Wikipedia.

in 1987. Similarly, the Mexican flag carrier, Aeromexico, was privatized in 2007. The European Union has been forceful in encouraging privatization. The largest carriers were privatized decades ago. EU member countries can only provide state aid to restructure an airline so it can successfully operate on a commercial basis without state interference. The largest EU airlines were successfully privatized in the latter part of the last century beginning with British Airways (BA) in 1987. A few remain in state ownership. The Italian government owns its newest airline, ITA, the successor to ever-struggling Alitalia. Portugal, having previously reduced its majority stake, saved TAP Portugal from liquidation in 2020, increasing its ownership to 72%. The Finnish government still owns 56% of its national carrier Finnair. The remaining state-owned European carriers are mostly small with governments attempting to shed or reduce their ownership shares. Outside Europe, the largest state-owned carriers include Aerolineas Argentinas, Air New Zealand, Singapore Airlines, Air China, and the Big Three Gulf airlines – Emirates, Etihad, and Qatar – that are fully owned by their respective governments.

The reasons for early state ownership of the nations' flag carriers were compelling, but state-owned airlines often operate inefficiently, perpetually lose money, and serve as a drag on the economy. Subsidies, though substantial, are routinely indirect and hidden in minimal or below-market airport fees, low-interest rates on loans, and below-market fuel prices. The carriers are politicized and pressured to provide cheap fares or free travel for government officials, patronage jobs that change with each new government, and services on unprofitable foreign and domestic routes. Unions become entrenched, usually demanding and obtaining a say in all management decisions, often resisting improvements in efficiency, and refuse layoffs. Staffs become bloated and unproductive. Frequent management changes and other government interference result in a lack of strategic direction, an overextended system with money-losing routes, and an aging fleet poorly matched to demand. Poor passenger service becomes endemic. Consequently, the typical bureaucratic management is unable to adapt and succeed in an increasingly competitive marketplace.

Faced with continuing losses and a liberalizing marketplace, many countries decided to privatize their state-owned airlines. Several considerations drove these decisions:

- a growing realization that free-market capitalism increases national wealth and the standard of living;
- a need to increase the efficiency and competitiveness of the airlines and, consequently, relieve the financial burden caused by continuing subsidies;
- a desire to create stakeholders with a vested interest in the financial success of the airline, including owners, management, employees, suppliers, and customers; and,
- restrictions on government subsidies to businesses, especially within the European Union.

11.5.2 Privatization

The steps to privatizing a state-owned airline are often difficult and protracted. The government and the unions must be committed to privatization. This usually occurs only when the airline is faced with insolvency and liquidation, a point at which privatization also becomes more problematic. State-owned airlines are invariably in debt to the government – these loans must be written off as few outside parties are willing to assume the debt. An infusion of funds is often necessary. When subsidies end, costs must be reduced to a competitive level by layoffs, wage cuts or freezes, increased flexibility in work rules, and management restructuring free from government interference. Finally, network and fleet rationalizations are necessary, including dropping unprofitable routes and reducing the number of fleet types.

When these preliminary steps are completed, several options are available to privatize the state-owned airline. The state may simply auction the airline to the highest bidder and keep the proceeds. Rather than an open auction, the state may negotiate the purchase with a single party. In some instances, the management and/or employees can arrange financing to purchase the airline. Employees may be granted stock, generally in return for concessions and efficiency improvements. Granting stock to employees aids in the transition to a competitive market. British Airways, for example, offered employees stock at a discount as it was privatized in 1987. If the country has a large, developed stock market, the state may issue an initial public offering of shares in the company. In such instances, the state may remain a minority owner.

11.5.3 Results

The results of privatization efforts are mixed with many examples of success and failure. In the early 1980s, the British government under Margaret Thatcher successfully privatized much of the UK-owned industry, including flagship British Airways. BA, of course, remains one of the world's premier airlines. Japan Airlines successfully privatized at the same time. Air France and Lufthansa were privatized later and more gradually than BA (Morrell, 2013). The Spanish national airline Iberia was fully privatized when it was acquired in 2011 by the International Airlines Group (IAG), the holding company formed to facilitate the merger of BA and Iberia. IAG, which also owns several other smaller airlines, added to its stable in 2015 with the acquisition of the Irish national airline Aer Lingus. Similarly, Air France/KLM and the Lufthansa Group have grown through acquiring previously state-owned airlines resulting in three large European airline groups – IAG, Air France/KLM, and Lufthansa – similar in many ways to the three largest US network carriers.

Elsewhere in Europe, other attempts at privatization suffered repeated failures before finally meeting with some success. Olympic Airways, the former national airline of Greece, traces its origin to 1930. Although initially privately owned, the Greek government purchased the carrier in 1973. It struggled under politicized management until 2009 when the remains were acquired by its competitor Aegean Airlines. The Hungarian government tried repeatedly to sell its flag carrier Malev, finally finding a buyer in 2007. Success was short-lived as the carrier was renationalized in 2010 and then ceased operations in 2012 when the European Commission ruled that continuing state subsidies were illegal. Italy failed at serval attempts to sell off its long-struggling national airline, Alitalia. Labor unions opposed efforts to privatize and restructure the airline. In 2015, Etihad Airways acquired 49% ownership in a promising but ultimately failed attempt at revitalization. Alitalia, again fully government-owned, officially operated its last flight in October 2021. The ghost of the airline, nonetheless, lives on in Italia Trasporto Aereo – ITA – Italy's newest government-owned airline. Alitalia is the case study for this chapter.

Although several small Caribbean airlines are state-owned, most Central and South American airlines are privately held. As in Europe, some have done well while others have not. One example of the perils of government ownership is Brazil's national airline, Varig, which was for many years pampered and protected by the government. As the Wall Street Journal stated in 2005,

> When politicians wanted flights, Varig was lavish in handing out courtesy tickets. As the government sought to weave together this massive country, Varig flew to the hinterlands, whether or not routes were profitable. When

Brasilia strengthened ties with Africa in the 1980s, Varig started flying to unprofitable destinations there.

(Samor, 2005, p. A18)

Following industry deregulation, Varig entered bankruptcy and was subsequently sold off piecemeal. One division was purchased by the low-cost carrier Gol.

Turning finally to Asia, India's flag carrier Air India is a surprising example of an airline that has come full circle. The airline was founded in 1932 as a division of the now huge Indian conglomerate Tata Group when it won a contract to carry mail for Britain's Imperial Airways (see Chapter 1). For a few years after WWII, it was a public company until the Indian government effectively nationalized the airline in 1952. As with many state-owned airlines, Air India's losses and debt grew along with its expansion in international markets. The government's attempts at privatization stretched over two decades ending with the sale back to the Tata Group in 2022.

Privatization may be the only salvation for loss-making state-owned airlines, but there is no guarantee of success. Of 32 passenger airlines that ended government ownership since 2000, only five are still flying ("List of Government-Owned Airlines," 2021).

11.6 Global Alliances

Membership in a global alliance allows an airline to offer destinations that cannot be served economically, such as international routes with relatively low demand, or legally because of restrictive air service agreements. For example, no single airline currently provides service between Albuquerque, New Mexico, and Cape Town, South Africa, but US carriers and their international alliance partners such as Delta/Air France, American/British Airways, United/Lufthansa do provide convenient flights between these cities. The coordination of schedules and other product features through a global alliance provides a competitive advantage over carriers without similar arrangements. Ultimately, the global alliance seeks to capture passengers from origin to destination anywhere in the world.

11.6.1 History of the Big Three Global Airline Alliances

Before deregulation, Pan American World Airways and TWA were the only US carriers to operate international routes. TWA had some domestic routes for passenger feed, but Pan Am had none. Pan Am was thus dependent on US domestic airlines to provide traffic from the US interior to its gateway cities in New York, Miami, and San Francisco. But US domestic carriers seeking to offer international destinations often entered into marketing agreements with foreign-flag airlines rather than with Pan Am or TWA; this practice, of course, weakened both US flag carriers. These various marketing agreements were the forerunners of today's global airline alliances.

US domestic alliances, another precursor of today's international alliance structure, date to the 1960s when Local Service Carriers petitioned the Civil Aeronautics Board (CAB) to withdraw from some small communities. The CAB responded by allowing commuter airlines, the predecessors to today's regional airlines, to provide substitute services. Allegheny Airlines, later part of US Airways, obtained CAB approval for Henson Airlines to serve several Allegheny cities. Other carriers later adopted this pattern and the trend accelerated after deregulation with industry vertical integration.

The beginning of modern global alliances, however, dates to the 1992 Open Skies agreement between the US and the Netherlands and KLM's equity stake in Northwest Airlines. Northwest and KLM were granted antitrust immunity permitting them to coordinate schedules and prices and to codeshare on flights through their "Wings" alliance.

The Northwest/KLM alliance demonstrated the benefits of international airline alliances with its early success. The alliance proved surprisingly successful in capturing traffic across the Atlantic between Northwest's hub in Detroit and KLM's Amsterdam hub. Traffic on the route grew 55% annually over the first five years of the partnership increasing nearly 10-fold from fewer than 63,000 passengers per year in 1992 to more than 572,000 by 1997. This dramatic increase shows the power of the dual-hub system established with the alliance. More than 60% of the traffic on the Detroit–Amsterdam flights did not originate in either Detroit or Amsterdam, but rather beyond each of those hubs in the spoke cities. Figure 11.2 shows a hypothetical itinerary from Indianapolis, Indiana, to Frankfurt, Germany. By connecting the two hubs and the many spoke cities, each airline marketed flights in 16,240 city pairs under its own code.

A passenger could travel from Indianapolis to Detroit on a Northwest flight, from Detroit to Amsterdam on either a Northwest or KLM flight, and then

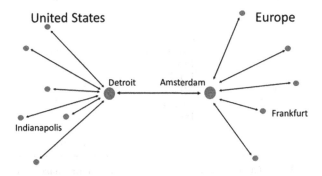

Figure 11.2 Northwest/KLM Alliance.

make a connection in Amsterdam for Frankfurt on a KLM flight. Through the code-share agreement, if the passenger booked this itinerary with Northwest, all flights would show as Northwest flights. The opposite, of course, would be true if the booking were with KLM. Coordinated flight schedules provided convenient connections at both hubs.

Other similar partnerships soon followed. In 1993, British Airways took a $300 million stake in US Airways from which BA sought valuable US domestic feed. This arrangement failed to meet BA's expectations and, in 1997, it wrote off the entire investment in US Airways. BA subsequently formed an alliance with American Airlines that is the foundation for today's Oneworld alliance (Ito & Lee, 2005).

By 2000, three international alliances emerged: Oneworld, Star, and Skyteam. Although membership has frequently shifted over the past 20 years, the shakeout has produced three competitive global alliances of comparable size as shown in Table 9.2. Each developed on an initial partnership between the US and an EU airline.

Until 2020, each alliance had at least one airline member from the US, Europe, South America, and Asia. Star and SkyTeam each have a member from Africa. Following the merger of LAN Chile and Brazil's TAM in 2011, LATAM joined Oneworld in 2013. Then Delta Air Lines purchased 20% of LATAM in 2020 forcing it to exit oneworld leaving the alliance without a South American member, a major blow to American Airlines.

Table 11.2 Global Airline Alliances

Alliances	Star	SkyTeam	Oneworld
Members	Lufthansa, United, Air China, Singapore, Avianca, and 21 others	Air France/ KLM, Delta, China Eastern, Aerolineas Argentina, Kenya Airways, and 13 others	British Airways, American Airlines, Japan Airlines, Cathay Pacific, and 10 others
Market Share	24%	20.8%	19%
Passenger per year	642 million	665 million	557 million
Countries	195	175	161
Destinations	1,360	1,062	1,016
Fleet Size	5000	3,937	3,560
Employees	432,603	481,691	382,913
Revenue (billions USD)	196	156	143
Daily Departures	18,521	16,323	14,296

Note: China Airlines, the largest Taiwanese airline, is easily confused with Air China, the Beijing-headquartered airline and one of China's big three network carriers. Status as of 2021.

Source: Airline Business, Wikipedia.

The three global alliances carry more than 50% of the world's passengers, but many airlines are not members of the big-three alliances. The growing number of low-cost carriers are not members mostly because their Spartan product is not compatible with the broader offering of the alliance members. Looking to reap the benefits of an alliance, nine Asia-Pacific low-cost airlines formed the Value Alliance in 2016 designed to "provide unparalleled access to more than 160 destinations from 17 hubs across Australia, North Asia, and Southeast Asia" (Value Alliance, n.d.). Typical of the instability of the early years of an alliance, Value now has five members with two having merged and two others succumbing to the pandemic.

11.6.2 Marketing and Revenue Benefits

Global alliances offer members many benefits that include both revenue enhancement and cost reduction. First, let's look at the potential revenue benefits.

An international or global alliance allows each member carrier to expand the destinations offered well beyond what it could do independently due to financial or air service agreement constraints. A member of any of the big-three alliances can offer the world to its passengers through its alliance partners promising a consistent level of service and convenience. If the alliance is successful in generating additional traffic, members can offer more flights that, in turn, improve the alliance's competitive advantage.

Through the various member sales and marketing organizations, distribution is expanded over a large geographical area, including all major business markets. Joint advertising and sales efforts similarly are expanded over the globe. Frequent flyers and other passenger loyalty programs become more attractive as passengers earn rewards, or rebates depending on point of view, on all member airlines.

Alliances strengthen the hubs of the largest members increasing market power and pricing flexibility. These airlines typically dominate their respective hubs so that partners may be able to share gates, customer service space, and even slots providing entry to airports otherwise not accessible.

Most of these marketing benefits should be familiar as they are essentially the same as those that arise from large, single-airline hub-and-spoke systems. Global alliances simply extend these benefits to a worldwide scale.

11.6.3 Operating Benefits

Alliances also provide potentially significant operating benefits. In the absence of an Open Skies agreement, an alliance allows members to access markets that would otherwise be unavailable. Even with Open Skies, alliances allow the marketing of routes not allowed under cabotage restrictions.

By combining facilities, expertise, and purchasing power, alliances can also reduce operating costs. Joint use of airport facilities reduces total space requirements and facilitates better staff utilization. London Heathrow's new Terminal 2, for example, provides common check-in and ground handling for 25 members of the Star Alliance. Maintenance of modern generation aircraft requires large capital investments. Participating carriers can specialize in specific aircraft types providing maintenance to all members. Likewise, pooling of replacement parts can reduce required inventory levels. Last, alliance members can engage in joint purchasing of aircraft and parts, which increases negotiating power and potentially reduces prices.

11.6.4 Antitrust Immunity

Reaping the benefits of global alliances requires immunity from antitrust laws. The United States, the European Union, and many other countries have laws that prevent firms from coordinating pricing (called price-fixing) and other business practices that might reduce competition. The US Congress passed a series of laws beginning with the Sherman Antitrust Act of 1890 that are collectively termed antitrust legislation. Without a grant of antitrust immunity, independent airlines are prohibited from practices such as fixing prices, schedule coordination, and jointly determining capacity. However, nothing prevents one carrier from contracting services from another as network carriers do with their regional partners.

Global airline alliances have the potential to reduce competition as previously competing airlines cooperate as alliance members. In granting antitrust immunity, the US, EU, and other governments must weigh the potential consumer benefits from improved service against the loss of some competition. As part of its Open Skies policy, the US has granted antitrust immunity to many foreign partners in the big-three alliances (US ATI is not granted to domestic airline partnerships). The EU has similarly granted immunity from its competition laws. With antitrust immunity, airlines may coordinate flight schedules, codeshare on routes, fix prices, pool revenues and costs, and set capacity at the level their analysis determines will be most profitable. These are all essential elements of a successful partnership.

11.6.5 Establishing an Alliance

Forming and maintaining an alliance is a complex management challenge. An alliance begins with a vision that capitalizes on each partner's strengths and develops common goals that allow all partners to benefit. The founding members of the global alliances jointly developed their respective alliance visions and objectives.

As the alliance expands, a top-level organization helps foster a common culture that values all partners. A fact-based decision process is required to build an integrated, seamless travel network. Teams representing the members set goals and monthly performance metrics, communicate effectively, and establish arbitration procedures to resolve disputes among members.

A dedicated staff sets prices and coordinates flights. Allocating revenues, costs, and profits across the membership is fraught as each member will want a larger profit. Successful alliances require extensive information sharing and compatible information technology systems. Integrating hardware and software is a long, complex, and expensive endeavor.

As the big-three alliances have added new members, the process for joining an alliance has become lengthy and expensive. New members must meet the alliance criteria that often require extensive modifications of existing business practices. Computer reservations systems, ticketing, and baggage processing must be compatible with those of other alliance members. Further agreements are needed on joint fares, joint sales and e-commerce ventures, reciprocal use of airport facilities, and *codesharing.* Associate or affiliate memberships offered to some smaller carriers require less accommodation (Dunn, Gubish, & Waldron, 2015).

Airline alliances and partnerships take many forms, essentially along a continuum as shown in the top section of Figure 11.3. A commercial alliance involves interline agreements and prorated pricing where each airline agrees to accept the other tickets and transfer baggage. Although interlining was once commonplace, many carriers have dropped interline agreements with other carriers as a cost-saving method. Of course, many LCCs such as Southwest, have never interlined. A codeshare agreement takes interlining a step further. Each partner places its airline code on some of the partners' flights marketing and selling these flights as their own. Strategic alliances are at the other end of the continuum. The joint venture is one step short of a merger in which two or more airlines closely coordinate their operations and frequently share revenues. In a merger, two airlines join to become a single company. As the degree of cooperation increases, the cost of exiting the relationship is ever more costly.

The bottom section of Figure 11.3 lists many of the steps toward expanded cooperation. The first stage focuses on revenue generation as the airlines

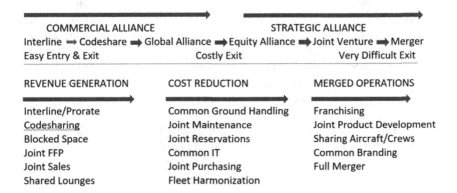

Figure 11.3 Continuum of Alliance Cooperation.
Source: Adapted from Doganis, 2006.

offer more destinations through the partnership. Establishing reciprocal frequent flyer programs, joint sales, and shared airport lounges are all intended to attract more passengers. These commercial alliances do not require substantial changes in either airline's business practices and are relatively easy to dissolve. Commercial alliances are common even between airlines in different global alliances.

At the next level of cooperation, airlines also seek cost reductions as they combine operations at common destination airports. Joint maintenance and purchasing offer potential savings. As the degree of cooperation increases through scheduling and capacity agreements, carriers give up some control and must modify their schedules and operating practices for the good of the alliance. As cooperation deepens, so does the cost of exit.

Joint ventures, one step below a merger, closely tie the airlines in most operations, such as schedule coordination, revenue sharing, and joint product development. Many alliance members have formed joint ventures that create tiers within the alliance, a potential source of friction ("Ten Joint Ventures," 2016). Except for the joint venture members, ties within alliances are not strong – airlines periodically change alliances.

In the last step, the airlines merge into a single airline, as in the US, or common ownership by a holding company, which is typical in Europe. While the benefits may be significant, dissolution of the agreement is like ending a marriage – expensive and probably bitter.

11.6.6 Passenger Benefits

Alliances can potentially either benefit or harm passengers. If international travel convenience is improved at lower fares, then the consumer benefits from alliance formation and expansion. Alliances can also be pro-competitive. If weaker competitors form an alliance, they are better able to compete with a more dominant airline, so competition is increased.

On the other hand, alliances may limit competition resulting in higher prices. If an alliance reduces competition on routes by combining carriers that previously competed, prices will probably rise and service may suffer as well. This may benefit the alliance but passengers suffer. Because US and EU regulators feared anti-competitive effects, American Airlines and British Airways were long refused antitrust immunity for the Atlantic routes in the Oneworld alliance. Regulators believed reduced competition would lead to less service and higher fares.

Most early academic studies show consumers benefit from alliances with improvements in service, including more flight frequency, shorter total trip time, easier booking and ticketing, more convenient service, and lower fares. Brueckner (2003) found that code sharing on international itineraries reduced fares by 8% to 17% while antitrust immunity reduced fares by 13% to 21%. More recent studies show more modest effects (Bilotkach, 2019).

11.6.7 *Alliance Instability*

The potential benefits of a global alliance notwithstanding, early alliances proved unstable. Some collapsed entirely and membership changes in the current global alliances are still common. There are many reasons for these failures. Of course, examining the reasons for failure also provides insight into the elements necessary for success, such as:

- Misaligned expectations. Partners sometimes don't have similar expectations over many aspects of operating the alliance. Expectations, objectives, and governance procedures were not fully explored before joining the alliance.
- Governance problems. No clear procedures are established for decision-making and resolving disputes. Governing by committee may leave some or all partners satisfied.
- Complex structures with two or three levels of membership.
- Flight schedule consolidation requires compromise – the optimal schedule for the alliance will not be optimal for one or more partners.
- Transfer pricing and revenue sharing. Disputes often arise over the allocation of revenues and expenses. These are difficult to solve analytically and require partners to compromise with a focus on the greater objective for the alliance over each airline's interest.
- Unwillingness to share information. For the alliance to operate seamlessly from the customers' view, airlines must share operating and customer information. Members may be unwilling to share information that could provide an advantage to other carriers or weaken an existing competitive advantage.
- Joint ventures between members in different alliances pose conflicts of interest and may confuse passengers.

Although the big-three alliances have matured and gained stability with only a handful of new members being added in recent years, members continue to shift and realign, especially in response to airline mergers and the rapid growth of the Gulf carriers. In 2013, Qantas joined Emirates in a codeshare agreement, a blow to its long-standing partnership with British Airways and the Oneworld alliance. At the same time, Qatar joined Oneworld (CAPA, 2012). In 2019, Delta Air Lines invested $1.9 billion to purchase a 20% stake in LATAM (see Equity Alliance below), South America's largest airline, a strategic move that surprised American Airlines. LATAM had long been a partner with American and a Oneworld Alliance member. Because Delta is a founding member of the SkyTeam Alliance, LATAM left Oneworld in 2020 but, as of early 2022, had not joined Delta in SkyTeam.

As Michael Porter (1990) noted, alliances are transitional devices that arise in industries undergoing structural change and increasing competition, certainly an apt description of the airline industry with spreading liberalization. Alliances are not just an airline industry phenomenon but are common in many industries. Consider, for example, the many partnerships and cross-equity holdings in the automobile industry.

11.6.8 Equity Alliance

There's a long history of large airlines purchasing an equity stake in other airlines to solidify commercial cooperation. Delta Air Lines has for years built equity stakes in foreign airlines with investments in Aeromexico and LATAM in the Americas; China Eastern and Korean in Asia; and Air France/KLM and Virgin Atlantic, in which it holds 49% stake, in Europe. The pandemic tested this strategy as Delta wrote off nearly $2 billion in 2020 from the bankruptcies of LATAM and Aeromexico plus a $200 million charge on its investment in Virgin Atlantic. In late 2021, Delta affirmed its strategy announcing it would invest $1.2 billion to maintain its stakes in LATAM, Aeromexico, and Virgin Atlantic (Clough & Jasper, 2021).

Etihad Airways' bold attempt at building a fourth global alliance through investments in several airlines fared even worse. By 2016, Etihad had purchased one-quarter to one-half ownership in six financially struggling airlines scattered across Europe, Asia, and Australia. The most ambitious was a 49% stake in long-struggling Alitalia with whom it has a substantial code-sharing agreement. With these equity stakes in airlines, Etihad expanded its network and gained feed traffic for long-haul flights across its hub in Abu Dhabi. Six years later, only one of the six, Air Serbia, had avoided bankruptcy, and three – Alitalia, Darwin Airlines, and Jet Airways – had ceased operations. The alliance folded and Etihad embarked on a new network strategy under a new leadership team.

The most sobering and cautionary tale comes from China's HNA Group, the once sprawling conglomerate that fell from the grace of China's government and collapsed under a mountain of debt. Hainan Airlines, founded in 1993 as a regional airline, is the foundation of the Group. It grew rapidly across China and then internationally becoming the fourth largest airline in China by 2003. Seemingly not satisfied with its internal growth opportunities, the HNA Group began buying other Chinese airlines in 2004 amassing an astonishing 16 additional Chinese airline subsidiaries as well as minority stakes in a few foreign airlines (CAPA, 2021). By 2020, Hainan Airlines and its subsidiaries operated nearly 1,800 domestic and international routes to 86 cities in Asia, Europe, North America, South America, and Oceania (Hainan Airlines, n.d.). As of the end of 2021, Hainan's fleet of 200 plus aircraft comprised Airbus A330, A350, Boeing B737, and B787s. The other HNA subsidiaries operated some 700 additional planes (Bailey, 2021; Hainan Airlines, n.d.).

As unwieldy as this collection of airlines is, it was not the direct cause of the HNA Group's failure. In 2016, HNA went on a $40 billion spending spree buying 1,400 hotels, businesses in construction and property management, airport management, and office towers. Two years later, its acquisitions spanned tourism, finance, media and entertainment, logistics, and technology. Having accumulated massive debt, it couldn't pay the interest on loans. The government of China's southern province of Hainan effectively took control of the group in February 2020. The airline division was sold and now operates as a form of equity alliance with an uncertain future.

History has not been kind to equity alliances, but the pandemic has laid the entire airline industry low. Delta Air Lines' highly regarded management retains faith that its investments in other airlines will earn long-run profits. In his analysis of equity alliances, Joe Pesek (2021) concludes that many airlines have considered equity stakes in other airlines to gain access to new markets by aligning operations and marketing strategies. "However, equity stakes don't guarantee influence, nor do they reduce risk in the event of an economic recession."

11.7 Consolidation – Mergers and Acquisitions

Airline analysts have long lamented that the world airline industry is too fragmented leading to chronic overcapacity, low fares, and marginal profitability. Citizenship requirements often prevent cross-border mergers and acquisitions, but, where allowed, consolidation through merger and acquisition seeks the same benefits as global alliances – seamless travel, revenue enhancement, cost synergies, and elimination of competition. Indeed, the new century has accelerated industry consolidation. European deregulation permitted citizens of any EU state to own EU-based airlines. This liberalization led to the merger of Air France and KLM in 2004. Although the airlines are under single management, both brands have been maintained. Further consolidation in the European Union began in 2011 as British Airways, Iberia, and Aer Lingus along with several smaller carriers were combined under the holding company International Airlines Group. The Lufthansa Group, the third of the European big three, has a turbulent history as several subsidiary airlines have been acquired, created, dismembered, and reassembled. The Group comprises Lufthansa, Swiss, Austrian, Brussels, Air Dolomiti, and Eurowings. In contrast to practice with the US and Latin American mergers, the European airline holding companies have maintained the independent branding of the airlines. A passenger flying on Swiss International Air Lines would likely not recognize that Swiss is part of the Lufthansa Group. Aaron Karp (2015) attributes this difference to Europe's long airline history and established brand loyalties.

In the US, consolidation of the largest airlines concluded with the merger of American Airlines and US Airways in late 2013 forming the world's largest carrier. This followed the mergers of Delta and Northwest in 2008, United and Continental in 2010, and Southwest and AirTran in 2011, leaving the US with four large airlines controlling about 80% of US domestic traffic. In contrast to European consolidation where merged airlines continue to operate under separate brands, mergers in the US result in a single surviving brand. When US Airways Flight 434 from San Francisco landed in Philadelphia on October 17, 2015, the US Airways website went dark, the airport kiosks and signs changed to American Airlines.

In Latin America, the Chilean flag carrier LAN completed its takeover of Brazil's TAM in 2012 to form LATAM, the region's largest airline. Earlier, Columbia's Avianca merged with El Salvador's Taca Airlines. The Avianca Group subsidiaries are all now under the Avianca nameplate. The integrated cargo carriers are much less fragmented than the passenger airlines with UPS, FedEx, and DHL controlling much of the market. A UPS agreement to purchase the Dutch carrier TNT Express would have placed the package delivery industry firmly within the grip of the big three, but in 2013, the European Commission refused to permit the $7 billion deal on antitrust grounds. UPS subsequently gave up its efforts to restructure the acquisition, but FedEx, seizing the opportunity, acquired TNT for $4.8 billion all-cash in 2016.

Large airline mergers and acquisitions slowed during the latter half of the 2010s, but the stress of the COVID-19 pandemic has rekindled interest. Korean Airlines and Asiana, South Korea's two largest full-service network carriers, were in the process of merging in early 2022 having received conditional approval from the South Korean government. The largest US ULCC, Spirit Airlines, agreed to an acquisition by JetBlue. The combined airline would be the fifth-largest in the US. Not to be left out, Allegiant Air, the other large US ULCC, and Mexico's Viva Aerobus applied for antitrust immunity to jointly schedule, operate, and market flights between the US and Mexico.

11.8 Summary

The airline industry remains under more economic restrictions than most other industries, especially with limits of foreign ownership that almost universally prohibit a foreign citizen from owning or exercising control of a domestic carrier, thus preventing the formation of truly multinational airlines. Faced with the likely failure of General Motors and Chrysler during the Great Recession of 2008, the US federal government bailed out both. The Italian automaker Fiat was enlisted to take control of Chrysler Motors Corporation. In 2021, Fiat Chrysler Automobiles merged with France's PSA Group to form the world's sixth-largest automaker. A similar arrangement, however, is not possible in the airline industry. Following the consolidation, US airlines enjoyed a decade of consistent profitability that had previously been unattainable. Multinational airlines might provide a more predictable, less turbulent industry than has been achieved through global alliances. International liberalization, however, has slowed in recent years suggesting that multinational airlines are not in the offing.

11.9 Case Study: Alitalia

Alitalia, Italy's former flag carrier, was founded one year after the end of World War II in 1946 jointly funded by the Italian government and British European Airways (later to become part of British Airways). By the late 1960s, it was Europe's third-largest airline and a favorite of the emerging jet set flying Italian and Hollywood stars between movie sets. However, like many airlines born in an era of economic regulation and government protection, Alitalia failed to adapt to industry deregulation that led to the emergence and fierce competition from low-cost carriers and consolidation among large legacy airlines. The spread of high-speed rail across Europe further eroded its markets. Yet, despite having only one profitable year since its founding, the Italian government repeatedly saved the airline from collapse spending an estimated €10 billion taxpayer in a series of bailouts since 2008. Financial distress led to a merger with Air France and KLM in 1997, a contentious and unstable partnership that lasted 10 years. With Alitalia on the verge of collapse in 2009, Silvio Berlusconi, then running a successful bid for prime minister, persuaded 16 companies to save the airline. Despite union opposition, the airline was privatized for the first time in its history. The new owners, with Air France/KLM retaining a minority stake, took the potentially profitable assets while dumping the airline's debt on the Italian taxpayers. The "new Alitalia" struggled; the government pumped in several billion euros to keep it running. A savior arrived in 2014 when Abu Dhabi's Etihad Airways acquired 49% ownership as part of its development of an equity alliance. Labor unions opposed Etihad's efforts at restructuring needed for the airline to prosper in the global aviation market and Alitalia returned to bankruptcy in 2017. European Union rules forbade further bailouts, so the government announced it would be auctioned. Delta, China Eastern, easyJet, Lufthansa, Ryanair, the Italian railway company, and others expressed interest in buying all or parts of the carrier, but all eventually dropped out. Having determined that it could not survive the COVID-19 pandemic, the Italian government again took full ownership in 2020 only to conclude Alitalia was doomed. Not willing to concede that Italy's once-proud flag airline would vanish, it determined to create yet another fully state-owned new airline, Italia Trasporto Aereo, ITA, in which the government would invest €1.35 billion during the first three years. Alitalia flew its last flight on October 14, 2021. ITA began flying the next day. Starting with 52 jets and 2,800 employees, ITA is less than half the size of Alitalia, but, of course, it plans to grow. ITA's core is still the old Alitalia. It may find it retains alitalia's legacy of high costs, mismanagement, and heavy political and trade union influence. Apparently aware of these risks, in late August 2022 the Italian government selected a consortium led by the private equity group Certares, Delta Air Lines, and Air France-KLM to begin exclusive negotiations to acquire ITA. The Italian state would retain 45% of the ITA shares.

Concepts: deregulation, competition, substitutes, state-owned airlines, bailouts, unions, equity alliances, privatization.

Sources: Air Transport World, Cranky Flier, New York Times, Wall Street Journal

References

Bailey, J. (2021, January 29). HNA Group bankrupt but Hainan Airlines to carry on flying. *Simple Flying*. https://simpleflying.com/hna-bankrupt/

Bilotkach, V. (2019). Airline partnerships, antitrust immunity, and joint ventures: What we know and what I think we would like to know. *Review of Industrial Organizations*, 54, 37–60. https://doi-org.ezproxy.libproxy.db.erau.edu/10.1007/s11151-018-9636-x

Brueckner, J. K. (2003). International airfares and the age of alliance: the effects of code-sharing and antitrust immunity. *The Review of Economics and Statistics*, 85(1), 105–118.

CAPA. (2012, September 6). Qantas and Emirates to codeshare in the first alliance shakeup of the season; next Qatar to oneworld.

CAPA. (2021). *Hainan Airline Group.* https://centreforaviation.com/data/profiles/airl ine-groups/hainan-airlines-group

Clough, R., & Jasper, C. (2021, December 13). Delta commits $1.2 billion to Virgin Atlantic, Aeromexico, Latam. *Bloomberg*. Delta Commits $1.2 Billion to Virgin Atlantic, Aeromexico, Latam

Doganis, R. (2019). The Airline Business 5nd ed.). Routledge.

Dunn, G., Gubisch, M., & Waldron, G. (2015, September). After the gold rush. *Airline Business*, 31(7), 30–33.

European Commission. (2015, December 7). *International aviation: an opportunity for growth and jobs in EU aviation sector.* http://europa.eu/rapid/press-release_MEMO-15-6145_en.htm

fBermuda Agreement. (2021, November 11). *In Wikipedia.* https://en.wikipedia.org/wiki/Bermuda_Agreement

Hainan Airlines. (n.d.) *About us.* www.hainanairlines.com/HUPortal/dyn/portal/DisplayPage?LANGUAGE=US&COUNTRY_SITE=US&SITE=CBHZC BHZ&PAGE=CHAP

Ito, H. & Lee, D. (2005). Domestic codesharing practices in the US airline industry. *Journal of Air Transport Management*, 11(2), 89–97.

Karp, A. (2015, August 10). Latin America's airline consolidation take a different path from Europe's [Web log post]. http://atwonline.com/blog/airkarp.

List of government owned airlines. (2021, December 15). *In Wikipedia.* https://en.wikipe dia.org/wiki/List_of_government-owned_airlines

Morrell, P. (2013). *Airline finance.* Ashgate.

Open Skies Partners. (2021, December 9). US Department of Transportation. www.tra nsportation.gov/policy/aviation-policy/open-skies-agreements-being-applied

Pesek, J. (2021, January 5). Airline alliances through equity stakes: Is the business model dead? *Airline Geeks.* https://airlinegeeks.com/2021/01/05/airline-alliances-thro ugh-equity-stakes-is-the-business-model-dead/

Porter, M. (1990): The Competitive Advantage of Nations. *Free Press.*

Ten airline joint ventures travellers ought to know about. (2016, December 6). *Business Traveller.*

Value Alliance. (n.d.). *About us.* www.valuealliance.com/#/about

Review Questions

1. What were the positions of the US versus most other nations at the Chicago Convention on regulating international air commerce? What was the outcome and why did most other nations not agree with the US position?
2. What aspects of the air transportation service between two countries are typically addressed in a bilateral agreement (such as Bermuda I)?

3. What does the US mean by "Open Skies?" Compare and contrast recent "Open Skies" agreements with earlier bilateral agreements.

4. In negotiations with the US to obtain a single air transportation agreement, the EU has proposed eliminating or significantly reducing the restrictions on citizenship and cabotage. What are these restrictions? Who might win and lose from such an agreement? What is the position of unions representing airline pilots? Why?

5. In most nations, the flag carrier has historically been state-owned. Why have states felt that public ownership was beneficial?

6. What are the characteristics of state-owned airlines that often render them uncompetitive with privately (or public stock) owned airlines.

7. What are the various ways that a state-owned airline can be privatized?

8. Provide an example of a domestic alliance and an international alliance.

9. What benefits do airlines attempt to obtain through an alliance? Why doesn't an airline simply add the new destinations it wishes to serve instead of entering into an alliance?

10. Why do alliances seek antitrust immunity? Under what conditions will the US grant such immunity?

11. Members of international alliances often leave or change alliances. What factors cause this instability?

12. Under some circumstances, alliances benefit the customer (pro-competitive), but in others, the customer may suffer. Explain.

13. Curiously, in the three largest airline markets, the US, EU, and China, there are three dominant full-service network carriers or airline groups. Who are these FSNCs in each market?

14. Name the three global airline alliances. Are the three largest network carriers in the US, EU, and China evenly divided among the alliances?

12 Looking Ahead

Upon completion of this module, you should be able to:

* List the important future challenges facing airlines.
* Explain why the airline industry is one of the most difficult to decarbonize.
* Discuss approaches to reducing airline emissions.
* Describe the threats to the growth in world trade and business travel.
* Characterize the potential pilot shortages by world region.
* Discuss the appeal and history of complex airline management structures.
* Assess the future of the global airline industry.

12.1 Introduction

The global airline industry is dynamic, constantly adapting to changes in passenger demand, technology, government regulation, and new entrant competition. Predicting the future of the industry is perilous; you will find predictions hedged with terms such as may or might. This concluding chapter looks ahead to the challenges facing the industry. It begins with the newest and likely most onerous challenges and then addresses the consequences of unsustainable profitability to the industry's frequent resorts to government assistance for survival.

12.2 Environmental Constraints

The gradual acceptance by world leaders that global warming presents an existential threat led to the United Nations Paris Agreement of 2015. One hundred eighty-nine members agreed to reduce greenhouse gas emissions to limit the projected global temperature increase by 2050 to 2 degrees Celsius (United Nations, 2015). In furtherance of this agreement, each nation is responsible for setting its own emissions goals. Most developed economies, led by the EU, pledged to attain net-zero carbon emissions by 2050. The US initially joined the agreement but withdrew under the Trump administration. In 2021, the new Biden administration rejoined committing the US to net-zero emissions by

DOI: 10.4324/9781003290308-12

2050. China, the world's largest carbon emitter, delayed stating its specific goal until 2020 when it agreed to achieve carbon neutrality by 2060. Whether the world will meet the UN goal is uncertain, but the transition has begun. Though estimates of the total costs and benefits vary greatly, all are in the hundreds of trillions of dollars. At its 2021 annual meeting, the International Air Transport Association (IATA) approved a resolution for the global air transport industry to achieve net-zero carbon emissions by 2050. This commitment, supported by most of the world's airlines, is an immense challenge because the aviation industry is among the most difficult to decarbonize.

Airlines produced only 2.8% of global CO_2 emissions in 2019 – 3.5% of global warming if non-CO_2 contributions are included (Cabera & de Sousa, 2022). Emissions have dropped by more than 70% per available seat kilometer since 1960; however, without further mitigation, the industry contribution to total emissions will rise to 15% by 2050. IATA proposed several steps to meet the industry's net-zero goal, including (IATA, 2021):

- Bringing large-scale, cost-competitive sustainable aviation fuels (SAF) to the market.
- Producing radically more efficient airframe and propulsion technologies.
- Providing the airport infrastructure to supply sustainable aviation fuel (SAF), at cost, and in a cost-effective manner.
- Eliminating inefficiencies in air traffic management (ATM) and airspace infrastructure.

The first two steps require huge, perhaps unattainable, leaps in technology and trillions of dollars of investment.

Several firms manufacture SAF from feedstock such as biomass, waste oil and fats, and municipal waste. These fuels, when mixed with conventional jet fuel, power the existing airline fleets with minimal modification, hence the term *drop-in fuels*. The cost, currently two to four times that of jet fuel, may fall with scale but will remain higher than for jet-A kerosene. A study by Bain and Company, a consultancy, estimated that a $1.3 trillion investment in SAF manufacturing would supply less than 25% of the 2050 fuel requirements (Harris, 2020).

Synthetic fuel, or eFuel, is hydrocarbon fuel made commercially rather than refined from fossil fuels. A new process can produce a type of eFuel using CO_2 and hydrogen. If the hydrogen is produced through the electrolysis of water (using "green electricity"), the eFuel product would be carbon-neutral. eFuels could be the holy grail of SAF, but the technology has not advanced beyond the lab. Rather than chemically converting hydrogen obtained from electrolysis into liquid fuel, liquefied hydrogen could also power future commercial aircraft. New aircraft designs are required to increase fuel tank capacity by some four times that of current aircraft. Bain concludes that "despite aggressive assumptions around technology and efficiency improvements over the coming decades, SAF production cost and supply availability will fail to meet aviation needs, falling short even with government incentives and mandates that spur early investments and adoption" (Harris, 2020).

Further improvements in engine technology offer a more promising medium-term solution for reducing emissions. Next-generation aircraft engines designed to run on jet fuel, SAF, and pure hydrogen could enter service by the mid-2030s. The new engines retain most features of existing engine designs relying on higher bypass ratios to obtain up to 25% better fuel efficiency and emissions reductions. CFM International, a joint venture between GE Aviation and Safran, envisions an open fan, hybrid electric engine compatible with SAF and hydrogen (CFM, 2021). Pure electric systems are also in the test stages. Battery weight and low energy density seem to limit this application to smaller regional aircraft operating short to medium-haul flights.

SAFs derived from biomass are compatible with existing delivery systems and airport infrastructure. Hydrogen, which boils at -253 °C, is naturally a gas. To be transported as a liquid, it must be cooled and compressed requiring a massive investment in delivery and distribution systems. Some of these investments may be government-funded, but airlines would likely bear most of the cost. Improving European ATM efficiency is an old goal stymied mostly by political and national sovereignty objections.

For now, meeting the IATA's commitment to net-zero goal emissions by 2050 does not seem technologically or economically feasible. Wide use of alternative fuels will raise airline costs that must be recouped in higher ticket prices. Fewer people will choose to purchase air travel at higher prices. If the goal is not revised or extended, a distinct possibility, airlines will have to purchase carbon offsets from third parties, further increasing costs. If the industry stumbles in meeting its net-zero goal, it will come under intense popular and political pressure, especially in Europe. Governments may resort to capping flights. While airline travel in 2050 may not be a luxury good, the real price of air travel may be substantially higher.

12.3 Fuel Expenses

Environmental constraints and fuel expenses are closely related. The price of jet fuel varies substantially often over short periods and defies reliable prediction. Nonetheless, investors, financial institutions, and governments are becoming averse to new fossil fuel extraction. Drilling in some world regions will not be affected – Saudi Arabia vowed to pump every last molecule – but the world petroleum supply may fall well before the reserves are deleted. If true, the long-run real price of jet fuel will increase. Many airlines use financial instruments to hedge against short-term changes in petroleum prices, but hedging is essentially a gamble that has lost favor after fuel prices plunged more than 70% in 2008 and again in 2019 leaving airlines with hedges paying higher than the market rate for fuel. With fuel prices again climbing sharply in 2022, airlines with fuel hedges have gained immensely. Air France and Southwest each gained over $1 billion. Based on its fuel hedges, British Airways parent company IAG's fuel bill went up only 45% despite a jet fuel increase of 150% (Longley & Kumar, 2022). In the future, airlines may continue to seek the protection hedging offers because ticket price increases in response to higher fuel prices are usually insufficient and too late to offset the increase in fuel cost. In the long run, however, there is no escaping higher fuel prices.

12.4 Diminishing World Trade and Business Travel

After decades of steady growth, world trade measured as a percentage of world GDP peaked in 2008 at 62% followed by an erratic 10-point decline by 2020 (World Bank, n.d.). A lively debate over the future of globalization ensued. Trade in services such as legal and data analyses has continued to grow but has not offset the fall in the trade of goods. Reemerging nationalism and protectionism, especially with the trade wars between the US and China, are partially responsible for the fall in trade in goods. The effects of the Great Recession of 2008, the 2020 pandemic, and the Russia-Ukraine war beginning in 2022 complicate the analysis. One dire possibility is a world that is split between a group of nations led by China and the developed world democracies. If globalization does recede, international business and, to a lesser extent, leisure air travel will be adversely affected.

During the COVID-19 pandemic, government-imposed air travel restrictions and widespread fear of infection caused business air travel to plummet. Many employees began working remotely, abandoning big-city corporate offices. Zoom and other platforms replaced in-person meetings that often required air travel. Teleconferencing as a substitute for in-person meetings now seems entrenched, leading some analysts to conclude that 40% of business travel will not return when the pandemic ebbs or becomes endemic (Baldanza, 2022).

A decline in globalization, balkanization of world trade, and teleconferencing instead of in-person meetings would reduce air travel demand.

12.5 Labor Shortage

A long-term pilot shortage threatens to constrain airline growth and increase labor costs. The US airlines first faced the shortage before the pandemic. Anticipating a lengthy recession, airlines offered early retirements and extended leaves to absences to all employees, including pilots. When demand roared back in the summer of 2022, however, they found themselves short of pilots for schedules they wished to fly. The regional airlines were hardest hit because their pilots are the major source of new pilots for the FSNCs who are often the regional airlines' commercial partners. That summer, the US lacked about 8,000 pilots, or 11% of the total supply (Murray, 2022).

If the shortage occurred only in the US, the effect on the worldwide industry would be limited. However, in a forecast updated in 2022, Oliver Wyman, a consultancy, warned that the shortage would expand to most of the industry over the following decade leaving airlines short about 80,000 pilots by 2032

(Murray, 2022). A combination of retirements from a workforce that is older than the average, a tough value proposition for new civilian pilots, and a limited training capability constrained by a lack of instructors and flight simulators, promises to quadruple the US shortage to 30,000. Rapid air travel growth in the Middle East will leave it 18,000 pilots short by 2032. Europe will enjoy a surplus until the middle of the decade, but then face a shortage of 19,000 pilots by 2032. Asia will not face a shortage until the end of the decade when demand growth resumes. Latin America and Africa will be mostly spared with only minor shortages expected.

Pilots, of course, are mobile with some willing to take jobs in other world regions for higher pay and benefits. Airlines will have to curtail some expansion plans and increase pilot wages for at least a decade. Passengers will experience the pilot shortage in fewer flights and higher fares. For aspiring pilots, the future has never looked better.

A shortage of other skilled airline labor, especially aircraft technicians, also looms but should be less constraining.

12.6 Complex Airline Structures

Despite rapid liberalization in many world regions, ubiquitous restrictions on foreign ownership and control remain. Unlike other global industries, airlines are unable to become truly multinational corporations. The result is a fragmented industry. Expansion-focused airlines attempt to overcome the citizenship restrictions by setting up subsidiaries outside their home countries, usually in partnership with a home-country airline or investment firm. The many airline subsidiaries of the AirAsia Group, now Capital A Berhad, stem from the various government restrictions on airline ownership. To comply with the laws, the subsidiary airlines must be majority owned by citizens of the home country. Consequently, the sponsoring airline may not exercise managerial control, a circumstance that can result in conflict. For example, between 2012 and 2020, AirAsia twice failed to establish a successful airline in Japan. It withdrew from its first partnership with All Nippon Airways after less than a year of operation. AirAsia's second partnership with a Japanese retailing conglomerate survived for several years before succumbing to the pandemic.

Similar management challenges arise, especially in Europe, when large FSNCs strive to offer a product tailored to every passenger segment, often in response to LCC competition. The Lufthansa Group is an example of a complex assemblage of several airlines and other businesses. As Lufthansa explains (Lufthansa Group, n.d):

> The Lufthansa Group is composed of the segments Network Airlines, Eurowings and the Aviation Services. The Network Airlines segment comprises Lufthansa German Airlines, SWISS, Austrian Airlines and Brussels Airlines. Lufthansa German Airlines also includes regional airlines Lufthansa CityLine and Air Dolomiti as well as Eurowings Discover, the new holiday airline from the Lufthansa Group which started operations in July 2021 and

focuses on the touristic segment. Eurowings focuses on short-haul traffic in European point-to-point traffic. Aviation Services particularly includes the Logistics, MRO and Catering segments. The Lufthansa Group also includes the Additional Businesses and Group Functions. This business segment includes in particular Lufthansa AirPlus, Lufthansa Aviation Training and Lufthansa Systems.

Although not unique to the airline industry, large holding companies are difficult to manage. The airlines within a holding company may work at cross-purposes rather than cooperate as each airline's management tends to push for its own growth and profitability even at the expense of the best interests of the group. Some complex airline groups, such as the International Airline Group, parent of British Airways, have succeeded. Lufthansa, on the other hand, continues to shuffle its stable of airlines in search of the right mix. The Group's newest edition, Eurowings Discover, is the latest of many changes.

While the management of airline holding companies has proved challenging, management of equity alliances seems even more daunting. The dominant airline does not hold a controlling stake in the other partners in the alliance, an ownership relation that may be further complicated by other cross-company equity holdings among the members. But, the very nature of the restricted ownership leads to management by committee. In most instances, one carrier will lead, but conflicting interests of the individual airlines may make consensus building difficult. The proliferation of joint ventures between airlines that sometimes extend across other existing alliances or groups further complicates governance. The collapse of the Etihad equity alliance and the bankruptcy of the HNA Group illustrate the perils of wide-ranging and complex groupings of airlines with other business ventures.

12.7 Struggle for Stable Profits

Irregular profits followed by period steep losses have historically plagued the world airline industry, a trend exacerbated by deregulation and liberalization. Demand for the airline product is derived from the travel needs of business and leisure passengers; demand is highly correlated with the global business cycle. When economic growth slows or becomes negative, as in a recession, firms reduce employee travel as part of cost reduction aimed at weathering the downturn. Airlines then suffer a loss of some of their highest-paying passengers. Leisure passengers, who often buy discounted tickets, find themselves with tighter budgets and less disposable income. This leads to cutbacks on family vacations and airlines scrambling to fill even the discounted seats. Cargo shipments tend to follow the same pattern. The result is large cyclical swings in world airline profits, as shown in Figure 12.1.

Airline consolidation is the most promising strategy for sustained profitability. Indeed, the consolidation of the US FSNCs during the first decade of the century resulted in several years of uninterrupted profits until upended by the pandemic. The world airline industry remains fragmented with many

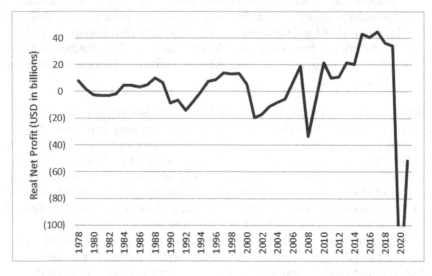

Figure 12.1 World Airline Real Profits.
Note: Real profits in 2021 USD. The 2020 loss of $139 billion extends off the chart.
Sources: Airlines for America (A4A); Federal Reserve Bank of St. Louis data (FRED).

opportunities for consolidation. In Asia, a merger between Korean Airlines and Asiana received regulatory approval while in South America, Avianca and GOL plan to merge. While consolidation promises better financial results, it also provides an opportunity for new entrants. In the summer of 2022, Spirit Airlines agreed to be acquired by JetBlue. Although the acquisition faces potential antitrust objections from the Department of Justice, it would create the fifth largest airline in the US and the 11th worldwide. Spirit will abandon its ultra-low-cost model as it is absorbed into JetBlue's higher-cost hybrid model creating opportunities for existing ULCCs – Frontier, Allegiant, and Sun Country. However, they face competition from new-entrant LCCs Avelo and Breeze which began flying in 2021. This trend is not limited to the US. New LCCs Flair Airlines and Lynx Air, seeing an opportunity in the Canadian market dominated by Air Canada and WestJet, began domestic flights in 2022. Also in 2022, Bonza entered the Australian domestic market. The pattern seems likely to continue. As established airlines seek higher profits through consolidation, new and existing LCCs see opportunities to enter and expand, so any reduction in competition in major markets is likely to be temporary.

12.8 Summary

Looking ahead, airlines face several daunting challenges, including carbon emission restrictions, labor shortages, and slowing growth in world trade and business travel. Airlines remain particularly susceptible to economic cycles, pandemics, war, and other events outside management's control. Structural reasons for cyclical profits and losses, including variable demand, volatile fuel prices, and strong unions, were explored in earlier chapters. Some LCCs are consistently profitable, but LCCs cannot meet all of the world's air travel needs. FSNCs typically struggle to earn a competitive return on investment with little room to accumulate reserves needed to withstand the next external shock. To survive, they depend on various forms of government support. Global network airlines are too important to the global economy to rely on bankruptcies and government bailouts for survival. Nonetheless, as the Centre for Aviation (CAPA) rather dismally concluded as the industry was emerging from the pandemic in April 2022, "With a handful of exceptions, the common feature is that airlines are commercially unviable" (CAPA, 2022). CAPA offers only a partial and unsatisfactory solution – let the world's airlines merge, but erect protective barriers against foreign investment and competition. Strong demand for air travel ensures that airlines will continue to fly, but the industry continues to search for business models that do require continuing government market interventions to achieve a modicum of stability and sustainable profitability.

References

Baldanza, B. (2022, April 1). Updated study estimates up to 40% of airline business travel may not return. *Forbes*. www.forbes.com/sites/benbaldanza/2022/04/01/updated-study-estimates-up-to-40-of-airline-business-travel-may-not-return/?sh=6c28bbe9146e

Cabrera, E., & de Sousa, J. M. (2022). Use of sustainable fuels in aviation – a review. *Energies, 15*(7). DOI:10.3390/en15072440

CAPA. (2022, April 8). Airlines, governments much fundamentally change air transport model. *Aviation Week Intelligence Network*.

CFM. (2021, June 14). *GE Aviation and SAFRAN launch advanced technology demonstration program for sustainable engines; extend CFM partnership to 2050*. [Press release]. www.cfmaeroengines.com/press-articles/ge-aviation-and-safran-launch-advanced-technology-demonstration-program-for-sustainable-engines-extend-cfm-partnership-to2050/

Harris, J. (2020, July 1). *Opinion: Inconvenient truths behind sustainable aviation fuel*. https://aviationweek.com/special-topics/sustainability/opinion-inconvenient-truths-behind-sustainable-aviation-fuel

IATA. (2021, October 4). *Net-zero carbon emissions by 2050*. www.iata.org/en/pressroom/2021-releases/2021-10-04-03/

Longley, A, & Kumar, D. K. (2022). Hedges set to save billions for airlines after $200 oil. *Bloomberg US Edition*. www.bloomberg.com/news/articles/2022-08-03/airline-fuel-hedges-set-to-save-billions-for-some-with-100-oil#xj4y7vzkg

Lufthansa Group. (n.d.) *Company portrait-business activities*. Retrieved August 4, 2022, from www.lufthansagroup.com/en/company.html

Murray, G. (2022). After COVID-19, aviation faces a pilot shortage. *Oliver Wyman*. www.oliverwyman.com/our-expertise/insights/2021/mar/after-covid-19-aviation-faces-a-pilot-shortage.html]

Samor, G. (2005, September 14). Brazil's not-so-favored airline; once a government dar-
    ling, Varig faces vagaries of competition. *Wall Street Journal* (Eastern edition). p. A18.
United Nations. (n.d.) *The Paris Agreement*. https://unfccc.int/process-and-meetings/
    the-paris-agreement/the-paris-agreement
World Bank. (n.d.). *Trade (% of GDP)*. https://data.worldbank.org/indicator/NE.TRD.
    GNFS.ZS

Review Questions

1. Why did IATA commit the airline industry to net-zero emissions by 2050?
2. Why is the airline industry one of the most difficult to decarbonize?
3. Can SAF produced from biomass replace conventional jet fuel? Why or why not?
4. Why would liquid hydrogen require new aircraft designs?
5. Can liquid hydrogen be delivered to aircraft using existing airport fuel infrastructure?
6. How is the price of petroleum determined and why is it so volatile?
7. Why might the world split into two competing economic blocks? Characterize both.
8. Why is there a risk of a substantial reduction in business travel growth?
9. Which world regions are most at risk of a pilot shortage? How will a shortage affect airline costs?
10. What are two types of complex airline structures? What management challenges do these complex structures present?
11. What are the airline industry's prospects for stable, competitive profits? Why?

Glossary

This glossary includes the definitions for all terms in the chapters that are printed in *italics* as well as other significant and/or commonly used terms in the airline industry whose definitions are important to a student's understanding. Terms are defined as used in the airline industry even when there are broader applications.

Air Mail Act of 1930 (aka the McNary-Watres Act) gave the postmaster more power to award mail contracts and changed the method used to pay carriers for carrying mail, basing the payment on the size of the aircraft and not the weight of mail carried. This change discouraged the carriage of junk mail and stabilized carriers' revenue. The Act also authorized the postmaster to offer longer-term contracts to the carriers.

Air Mail Act of 1934 imposed new rules on the awarding of contracts to private air mail carriers and allowed carriers to re-bid on routes canceled by the Spoils Conference. The Act returned the mail to private carriers after temporarily being flown by the Army Air Corps' AACMO.

Air Service Agreements (ASA) (aka Bilateral Agreements) are official agreements between two countries to authorize and regulate air travel between the countries. ASAs are generally developed by the State Departments and designate specific airlines, destinations, aircraft size, and frequency of flights.

Aircraft Utilization is the time an aircraft spends flying during a 24-hour day. Generally, low-cost carriers have high aircraft utilization – sometimes 12–14 hours a day. Network carriers might have slightly less utilization due to more complex schedules with more aircraft slack time built into their schedules.

Airline Deregulation Act (1978) lifted the CAB's restrictions on fares and routes and began the phase-out of the CAB. Under CAB control, airline fares were high, load factors were low and airlines were rewarded for inefficient operations. The economic liberalization allowed airlines to fly routes and charge fares that the market would support.

Airline Operations Control Center (AOCC) is one of several names for an airline's operations command center. An AOCC will usually incorporate dispatchers, schedulers, weather experts, air traffic control experts, chief pilots, and other essential personnel needed to make decisions concerning the airline's tactical control.

Airport & Airway Trust Fund   (AATF) originated in 1970 to help the government finance aviation projects. The AATF's source of funds is from taxes on passenger fares and cargo fees, as well as airline fuel purchases. This tax, a percentage of the fare, provides the majority of funding for the FAA's operation.

All Cargo Carrier   operates freighter aircraft and carries only cargo. They do not provide door-to-door services and are often under contract from large freight forwarders.

Amendments to the CBA   (also called side letters) are agreements between airline management and unions that modify the terms of the existing CBA.

Ancillary Revenue   is revenue earned by an airline from sources other than the sale of passenger seats. Ancillary revenue includes fees for checking luggage, printing a boarding pass, assigned seats, or carry-on luggage. Additionally, ancillary revenue could come from a credit card company for their issue of frequent flier miles.

Antitrust Immunity   permits an airline coordinate product offerings with its alliance partners. To protect consumers, antitrust laws prohibit collaboration between competitors. With antitrust immunity, airlines may collaborate with their alliance partners to coordinate flight schedules, fix prices, and set capacities at profitable levels.

Army Air Corps Mail Operation   (AACMO) was formed to carry mail after air mail contracts were canceled following a Congressional investigation of the Spoils Conference. The operation was a failure as military pilots were not trained to fly in poor weather conditions.

Atypical Employment   refers to employees considered self-employed rather than direct-employed. (Also see Bogus Self-Employment).

Available Seat Mile   (ASM) (or ASK for kilometers) is the basic unit of airline production. It is defined as one seat flown one mile (or kilometer) regardless of whether the seat is occupied or not. An airline's annual total ASMs is also called its capacity.

Bank   (aka Wave) is a group of aircraft arriving at a hub airport or departing from a hub airport. A hub complex starts with a bank of aircraft arriving from spoke cities and ends with a bank of aircraft departing for spoke cities.

Bermuda I   was the first Air Service Agreement negotiated between the United States and the United Kingdom in 1946. It allowed restricted air service between the US and the nations then in the UK including some points in Latin America, the British colonies, and Singapore.

Bermuda II   was the first revision of the Bermuda I Air Service Agreement between the US and the UK in 1977. It further restricted flights at rapidly growing Heathrow Airport and remained in effect until replaced by Open Skies in 2007.

Bid Line Process is a traditional crew scheduling system in which pairings are assembled into nameless bid lines that typically cover a calendar month. Crewmembers bid on the monthly schedules which are awarded in seniority order.

Big-Three alliances are the Star Alliance, SkyTeam, and Oneworld.

Bilateral Agreements (aka Bilaterals) (See Air Service Agreements).

Block Hour is the time an airliner is in operation. Block time starts when the aircraft begins pushback and ends when the aircraft comes to a final stop at the destination. Start and end are usually at a jetway.

BOAC (British Overseas Airways Corporation) was a British-owned airline formed to operate during WWII. After the war, BOAC continued to operate air services for Britain destined outside Europe. BOAC merged with BEA (British European Airways) in 1974 under the brand British Airways.

Bogus Self-Employment (also called False Self-Employment) refers to employees hired at airlines as "self-employed" although they do not fit the traditional definitions of self-employment. Bogus self-employment can be advantageous for airlines because of less restrictive employment regulations.

Breakeven Load Factor (BLF) is the average percentage of seats that must be sold for the airline to break even. Breakeven load factor can be determined by dividing the airline's cost per available seat mile by the yield (CASM ÷ yield).

Brick-and-Mortar Travel Agency is a traditional travel agency with an office that travelers can visit or call to make travel plans. Brick-and-mortar travel agencies use one or more Global Distribution System to make reservations.

Cabin Crew is the term used when referring to flight attendants.

Cabotage occurs when an airline carries passengers between two points in a foreign country. To protect their domestic airlines from foreign competition, most countries do not allow cabotage. As an example, Air France carrying passengers from New York to Los Angeles is cabotage.

Capacity (See Available Seat Miles).

Capacity Discipline is the restraint of capacity during a booming business period. Raising capacity during a boom period results in lower average fares. Lower demand during economic recessions results in capacity reductions during economic down cycles perpetuating the cyclic nature of the industry.

Capacity Purchase (See Fee for Departure).

Capital Lease is an aircraft lease similar to a rent-to-own agreement. A capital lease is used for long-term aircraft purchases. At the end of the capital lease, the aircraft belongs to the airline.

Catchment Area is the area from which an airport can reasonably expect to draw commercial air service passengers.

Chicago Convention of 1944 brought together countries to draft international aviation protocols. The International Civil Aviation

Organization was founded at the Chicago Convention. Attending countries agreed that they would use a system of bilateral Air Service Agreements to regulate international air commerce.

City Pair is a routing between any two cities. A city pair includes an origin (O) and destination (D) and may include intermediate connections between the O&D.

Civil Aeronautics Act (CAA) (1938) required the US government to ensure aviation safety. The Civil Aeronautics Authority was established to investigate airline accidents. The Act also gave the CAA power to regulate airline fares and routes. In 1940, the CAA was split into the Civil Aeronautics Administration and the Civil Aeronautics Board.

Civil Aeronautics Board (CAB) was initially part of the Civil Aeronautics Authority (See Civil Aeronautics Act). Changed to the CAB in 1940, it became responsible for safety rulemaking, accident investigation, and economic regulation of airlines. As part of economic regulation, the CAB controlled three broad categories of airline operations: initial approval to operate, routes flown, and fares charged.

Codeshare is a form of airline cooperation whereby airlines share identification codes within the reservations system. Code sharing allows passengers to easily book multiple leg flights that require two or more different airlines to travel from the origination to the destination. Using a codeshare, a ticket might be purchased from airline A, but travel might include flights on airline A and airline B.

Co-Hosting occurs when an airline owning a computer reservations system (host) gives host status to another airline. A co-hosting airline enjoys the same status as the host and is not subject to display bias.

Collective Bargaining Agreement (CBA) is a contract negotiated between an airline and a labor union that establishes detailed terms and conditions of employment.

Collective Labor Agreement (CLA) (See Collective Bargaining Agreement).

Complex is the sequence of aircraft arrivals, passenger connections, and departures from a hub airport. In a typical complex, many aircraft arrive from spoke cities at the hub airport in a close time window, passenger and luggage transfer to outbound flights, and then aircraft depart for spoke cities. A large airline will conduct several daily complexes (eight to 10) at each hub throughout the day.

Computer Reservation System (CRS) are computer systems used to manage airline seat reservations. In early airline history, seat sales were tediously managed by hand, then slowly, computer systems developed to automate the process. Eventually, CRSs grew to include other areas of travel and are now called Global Distribution Systems.

Connecting Flight is one to which a passenger connects, usually at a hub, to reach his or her destination.

Consumer Surplus is the difference between what a consumer(s) is willing to pay and the price.

Contract Air Mail Act of 1925 (aka the Kelly Act) provided for the award of government contracts to private airlines to carry US Mail. Previously, air mail was carried by the Post Office using Post Office planes and pilots. The Kelly Act is often cited as the "birthplace of the airline industry."

Contract Air Mail Routes (CAM routes) were developed by the Post Office and contracted to private companies after the Kelly Act. Designated by a number, the initial 12 CAM routes were awarded following the passage of the Contract Air Mail Act. As the air mail route system grew, so did the number of CAM routes, eventually forming a network covering the entire US.

Cost Leadership Strategy is a business strategy that focuses on low production costs and promotes a product based on low price, generally in high volume. In the airline business, a cost-leadership airline provides no-frills service at lower fares than competitors to appeal to cost-conscience customers.

Cost Per Available Seat Mile (CASM) (aka Unit Cost) is the cost an airline incurs to produce one seat flown one mile. CASM is calculated by dividing the total cost by the total available seat miles (total cost ÷ ASM). In the US, CASM is measured in cents per mile.

Cost-Based Pricing is based on a seller's cost plus a markup (profit).

Craft Unions are formed by workers of the same craft. For example, the Association of Flight Attendants includes flight attendants from different airlines.

Crew Pairings (See Pairings).

Crew Schedulers track individual crewmembers as they move through the airline's route network, maintaining up-to-date status, calling in reserve crewmembers or readjusting crewmember schedules as necessary when schedule disruptions occur. Crew schedulers are often located at the Airline Operations Control Center.

Cumulative Distribution Function (CDF) is the plot of the area under a normal distribution curve. It graphically represents the probability the sample data will be equal to or less than a selected amount.

Debt Financing is funding obtained from a loan. Debt financing is similar to an individual obtaining a bank loan to purchase a car, except, collateral for an airline is its aircraft.

Defined Benefit Plan is a pension plan that provides the retiree with a fixed monthly payment. Airlines offering these plans invest money throughout a person's employment to fund retirement payments. Defined benefit retirement funds remain under company control. If investment returns are insufficient, the plan is underfunded. In bankruptcy, plans can be terminated.

Defined Contribution Plan is a retirement plan, such as a 401k, where an employee makes contributions to their account and the company matches the contributions to a predetermined level. Defined contribution plans grow tax-deferred and are held separately from the company making them safe if the company fails. The employee bears the risk of investment returns.

Demand-Related Pricing   is a revenue management tool that bases fares on demand (as opposed to cost-based pricing). Demand-related pricing is used to exploit the passengers' willingness to pay.

Denied Boarding   occurs when a passenger is present for a flight but is denied boarding because the flight is oversold. Denied boarding is involuntary.

De-Peaking   (See Rolling Hub).

Deregulation   (See Airline Deregulation Act).

Derived Demand   describes a situation where a customer does not purchase a product for its own sake, but rather for a purpose derived from it. Airline travelers do not purchase tickets to go for an airplane ride; rather, they purchase a ticket to travel to a different location. As a result, demand for air travel is derived from the need to be somewhere else.

Differentiation Strategy   is a business strategy in which a company creates a unique product tailored to each targeted consumer segment. Airlines that employ a differentiation strategy produce a variety of service offerings, some for cost-conscience customers (like coach fares) and some for travelers wanting higher levels of service.

Direct Flight   is a flight from one point to another with a stop in an intermediate airport, but no aircraft change.

Direct Operating Costs   (DOCs) (aka Variable Operating Costs) are those resulting directly from the operation of the aircraft (no DOCs are incurred when the aircraft is not in operation). For an airline, fuel is the largest direct operating cost.

Direct Sales   are airline ticket sales directly from the airline. Direct sales can be through the airline's internet site, a reservations agent via telephone, or an airline's ticket office.

Directional Hub   is a complex in which flights arrive from spoke cities on one geographical side of the hub and then depart to spoke cities on the opposite geographical side. For example, a complex at a directional hub in the center of the US might begin with arrivals from the East Coast spokes and end with departures to the West Coast spokes.

Dispatchers   are FAA-licensed professionals responsible for flight planning, issuing flight plans to captains, and following each flight's progress. The dispatcher and captain are legally jointly responsible for the safe operation of the flight. Dispatchers are usually located at the Airline Operations Control Center.

Display Bias   (See Screen Bias).

Distribution Cost Charge   is added to tickets by some airlines when a customer uses a GDS system instead of the airline's reservation system.

Domicile   is the term for the home base of a pilot or flight attendant. Medium to large airlines have several domiciles while small airlines might have only one. Pilot and flight attendant schedule pairings will all start and end at their domicile.

Drop-In Fuels   are alternative fuels (like sustainable aviation fuel) that, when mixed with conventional jet fuel, can power the existing airline fleets with minimal modification.

Dynamic Pricing falls under the umbrella of revenue management. Dynamic pricing uses computer algorithms to calculate and adjust prices to maximize revenue.

Dynamic Scheduling is a process of rescheduling aircraft to better match passenger demand. After an initial aircraft assignment, the airline might re-assign a different-sized aircraft to better match demand. Dynamic scheduling minimizes spill and spoil.

Enterprise Unions are unions made up of workers at the same airline. For example, the Southwest Airlines Pilots Association.

Equity Alliance is an alliance created by one airline purchasing an equity stake in another airline(s).

Escapable Cost is the cost avoided by not operating a flight. Escapable costs increase with time, for example, employees can be furloughed relatively quickly but leased aircraft can be returned to the lessor only when the leases expire.

E-Ticket is issued electronically without a paper copy. E-tickets may have bar codes or QR codes that can be displayed on smartphones or other devices (or they can be printed) and used to board aircraft.

Expected Marginal Seat Revenue (EMSR) is the revenue an airline expects from selling a particular seat on the aircraft. EMSR is determined by multiplying the fare by the probability of selling that seat.

Fair Assignment is a crew assignment process based on a no-bid, equitable distribution of workload. This process attempts to equalize block hours, days off, and various other aspects of the workload to maximize both crewmember utilization and satisfaction.

Fare Aggregators (See Metasearch Engines).

Fare Buckets are a figurative way to categorize seat allocation when fare nesting. As reservations for a particular flight are received, seats are removed from the respective fare buckets. The highest fare bucket would contain all of the seats on the aircraft.

Fare Nesting is a process of ensuring that seat allocation restrictions never deny the sale of a higher-priced seat while allowing the sale of a lower-priced seat.

Federations are groups of unions. An example would be the Coalition of Airline Pilots Associations.

Fee for Departure (aka Capacity Purchase) is a contractual arrangement between a network carrier and its regional carrier(s) in which the major carrier agrees to pay the regional carrier a fixed fee to operate the flight regardless of the number of passengers paying. The major carrier assumes all the financial risks of the operation. As the name suggests, the major is simply buying capacity on the regional airline whether it's filled or not.

Fixed Costs are those incurred even if the airline's aircraft are not flown. Fixed costs for an airline might include insurance or hangar rent.

Fixed Scheduling is a flight crew scheduling process with a repeating on/off schedule pattern allowing crewmembers to know their workdays and days off well in advance.

Fixed Short-Term Capacity   means the capacity of an airline flight becomes more and more fixed as the departure time approaches.

Fixed-Fee Contract   is a contract between a regional carrier and a FSNC where the regional airline is paid a fixed fee for each departure plus an additional fee per passenger.

Flag of Convenience   is a scheme where an airline is owned and operated in one country, but holds an operating certificate in another country to obtain less restrictive regulatory standards.

Fleet Rationalization   is a reduction in the number of aircraft types flown by an airline. Although a fleet of diverse aircraft is sometimes required to meet the needs of a major carrier, minimizing the fleet diversity increases efficiency and decreases cost.

Flight Crew   (aka Cockpit Crew) are the pilots operating a flight.

Flight Crewmembers   are the pilots and flight attendants assigned to a flight.

Focus City   is not a traditional hub in a hub-and-spoke system but from which an airline operates many flights often allowing for some passenger connections.

Focus Strategy   is a business strategy (aka Niche Strategy) of a company that offers a highly tailored product to a small market segment. In the airline business, a focus airline might include former carriers Eos and MAXjet.

Fortress Hub   is an airport where the dominant airline exercises considerable influence over airport operators at a hub such that it can impede competition through control of airport facilities.

Freight Forwarders   are intermediaries that manage all aspects of air cargo, usually including ground transport, except the flight portion which is contracted to one or more airlines. Freight forwarders handle the complexities of air freight, customs, and other complex international shipping issues.

Frequency   is the number of daily flights in a city pair. With a given demand, an airline might fly several smaller aircraft with high frequency, or only one or two flights in a large aircraft.

Fuel Hedging   is a financial means to protect an airline from the volatility of fuel prices. Hedging instruments lock in the price of an airline's fuel for the duration of the hedge, providing an insurance policy against catastrophic fuel price increases.

Full-Service Network Carrier   (FSNC) is an airline offering a wide range of products from first class to budget coach class. FSNCs also usually offer connections to alliance partners further expanding their networks. Examples of FSNCs include Delta, British Airways, Lufthansa, and Japan Airlines.

Generic Business Strategies   are three strategies developed by Harvard Business School Professor Michael Porter. They include cost leadership, differentiation, and focus.

Global Distribution System   (GDS) (aka Computer Reservation System) is a third-party integrated system to sell airline tickets and other travel needs. As an analogy, the GDS is to airline tickets as Ticketmaster® is to concert seats – a middleman who facilitates sales from the producer (the airline) to the consumer (the traveler).

Global Logistics Suppliers provide transportation services from the manufacturer's warehouse to the customer's door.

Globalization broadly refers to the increasing integration of world economics and societies. Long-distance trade existed for years but accelerated after WWII as firms took advantage of relaxed restrictions on trade.

Gravity Model is a demand estimating tool used to predict passenger traffic between two cities. Demand between two cities is hypothesized to vary directly with the product of the populations and inversely with the square of the distance between cities. The name came from the similarity to Newton's law of gravity.

Halo Effect was a screen bias in the early CRS systems that came about because travel agents would tend to book flights on the host airline because the information for the host was generally more accurate, timely, and reliable.

Hidden City Ticketing (aka Point Beyond Ticketing) is a way passengers can circumvent an airline's segmentation devices by planning a trip to a destination with the intent to get off the aircraft at an intermediate stop.

High-Density Rule (HDR) Airports (aka Slot Controlled Airports) limit the number of takeoff and landing slots. The HDR in the US began in 1968, initially limiting takeoffs and landings at five highly congested airports in the US: three in the New York City area (LaGuardia, Newark, and Kennedy), Washington National, and Chicago O'Hare.

Homeland Security Fees were added to passenger fares after 9/11 to cover the cost of civil aviation security provided by the Transportation Security Administration. These fees are flat rates added to the cost of one-way tickets.

Hub and Spoke Route Structure (H&S) became the industry standard after deregulation. In a simple H&S system, passengers from outlying spoke cities all fly to a common hub airport where they transfer to another flight to their destination – another spoke city. H&S networks offer service from anywhere in their network to anywhere else in the network with the minimum number of aircraft.

Hub Dominance results as a hub grows and the predominant carrier establishes the highest market share at the hub. Hub dominance brings a degree of market power enabling the carrier to charge premium fares on flights to and from the hub. (See S-Curve)

Hybrid Airline is a term for an airline that fits between a low-cost carrier and an FSNC exhibiting traits of both types. Their inflight product might be better than a typical LCC, but they do not have the broad international networks typical of FSNCs. Examples of hybrid airlines would include Alaska, JetBlue, and Southwest.

Hybrid Route Structure is a route structure that uses a combination of the three main route structures: point-to-point, linear, and hub and spoke.

Most larger carriers operate hybrid route systems offering passengers a variety of hub and spoke connections as well as some point-to-point and linear routings.

Industrial Unions   are comprised of workers in the same industry. An example might include the Transportation Workers of America where membership would include workers in any transportation industry – airline workers, bus company workers, etc.

Integrated Cargo Carrier   operates door-to-door services using aircraft of all sizes and ground vehicles to deliver cargo. Examples are FedEx, UPS, and DHL.

Interline   is a process where airlines accept each other's tickets and transfer checked luggage between them.

Internal Financing   is a method airlines use to finance purchases using internally generated funds. Internal financing is analogous to using your savings to buy a car.

Internal Reservations System   (RES) is a computer system used to manage an airline's own reservations. RES setups can be developed by external vendors (possibly Global Distribution System companies) or by the airline.

International Civil Aviation Organization   (ICAO) is the international organization charged with developing and promulgating international aviation standards.

Involuntarily Denied Boarding   (See Denied Boarding).

Kelly Act   (See Contract Air Mail Act of 1925).

Labor Protective Provisions   (LPPs) are laws in the US designed to ensure that employees involved in an airline merger are treated fairly. The McCaskill-Bond Statute codified the LPPs into law in 2007.

Law of Demand   is the inverse relationship between price and the quantity of goods sold.

Linear Route Structure   was extensively employed in early airline history. Much like a train, an aircraft originates in a city and makes several stops on the way to its destination city. Passengers can travel between any two cities on the route, sometimes with several intermediate stops.

Load Factor   (LF) is the ratio between ASM and RPM, in other words, the percentage of the airline's production that is sold. The load factor is computed by dividing revenue passenger miles by total available seat miles (RPM ÷ ASM) and expressed as a percentage. LF can be calculated for an entire airline over a period of time, or for a single flight.

Local Service Carriers   were those airlines approved by the CAB for air service between small communities or from small communities to trunk airline cities. Sometimes called "feeders," the Local Service Carriers were first approved by the CAB in 1945. Initially, there were 20 companies allowed to fly in 45 states.

Low-Cost Carrier   (LCC) is an airline that offers no-frills flights. As compared with full-service network carriers that offer a wide range of products, LCCs typically follow a simpler business plan. Some traits associated with LCCs include: open seating, no interlining, distribution via website only, single class seating, lower fares, shorter flights, minimal inflight service, and a single aircraft type. Examples of LCCs include Ryanair and EasyJet. Also, see Ultra-Low-Cost Carrier.

Maintenance Controllers coordinate with line mechanics for aircraft maintenance, especially when malfunctions occur, ensuring that required parts and mechanics are available when and where needed. Maintenance controllers work closely with dispatchers and are usually located at the Airline Operations Control Center.

Marginal Cost is the cost of carrying one more passenger on a flight. Airline marginal costs are low because the added weight of one passenger is almost negligible compared to the weight of the aircraft.

Market Share is the percentage of total demand earned by a particular company over a specified time period. Market share is calculated by dividing a company's sales by the total industry sales over the same period and is a measure of consumers' preference for one company over another for a similar product.

Marketing Concept is the idea that a company must first determine customers' needs and wants and then offer a product satisfying those desires at a price that yields a profit.

Marketing Mix refers to the four variables that make up an airline's marketing plan: product, price, promotion, and place.

McNary-Watres Act (See Air Mail Act of 1930).

Mega-City is a term used by Airbus in their Global Market Forecast defined as an area of urbanization and wealth creation capable of supporting 10,000 or more daily long-haul passengers. Mega-cities are sometimes categorized into three levels: greater than 10,000 passengers a day, greater than 20,000 per day, and greater than 50,000 per day.

Menu Pricing (aka Unbundling) is an airline pricing scheme that allows passengers to specifically choose which of the airline's services they want included in their total fare. With menu pricing, the basic fare might include only the transportation. Travelers can then add in fees for additional services like checked baggage, early boarding, or even sometimes carry-on baggage.

Metasearch Engines (aka Fare Aggregators) are systems that search airline internet sites for fares and display flight and fare data for purposes of comparison. Metasearch engines typically do not sell tickets, but rather earn revenue by selling internet advertising on their websites. They provide the traveler the comparison information, then the traveler makes a reservation on the airline website.

Narrow Body Aircraft is a passenger airliner with a single aisle.

NDC Aggregator is a company that pulls together multiple NDC connections and offers that access as a service to travel agents.

Net Profit Margin is the percentage of total revenue that remains after paying all expenses. Net profit margin is calculated by dividing after-tax net income (profit) by total revenue. Net profit margin is commonly used to compare the earnings of firms of varying sizes.

Network Allocation   is the process of allocating seats on a particular flight based on all of the possible passenger connections available. Network allocation might limit the seats available on a short leg to accommodate passengers connecting to longer, higher-revenue segments.

Network Inventory Allocation   (aka O&D Revenue Management) is a revenue management tool used to manage seat inventory not just for a single flight, but for all flights connecting the passenger's origination to their destination.

Network Revenue Management   is the process of revenue management considering the potential revenue from connecting passengers. Also see Network Allocation.

New Distribution Capability   (NDC) is an upgrade to the original global distribution system software that allows more options for the user. An NDC display is much more like an airline's website, allowing the GDS to display more options, like checked baggage sales, to the travel agent booking the flight.

Niche Strategy   (See Focus Strategy).

No Frills   is the term sometimes given to airlines that offer only basic transportation with limited amenities and services. Low-cost carriers are also termed no-frills airlines.

Non-Stop Flight has no intermediate stops between origin and destination (O&D).

Non-Transferable Tickets   are tickets that cannot be transferred from one traveler to another.

No-Show Rate   is the percentage of passengers who make reservations for a flight but then do not show up or cancel.

O&D   (See Origination and Destination).

Official Airline Guide   (OAG), originally a print publication, is now an electronic listing of all scheduled airline flights.

Online Travel Agency   (OTA) offers similar services as a brick-and-mortar travel agency via the internet, but has no physical retail office.

Opaque Travel   sellers sell airline flights without disclosing the airline name or flight information until the sale is finalized.

Open Skies   is a system of less restricted air commerce between countries. Open Skies agreements are replacing the more-restrictive Air Service Agreements and allow for airline competition with a minimum of government interference.

Operating Lease   for commercial aircraft is similar to a car lease. When the lease is over, the aircraft is returned to the lessor. Operating leases vary in duration from months to five or more years.

Origination and Destination   (aka O&D) is the city pair where a passenger begins travel (origination) and ends travel (destination). A traveler flying from JFK to DFW with a stop/transfer in ATL, is considered an O&D passenger for JFK and DFW, but not ATL.

Overbooking occurs when airlines allow customers to make more reservations than there are seats on an aircraft. Overbooking refers to reservations only. If more people show up for a flight than there are seats, it is called overselling (See Oversale).

Oversale occurs when more passengers show up for a flight than there are seats on the aircraft. Overselling is a possible result of overbooking.

Pacing Spokes within a hub-and-spoke route system that are farthest from the hub. Flights to and from pacing spokes will generally set the hub complex timing. Aircraft operating from closer spoke cities wait for the return of those aircraft from these more distant spokes.

Pairings are work schedules consisting of a sequence of flights that begin and end at the crewmember's domicile.

Paris Convention (1919) was the first international conference to address the conflicting claims of nations concerning the sovereignty of airspace. Conferees agreed that nations had sovereignty over the airspace over their territorial land and waters and could restrict flights through that airspace; however, they also encouraged as much freedom as possible.

Passenger Facility Charges (PFC) are collected by the airlines (added to the fare) and remitted to airports to fund federally approved improvement projects. These funds are used for both airside (taxiway and runway) and ground-side (terminal) improvements. PFCs are a fixed fee per flight segment with a limit of two per one-way trip.

Passenger Revenue per Available Seat Mile (PRASM) is a refinement of revenue per available seat mile (RASM) excluding revenues from cargo and other sources.

Passenger Segmentation is an element of airline revenue management. Potential passengers are assigned to categories based on their buying or travel needs. Airlines then devise fare restrictions (See Segmentation Devices) to prevent passengers who are willing and able to pay higher fares from purchasing lower-fare tickets. Passengers are broadly segmented by purpose of travel, for either business or leisure.

Passenger Service System (PSS) is a computer system that electronically stores the airline's flight schedule. The PSS is developed and customized by each airline; individual systems vary in sophistication and capability.

Passenger-Name Record (PNR) is passenger information from a flight reservation. A PNR associates the passenger's name (not just a number) with a reserved seat.

Payload is the weight of passengers and cargo carried on the aircraft.

Perimeter Rule is a federal law that limits the distance an airline can fly from certain airports. Washington Reagan and New York LaGuardia airports have perimeter rules. Reagan's perimeter rule limits flights to 1,250 miles and LaGuardia's perimeter rule extends to 1,500 miles. The perimeter rules were instated when nearby larger airports (Dulles and Kennedy, respectively) opened in an attempt to steer airlines and travelers to the new airports. Exceptions are on a case-by-case basis.

Perishable Inventory   is product inventory with little or no shelf-life.

Point Beyond Ticketing   (See Hidden City Ticketing).

Point-to-Point Route Structure   is the simplest means to connect two cities. Passengers board at the origination city and deplane at the destination. In a pure point-to-point system, passengers do not connect to any other flights. Point-to-point provides the least travel time from origin to destination and is often the preferred routing for passengers.

Positive Network Externalities   refers to benefits gained by a consumer's familiarity with the airline's product. Familiarity lessens uncertainties and increases loyalty, particularly when linked to loyalty programs.

Power-by-the-Hour   is an agreement between an airline and a third party such as the engine manufacturer whereby the airline pays a fixed fee per hour of engine use. Engine ownership remains with the third party. Power-by-hour agreements can cover other aircraft components such as auxiliary power units.

Preferential Bidding System   is a crew scheduling system that uses sophisticated computer software to develop individualized monthly flight schedules based on crewmembers' preferences for a variety of options, including workdays, overnight cities, and start and end times.

Price Discrimination   occurs when the same product is sold for different prices based on a customer's willingness to pay.

Price Diversion   occurs when a customer who is segmented into a particular fare category circumvents the segmentation devices and purchases a cheaper fare. An example is a business passenger who spends a weekend at their destination to get a lower fare.

Price Elasticity   is a measure of consumer price sensitivity defined as the percentage change in quantity demanded divided by the percentage change in price. If the quantity demanded falls by 5% when the price increases by 5%, the elasticity is 1. An elasticity greater than 1 means that changes in price have a relatively large effect on the quantity demanded. An elasticity of less than 1 means that a change in price will have little effect on the quantity demanded.

Privatization   is the process of turning a state-owned airline into a private airline. Privatization can occur with the state simply auctioning off their airline, or the state can negotiate a purchase. In some cases, employees have arraigned financing to purchase the airline.

Probability Density Function   is the plot of sample data that includes every possible outcome. Airline seat sales fit a normal distribution, that is, a standard bell curve. A normal distribution can be plotted when the mean and standard deviation of the data set are known.

Product Scope   refers to the range of products an airline offers. The range might extend from a no-frills economy seat to a first-class premium service seat. Low–cost carriers, such as Europe's Ryanair, offer a narrow product scope while full-service network carriers have a broad scope.

Productivity Index is a measure of aircraft productivity based on available seat miles per year. It is the product of three factors: seat capacity, speed, and aircraft utilization. A small, slow aircraft has a very low productivity index, while a large, fast aircraft has a large productivity index.

Pro-Rate Agreements are a type of contract between major carriers and their regional partners. With a pro-rate agreement, the regional carrier and major airline split the passenger fares. In this way, both the regional and the major airline share in the risk associated with the operation. Large regional airline contracts are usually either pro-rate agreements or capacity purchase (See Fee for Departure) agreements.

Quality Function Deployment is a concept of designing quality into a product rather than something obtained after production by inspection and correction of defects. The basis for this concept is the conviction that high-quality products are ultimately cheaper to produce and sell.

Re-Banking is the process of turning a rolling hub back into a peaked hub. The opposite of de-peaking, re-banking adds cost to the airline's operation but regains revenue lost from poor connections associated with rolling (de-peaked) hubs.

Regional Airline (RA) is an airline that flies aircraft smaller than mainline jets to provide service to cities with insufficient demand to be profitable with FSNC service.

Rejected Demand (See Spill).

Request and Reply System was an early airline ticketing system that required an agent to call the airline reservation office to request a seat for a customer, and then wait for the airline to reply to the request.

Reserve Crews are pilots and/or flight attendants assigned to short-notice flight assignments.

Return on Invested Capital (ROIC) is the ratio of profit to all invested capital – both equity and debt-financed capital. ROIC is a good measurement of how efficiently a company puts its capital to work to earn profits.

Revenue Management (aka Yield Management) is a pricing method used by airlines and some other retailers to maximize revenue by setting prices that exploit customer preferences and ability and willingness to pay.

Revenue Passenger Mile (RPM) (or RPK for kilometers) is a measure of an airline's sales defined as one occupied seat flown one mile.

Revenue Per Available Seat Mile (RASM) is the revenue an airline earns from each unit of production. RASM is calculated by dividing total revenue by the total available seat miles (Revenue ÷ ASM). In the US, RASM is reported in cents per seat mile.

Revenue-Sharing Arrangement is a contract between a regional carrier and an FSNC whereby the regional airline receives a percentage of the ticket revenues.

Rolling Hub   (aka De-Peaking) is an approach used to reduce costs and increase the efficiency of assets and personnel at hubs. Working tight connections between arrival and departure banks is labor and infrastructure intensive, yet personnel and equipment sit idle between complexes. Rolling hubs spread out the arrivals and departures to de-peak the spikes in workload during the complex. Rolling hubs also improve aircraft and flight crew utilization, but passenger convenience can suffer with increased connection times.

Schedule of Services   is the airline's flight schedule.

Scope Clause   is a pilot contract provision that limits the size of aircraft a regional airline can operate under contract to the major carrier. A major airline with less restrictive or no scope can contract with regional airlines to operate larger, more efficient, regional jets (and lower-paid regional pilots) on routes that would otherwise be flown by mainline aircraft and pilots. Major airline pilot unions negotiate scope clauses to protect their members' jobs from being outsourced to regional airlines.

Screen Bias   (aka Display Bias) is built into computer reservation systems to favor the airline that owned the system (host airline). Originally, computer reservation systems were developed by one airline, but used by many. The host airline would bias CRS displays by listing its flights first.

Screen Padding   is a computer reservation system biasing scheme that lists one flight under several flight numbers to increase visibility on reservation screens.

S-Curve   is a graphical representation of a phenomenon in which an airline with greater flight frequency in a market captures a disproportionately higher market share measured in revenue or total passengers. As an example, if one airline in a market city offers 70% of the total capacity from that city, it will gain more than 70%, maybe even 80% of the passengers.

Seat Allocation   is the process of allocating seats in an aircraft to different fare categories, then offering seats for several different prices by controlling the seat inventory and restrictions to purchase.

Segment Length   (aka Stage Length) is the length of a flight leg in miles. Segment length is averaged over a period of time for an airline's entire operation or the operation of a single type of aircraft. Segment length determines whether an airline is considered a long- or short-haul airline.

Segmentable Market   refers to a customer base than can be categorized by purchasing habits. Although several others can be identified, the two most recognized passenger segments are business passengers and leisure passengers.

Segmentation Devices   are restrictions an airline sets on seat reservations to force passengers into a predetermined group. Typical segmentation devices might include a required weekend stay at your destination, advance purchase restrictions, or non-refundable.

Selling Up   refers to a situation where a passenger fits into a reduced fare segment, but instead purchases a higher fare. Selling up is, in a way, the opposite of price diversion. As an example, a leisure passenger meets all of the segmentation devices to purchase a deeply discounted seat, but instead purchases a higher fare, probably because the deeply discounted seat it already sold out.

Side Letters   (See Amendments to the CBA).

Single-Class Seating is an aircraft cabin seat configuration with all passenger seats in a single class. In single-class seating, there are no first-class or business-class seating sections. Single-class seating is a business model characteristic of low-cost and ultra-low-cost airlines.

Slack Time is built into flight schedules to mitigate irregular operations. Slack time makes it easier to recover from irregular operations but is costly to the airline.

Slot-Controlled Airports   (See High-Density Airports).

Southwest Effect describes the increase in originating travel resulting from a low-cost carrier entering a market. The lower fares and additional capacity offered by the low-cost carrier force the incumbent airlines to lower fares to remain competitive. The result is a rise in sales for all carriers.

Spare Aircraft are kept in reserve to substitute for scheduled aircraft that must be taken out of service due to mechanical problems. If not used, spare aircraft result in a cost to the airline. Aircraft sitting on the ground are not earning revenue.

Spill (aka Rejected Demand) is the loss of a potential customer due to insufficient capacity. Spill is also defined as total unconstrained demand for a flight minus the total capacity of the flight. Because total demand is not precisely known, spill can only be estimated. Researchers have developed spill models based on normal distributions of demand.

Spoil is airline seats that were produced but left empty. It's much like the term used for a grocer. Food that is produced but not purchased eventually spoils. Airline seats are produced when the airline schedules a route and sets a desired capacity. The product (passenger seat) spoils when the aircraft departs with unused capacity (empty seats).

Spoils Conference (1930) was a series of closed-door, semi-secret meetings held by Postmaster General Walter Folger Brown to re-distribute air mail contracts. Brown, using new powers gained from the McNary-Watres Act, set out to re-award contract mail routes to only the largest and most established carriers. His goal was to make the carriers more profitable and end government subsidies.

Stage Length   (See Segment Length).

State-Owned Airline is owned and controlled by the national government. Although the United States never had state-owned airlines, many worldwide airlines are either state-owned or were at one time state-owned and now privatized.

Stochastic Demand   (See Uncertain Demand).

Survivor Function is the plot of complement of the cumulative distribution function (1-CDF). The plot resembles a spread-out number 2, starting at 100% and dropping to 0%, passing through 50% at the mean.

System Board of Adjustment (SBOA) is a panel consisting of an equal number of airline management and union representatives (and sometimes an arbitrator provided by the National Mediation Board). An SBOA is set up to oversee a minor grievance that could not be resolved by negotiation.

Total Quality Management is a management approach that focuses on delivering services with the highest quality to maximize customer satisfaction and meet regulatory standards. Total quality management is an organization-wide effort for continuous improvement.

Travel Agent Commission Override (TACO) is an incentive paid to travel agencies, in addition to the base commissions, for exceeding predetermined sales targets.

Travel Management Company (TMC) is a travel agent business that specializes in corporate travel management. TMCs provide also provide consulting, executive travel services, and meeting or conference support logistics.

Trunk Carriers were those airlines approved by the CAB for long-distance flights. Sixteen carriers were grandfathered as trunk carriers when the CAB began in 1938 (See Civil Aeronautics Board and Civil Aeronautics Act). No new trunk carriers were added in the CAB's 40 years of control (although there were 79 applications), however, six trunk carriers either ended business or merged.

Ultra-Low-Cost Carrier is a spin-off of the typical LCC that offers even fewer amenities than LCCs, typically with unbundled pricing. In the US, there are three ULCCs – Spirit, Allegiant, and Frontier.

Unbundling (See Menu Pricing).

Uncertain Demand (Also called stochastic demand) is variable and can only be estimated for a particular flight – the exact demand cannot be known. Scheduled airlines face uncertain demand whereas charter airlines know the exact passenger demand per flight.

Unit Cost (See Cost Per Available Seat Mile).

Variable Costs (See Direct Operating Costs).

Variable Scheduling is a flight crew scheduling process that varies workdays from month to month with no particular sequence. With variable scheduling, a crewmember receives the new month's schedule a few weeks before the end of the current month.

Virtual Nesting is a technique to solve network revenue management problems by creating virtual buckets that correspond to O&D revenue values instead of single-leg revenue values.

Visiting Friends and Relatives (VFR) refers to traveling to visit or reunite with relatives or friends.

Warsaw Convention (1929) established the first rules of liability for international airlines carrying people cargo, luggage, or other goods. The convention recognized the right to compensation for the loss of luggage and cargo, as well as injury or death of passengers, but limited the airlines' liability.

Wave   (See Bank).

Weighted Average Cost of Capital   (WACC) is the return investors would receive if their money were invested elsewhere at similar risk. The return on invested capital (ROIC) is often compared to the WACC to determine how well the airline is performing relative to an industry average.

Wet Lease   is an operation by an airline, often a specialist in the business, that provides an aircraft, crew, maintenance, and insurance (aircraft, crew, maintenance, and insurance or ACMI) under contract to another airline. The lessee provides passenger and ground handling and, critically, fuel. A wet lease provides the airline with immediate capacity without long-term obligations

Wide Body Aircraft   is a passenger airliner with two (or more) isles.

Yield   is the average dollar amount each passenger pays to fly one mile. Yield is calculated by dividing total revenue by the revenue passenger miles (Revenue ÷ RPM). In the US, yield is measured in cents.

Yield Management   (See Revenue Management).

Index

Note: Page numbers in *italic* refer to figures. Endnotes are indicated by the page number followed by "n" and the note number e.g., 171n1 refers to note 1 on page 171.